Cetacean Paleobiology

Books in the **Topics in Paleobiology** series will feature key fossil groups, key events, and analytical methods, with emphasis on paleobiology, large-scale macroevolutionary studies, and the latest phylogenetic debates.

The books will provide a summary of the current state of knowledge and a trusted route into the primary literature, and will act as pointers for future directions for research. As well as volumes on individual groups, the Series will also deal with topics that have a cross-cutting relevance, such as the evolution of significant ecosystems, particular key times and events in the history of life, climate change, and the application of new techniques such as molecular paleontology.

The books are written by leading international experts and will be pitched at a level suitable for advanced undergraduates, postgraduates, and researchers in both the paleontological and biological sciences.

The Series Editor is *Mike Benton*, Professor of Vertebrate Palaeontology in the School of Earth Sciences, University of Bristol.

The Series is a joint venture with the *Palaeontological Association*.

Previously published

Amphibian Evolution
Rainer R. Schoch
ISBN: 978-0-470-67178-8 Paperback; May 2014

Dinosaur Paleobiology
Stephen L. Brusatte
ISBN: 978-0-470-65658-7 Paperback; April 2012

Cetacean Paleobiology

Felix G. Marx, Olivier Lambert, and Mark D. Uhen

WILEY Blackwell

Library of Congress Cataloging-in-Publication Data

Names: Marx, Felix G., author. | Lambert, Olivier, (Paleontologist), author. |
   Uhen, Mark D., author.
Title: Cetacean paleobiology / Felix G. Marx, Olivier Lambert and Mark D. Uhen.
Description: Chichester, UK; Hoboken, NJ : John Wiley & Sons, 2016. |
   Includes bibliographical references and index.
Identifiers: LCCN 2015047431 (print) | LCCN 2016005795 (ebook) |
   ISBN 9781118561270 (cloth : alk. paper) | ISBN 9781118561539 (pbk. : alk. paper) |
   ISBN 9781118561362 (Adobe PDF) | ISBN 9781118561553 (ePub)
Subjects: LCSH: Cetacea, Fossil. | Evolutionary paleobiology.
Classification: LCC QE882.C5 M37 2016 (print) | LCC QE882.C5 (ebook) |
   DDC 569/.5–dc23
LC record available at http://lccn.loc.gov/2015047431

Contents

Series Editor's Preface

Paleobiology is a vibrant discipline that addresses current concerns about biodiversity and about global change. Furthermore, paleobiology opens unimagined universes of past life, allowing us to explore times when the world was entirely different and when some organisms could do things that are not achieved by anything now living.

Much current work on biodiversity addresses questions of origins, distributions and future conservation. Phylogenetic trees based on extant organisms can give hints about the origins of clades and help answer questions about why one clade might be more species-rich ('successful') than another. The addition of fossils to such phylogenies can enrich them immeasurably, thereby giving a fuller impression of early clade histories, and so expanding our understanding of the deep origins of biodiversity.

In the field of global change, paleobiologists have access to the fossil record, and this gives accurate information on the coming and going of major groups of organisms through time. Such detailed paleobiological histories can be matched to evidence of changes in the physical environment, such as varying temperatures and sea levels, episodes of midocean ridge activity, mountain building, volcanism, continental positions and the impacts of extraterrestrial bodies. Studies of the influence of such events and processes on the evolution of life address core questions about the nature of evolutionary processes on the large scale.

As examples of unimagined universes, one need only think of the life of the Burgess Shale or the times of the dinosaurs. The extraordinary arthropods and other animals of the Cambrian sites of exceptional preservation sometimes seem more bizarre than the wildest imaginings of a science fiction author. During the Mesozoic, the sauropod dinosaurs solved basic physiological problems that allowed them to reach body masses 10 times larger than those of the largest elephants today. Furthermore, the giant pterosaur *Quetzalcoatlus* was larger than any flying bird, and so challenges fundamental assumptions in biomechanics.

Books in the Topics in Paleobiology series will feature key fossil groups, key events and analytical methods, with emphasis on paleobiology, large-scale macroevolutionary studies and the latest phylogenetic debates.

The books will provide a summary of the current state of knowledge and a trusted route into the primary literature, and will act as pointers for future directions for research. As well as volumes on individual groups, the Series will also deal with topics that have a cross-cutting relevance, such as the evolution of significant ecosystems, particular key times and events in the history of life, climate change and the application of new techniques such as molecular paleontology.

The books are written by leading international experts and have been pitched at a level suitable for advanced undergraduates, postgraduates and researchers in both the paleontological and biological sciences.

Michael Benton
Bristol
November 2011

Preface

All the fossil whales hitherto discovered belong to the Tertiary period, which is the last preceding the superficial formations. And though none of them precisely answer to any known species of the present time, they are yet sufficiently akin to them in general respects, to justify their taking rank as Cetacean fossils.

Detached broken fossils of pre-adamite whales, fragments of their bones and skeletons, have within thirty years past, at various intervals, been found at the base of the Alps, in Lombardy, in France, in England, in Scotland, and in the States of Louisiana, Mississippi, and Alabama. Among the more curious of such remains is part of a skull, which in the year 1779 was disinterred in the Rue Dauphine in Paris, a short street opening almost directly upon the palace of the Tuileries; and bones disinterred in excavating the great docks of Antwerp, in Napoleon's time. Cuvier pronounced these fragments to have belonged to some utterly unknown Leviathanic species.

But by far the most wonderful of all Cetacean relics was the almost complete vast skeleton of an extinct monster, found in the year 1842, on the plantation of Judge Creagh, in Alabama. The awe-stricken credulous slaves in the vicinity took it for the bones of one of the fallen angels. The Alabama doctors declared it a huge reptile, and bestowed upon it the name of *Basilosaurus*. But some specimen bones of it being taken across the sea to Owen, the English Anatomist, it turned out that this alleged reptile was a whale, though of a departed species. A significant illustration of the fact, again and again repeated in this book, that the skeleton of the whale furnishes but little clue to the shape of his fully invested body. So Owen rechristened the monster *Zeuglodon*; and in his paper read before the London Geological Society, pronounced it, in substance, one of the most extraordinary creatures which the mutations of the globe have blotted out of existence.

When I stand among these mighty Leviathan skeletons, skulls, tusks, jaws, ribs, and vertebrae, all characterized by partial resemblances to the existing breeds of sea-monsters; but at the same time bearing on the other hand similar affinities to the annihilated antichronical Leviathans, their incalculable seniors; I am, by a flood, borne back to that wondrous period, ere time itself can be said to have begun.

—Herman Melville's account of the cetacean fossil record, from "The Fossil Whale," *Moby Dick*

In what is maybe his most famous novel, Herman Melville provides an excellent account of the state of the cetacean fossil record in the mid-19th century. When *Moby Dick* was published in 1851, just a few years before Darwin's *On the Origin of Species*, whales were still among the most mysterious of all animals. How had a group of air-breathing, warm-blooded mammals come to live in the sea? As we shall see in the book, surprisingly little changed following Melville's early account. As late as 1945, great paleontologists like George Gaylord Simpson were still baffled by the origins of these seemingly "peculiar and aberrant" creatures. Since then, however, new finds and scientific approaches have led to a series of breakthroughs, and increased our knowledge of whale evolution to the point where it can no longer simply be summarized in a few paragraphs. Our goal here is to introduce our readers to this fascinating subject, and hopefully spark further interest in different areas of fossil cetacean research. Each chapter includes an extensive bibliography from which we have drawn the facts and hypotheses presented, as well as a list of suggested readings. We sincerely hope that you find the evolution of whales as interesting as we do, and that you will enjoy reading about it in this book. Afterward, be sure to keep your eye out for further developments in the field, as new information is unearthed around the globe in the future.

Felix G. Marx, National Museum of Nature and Science, Tsukuba, Japan

Olivier Lambert, Institut royal des Sciences naturelles de Belgique, Brussels, Belgium

Mark D. Uhen, George Mason University, Fairfax, Virginia, USA

Acknowledgments

Writing a book like this inevitably means to stand on the shoulders of giants. None of us could have begun to conceptualize such a project without being able to draw on the work of our colleagues and predecessors. To all of them, and to the many other people who have been involved in the discovery, preparation, curation, and study of fossil cetaceans, we extend our sincere thanks. Writing this book has been both a joy and a journey—not just for us but also for our families, who patiently endured the many hours we spent on this project. Brian, Catherine, Ikerne, and Matthias: we greatly appreciate your loving support, and dedicate this book to you. Much of what we know we owe to our former mentors and teachers. Our special thanks thus go to Michael J. Benton, R. Ewan Fordyce, Philip D. Gingerich, Pascal Godefroit, James G. Mead, and Christian de Muizon, all of whom have shaped our careers, never ceased to provide advice and help when needed and, more than anything, have become great friends.

Finally, we wish to thank the many people who helped to advance this book indirectly through insightful discussions, or went out of their way to help us out with photographs and information. Their help has been invaluable, and any errors in interpreting their contributions are entirely ours. Many thanks to: Lawrence G. Barnes, Ryan M. Bebej, Annalisa Berta, Giovanni Bianucci, Michelangelo Bisconti, Robert W. Boessenecker, David J. Bohaska, Mark Bosselaers, Mark D. Clementz, Lisa N. Cooper, Thomas A. Deméré, Erich M. G. Fitzgerald, R. Ewan Fordyce, Jonathan H. Geisler, Philip D. Gingerich, Stephen J. Godfrey, Pavel Gol'din, Oliver Hampe, Toshiyuki Kimura, Naoki Kohno, Lori Marino, James G. Mead, Ismael Miján, Christian de Muizon, Maureen A. O'Leary, Mary Parrish, George Phillips, Klaas Post, Nicholas D. Pyenson, Rachel A. Racicot, J. G. M. Thewissen, Mario Urbina, William J. Sanders, Frank D. Whitmore Jr., and Tadasu Yamada. Finally, we wish to thank Michael J. Benton for suggesting that we write this book, and Delia Sandford and Kelvin Matthews for their support and guidance during the writing process.

1 Cetaceans, Past and Present

1.1 Introduction and scope of the book

Cetaceans (whales, dolphins, and porpoises) are some of the most iconic inhabitants of the modern ocean. They are, however, also one of its most unlikely. This point was beautifully made by the famous paleontologist George Gaylord Simpson when he described cetaceans as "on the whole, the most peculiar and aberrant of mammals" (Simpson, 1945: p. 213). Living cetaceans are the result of more than 50 million years of evolution, which transformed a group of small, four-legged landlubbers into the ocean-going leviathans of today. As far back as the fourth century BC, the Greek philosopher Aristotle recognized in his *Historia Animalium* that whales and dolphins breathe air, give birth to live offspring, show parental care, and suckle their young. Along with their warm-bloodedness, these traits betray the terrestrial mammalian ancestry of cetaceans, and often present them with a considerable challenge. Put into water, most land mammals would struggle to swim for any length of time, breathe, cope with ingested saltwater, or maintain their body temperature. Yet cetaceans have managed to clear all of these hurdles, alongside many others. They can find prey even in murky water where eyes cannot see. Their air-breathing calves are born underwater, yet do not drown. They move around fast in three dimensions, yet avoid becoming dizzy. They dive deep beneath the surface, yet do not suffer from the bends.

For a long time, the story of how cetaceans managed to leave behind the shore and adapt so completely to life in the sea remained largely in the dark. Fossils of ancient cetaceans have been known since the early 19th century, but most of them were too fragmentary, or too similar to the living forms, to illuminate the morphological and ecological transition back into the water. This all changed in the early 1990s, when the first of a string of spectacular new fossil finds started to rewrite our understanding of how, when, and where the first cetaceans evolved. Over the following 25 years, further discoveries coincided with the emergence of an ever-more sophisticated array of analysis techniques, such as molecular phylogenetics, stable isotope analysis, **computed tomography** (CT) scanning, and molecular divergence time estimation. Together, these developments allowed unprecedented insights into not only the origin and evolutionary relationships of cetaceans, but also their ecology and functional biology.

In this book, we aim to provide an overview of the study of cetacean evolution from their first appearance to the present day. We start with a description of basic principles, including a brief summary of the ecology of living whales and dolphins, cetacean taxonomy, and an explanation of the main techniques and

Cetacean Paleobiology, First Edition. Felix G. Marx, Olivier Lambert, and Mark D. Uhen.
© 2016 John Wiley & Sons, Ltd. Published 2016 by John Wiley & Sons, Ltd.

concepts used to study extinct species (Chapter 1). This is followed by more detailed summaries of the cetacean fossil record (Chapter 2) and a description of their anatomy, phylogenetic relationships, and diversity (Chapters 3 and 4). Finally, Chapters 5–8 are devoted to particular topics and case studies of cetacean paleoecology, functional biology, development, and macroevolution.

1.2   What is a whale?

Whales and dolphins are the only mammals besides sea cows (sirenians) that have completely adapted to life in the ocean. Unlike the other major group of marine mammals, the pinnipeds (seals, sea lions, and walruses), cetaceans sleep, mate, give birth, and suckle their young in the water. Instead of hair, they rely on a thick layer of insulating blubber to maintain their body temperature. Their overall shape is extremely streamlined, with no external projections such as ears or genitals that could produce drag. Their forelimbs have turned into flippers and, having all but lost their original function in locomotion, are merely used for steering. To propel themselves through the water, they instead rhythmically beat their massive tail, which ends in a pair of characteristic horizontal flukes.

Given their distinctive anatomy, the question of how to define a cetacean may seem obvious to the modern observer. However, the issue becomes more vexed when fossils are taken into account. Taxonomically, cetaceans fall into three major groups: ancient whales (**archaeocetes**), baleen whales (**Mysticeti**), and toothed whales (**Odontoceti**), each of which comprises a range of families (Chapter 4). Broadly speaking, archaeocetes are defined by their retention of archaic morphologies, such as (1) well-developed hind limbs; (2) a small number of morphologically differentiated (heterodont) teeth, which are replaced once during life (diphyodonty); and (3) relatively close ties to land (e.g., to rest or give birth) (Figure 1.1). By contrast, mysticetes and odontocetes are completely aquatic, with no trace of an external hind limb, and they are unable to move or support their weight on land. Both groups furthermore underwent a pronounced reorganization of their facial bones—a process commonly known as **telescoping**—to facilitate breathing (section 3.2). Besides these shared features, modern odontocetes in particular are recognizable by (1) having a single blowhole; (2) having a variable but often large number of greatly simplified, conical teeth (i.e., they are both polydont and homodont); and (3) their ability to **echolochate** (i.e., use sound to navigate and detect prey). In contrast, mysticetes (1) are often extremely large, (2) have lost any trace of teeth as adults, and (3) possess a series of keratinous, sieve-like **baleen** plates suspended in two rows from their upper jaw (section 5.2.1). Incidentally, note that the term *whale* carries little biological meaning in this context, except when understood to mean *all cetaceans*. In common parlance, the word is usually applied only to large-sized species and their (presumed) relatives—including, ironically, some members of the dolphin family (e.g., the killer whale, *Orcinus orca*).

The morphological similarity of the oldest whales to terrestrial mammals can make it difficult to recognize their true evolutionary affinities. Potentially diagnostic features mostly relate to details of the morphology of the skull, such as incipient telescoping and the shape and arrangement of the teeth—in particular, the anteroposterior alignment of the tooth row and the absence of crushing basins on the check teeth (Thewissen *et al.*, 2007; Uhen, 2010). However, many of these are difficult to recognize across Cetacea as a whole or also occur in other, non-cetacean mammals. The clearest trait uniting all cetaceans is a marked increase in the thickness and density (**pachyosteosclerosis**) of the medial wall of the **tympanic bulla**, one of the two main ear bones located at the base of the skull (Figure 1.2) (section 3.2.5). A pachyosteosclerotic bulla was long thought to be unique to cetaceans, until a similar morphology was described for a group of extinct artiodactyls (even-toed ungulates) known as raoellids (Thewissen *et al.*, 2007). This wider distribution is, however, largely unproblematic, since raoellids are now known to be more closely related to cetaceans than to any other extant or extinct artiodactyls and, although never formalized as such, could therefore be seen as de facto cetaceans (sections 4.1 and 5.1.1) (Geisler and Theodor, 2009; Thewissen *et al.*, 2007).

Figure 1.1 Overview of the three main subdivisions of Cetacea: (a) archaeocetes (archaic whales), (b) Mysticeti (baleen whales), and (c) Odontoceti (toothed whales, including dolphins). Life reconstructions © C. Buell.

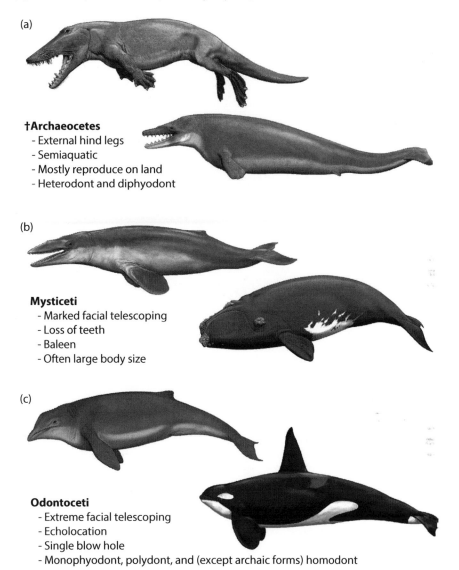

(a)

†Archaeocetes
- External hind legs
- Semiaquatic
- Mostly reproduce on land
- Heterodont and diphyodont

(b)

Mysticeti
- Marked facial telescoping
- Loss of teeth
- Baleen
- Often large body size

(c)

Odontoceti
- Extreme facial telescoping
- Echolocation
- Single blow hole
- Monophyodont, polydont, and (except archaic forms) homodont

1.3 Diversity, distribution, and ecology of modern cetaceans

Modern whales and dolphins form an essential part of the ocean ecosystem as **top predators**, as large-scale **nutrient distributors**, and as a **food source** for many deep-sea organisms (Croll *et al.*, 2006; Nicol *et al.*, 2010; Smith and Baco, 2003; Willis, 2014; Wing *et al.*, 2014). Their ranks include the holders of several world records, most of which are related to their often gigantic size: the blue whale *Balaenoptera musculus*, which at up to 190 tonnes is the Earth's heaviest animal (Tomilin, 1957)—and at least one-third again as heavy as the largest known dinosaur (Carpenter, 2006); the sperm whale *Physeter macrocephalus*, owner of

Figure 1.2 The pachyosteosclerotic tympanic bulla (highlighted in gray) characteristic of all cetaceans, as developed in (a) the early archaeocete *Pakicetus* and (b) the archaic mysticete *Aetiocetus*. Drawing of *Pakicetus* adapted from Gingerich *et al.* (1983) and Luo and Gingerich (1999).

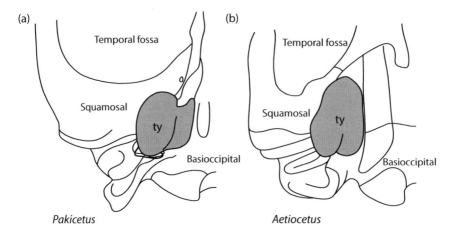

the world's largest brain (up to 8 kg) (Marino, 2009); the right whales of the genus *Eubalaena*, which possess the more dubious accolade of having the world's largest testes (approximating 1 tonne) (Brownell and Ralls, 1986); and the longest lived of all mammals, the bowhead whale *Balaena mysticetus*, which may reach a venerable age of more than 200 years (George *et al.*, 1999).

All extant species are either mysticetes or odontocetes, with archaeocetes having become extinct around 25 Ma (section 4.2). The Society of Marine Mammalogy currently recognizes 90 living species, 84% of which are odontocetes (Committee on Taxonomy, 2014). On the whole, the modern cetacean fauna is heavily biased toward three families in particular: the **rorquals** (Balaenopteridae), representing around 60% of all living mysticetes; and the **oceanic dolphins** (Delphinidae) and **beaked whales** (Ziphiidae), accounting for roughly 50% and 30% of all living odontocetes, respectively. Even more strikingly, nearly all balaenopterids and roughly two-thirds of all ziphiids each belong to a single genus (*Balaenoptera* and *Mesoplodon*). This skewed taxonomic distribution is probably an indicator of relatively recent radiations, possibly driven by the evolution of enlarged brains or particular feeding and mating strategies (sections 6.1, 6.5, and 7.5). Cetacean taxonomy remains in flux, and discover-

ies of new species (even large-sized ones) are still relatively frequent. Thus, a new beaked whale was reported as recently as 2014, and at least one new rorqual is currently awaiting formal description (Dalebout *et al.*, 2014; Sasaki *et al.*, 2006).

Living cetaceans range in size from about 1 m to more than 30 m, and they inhabit all parts of the world's oceans and seas. Geographically, modern diversity is highest at intermediate latitudes and sea surface temperatures of roughly 21 °C (Whitehead *et al.*, 2008). Mysticetes undergo long-distance migrations between low-latitude breeding and high-latitude feeding areas (Stern, 2009). Cetacean feeding strategies can broadly be divided into (1) **filter feeding**, which targets vast quantities of small-sized prey en masse and is characteristic of mysticetes; and (2) the targeting of individual prey items via **suction**, **raptorial feeding**, or a combination of the two, as seen in odontocetes (section 6.1) (Pivorunas, 1979; Werth, 2000). For their diet, most species rely on fish and cephalopods. Exceptions are the mysticetes, which also feed on tiny crustaceans (mostly copepods and krill), and the killer whale *Orcinus*, which regularly preys on other marine mammals and, occasionally, even turtles and sea birds. The false and pygmy killer whales, *Pseudorca* and *Feresa*, may also target other marine mammals, but tend to do so much less frequently (Werth, 2000). Feeding takes places at a range of depths. Sperm whales and beaked whales dive both the deepest (more than 2.9 km in the case of *Ziphius*) and the longest, with routine dives lasting 40–70 minutes (Aoki *et al.*, 2007; Hooker and Baird, 1999; Schorr *et al.*, 2014). By contrast, shorter (up to 10 min) and shallower (100–150 m) dives are characteristic of many dolphins and porpoises, as well as mysticetes (Stewart, 2009).

Nearly all living odontocetes are highly **gregarious**. Some species, such as the sperm, killer, and pilot whales, form matrilineal family groupings, whereas others are organized in less stable fission–fusion societies. Living in groups may help to guard against predators (e.g., in the case of sperm whales), facilitate cooperative feeding and serve mating purposes (Trillmich, 2009). Older killer and pilot whale females experience menopause, which may free them to support their descendants through day-to-day assistance and/or allomaternal care (Foster *et al.*, 2012; Marsh and Kasuya, 1986). In contrast to their tooth-bearing cousins, mysticetes are comparatively solitary creatures, but they aggregate during migration, in breeding areas and to engage in cooperative feeding (Brown and Corkeron, 1995; Weinrich, 1991). Relatively large groups of pygmy right whales have been observed at sea (Matsuoka *et al.*, 1996), and there is evidence of individual humpbacks forming long-term associations across several feeding seasons (Ramp *et al.*, 2010). Both mysticetes and odontocetes show signs of **culture** and engage in **complex social interactions**. These require flexible communication and sophisticated cognitive abilities, and likely explain both the intricate vocalizations of some taxa (May-Collado *et al.*, 2007) and the **enlarged cetacean brain** (sections 3.4.4 and 7.5) (Marino *et al.*, 2007; Rendell and Whitehead, 2001).

1.4 How to study extinct cetaceans

1.4.1 Comparative and functional anatomy
Anatomical observation has long been the mainstay of paleobiological inquiry, and it still plays a major role in (1) defining and classifying species; (2) establishing evolutionary relationships and certain measures of biological diversity (Slater

et al., 2010; Wiens, 2004; Wills *et al.*, 1994); (3) determining stages of physical maturity (Walsh and Berta, 2011); (4) gaining insights into developmental processes, such as heterochrony and vertebral patterning (Buchholtz, 2007; Galatius, 2010); and (5) reconstructing the feeding strategies, brain size, reproduction, sensory capabilities, and modes of locomotion of extinct taxa (Deméré *et al.*, 2008; Ekdale and Racicot, 2015; Montgomery *et al.*, 2013; Racicot *et al.*, 2014). **Anatomical descriptions** rely on specialized terminology relating to particular structures, locations, and motions (Figure 1.3). The sheer bulk of anatomical vocabulary may sometimes appear overwhelming, but it is hard to avoid given the complexity of biological systems and the need to ensure consistency. Luckily, there are some excellent summaries that help to navigate the jungle of jargon, especially with regards to the highly modified body of cetaceans (e.g., Mead and Fordyce, 2009).

Descriptive osteology forms the basis for phylogenetic analyses (section 1.4.2) and can be used to assess morphological disparity, or variation in body shape, through time (section 7.3). In addition, functionally relevant observations, such as the range of motion allowed by a particular

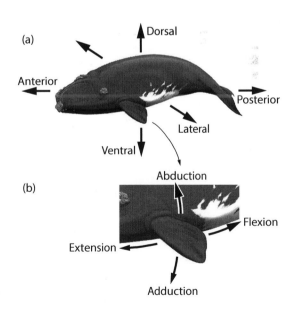

Figure 1.3 Standard anatomical terms of (a) location and (b) motion. Life reconstructions © C. Buell.

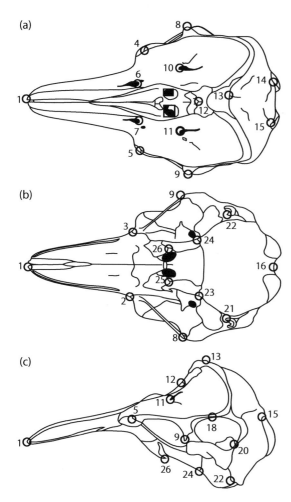

Figure 1.4 Example of a three-dimensional set of landmarks, based on the skull of a porpoise. (a) Dorsal, (b) ventral, and (c) lateral views. Reproduced from Galatius (2010), with permission of the Linnean Society of London.

data, but usually it can only be applied to largely complete, undistorted fossil specimens. Besides quantifying shape, direct measurements of particular parts of the skeleton are used to estimate the total body size of incompletely preserved fossil specimens (Lambert *et al.*, 2010; Pyenson and Sponberg, 2011).

Soft tissues are also a rich source of information on evolutionary relationships, ecology, life history, and functional anatomy, but, unlike bones, they are prone to rapid decay following death. With very few exceptions, details on the external anatomy, musculature, and inner organs of fossil organisms are thus invariably lost. Sometimes, however, soft tissues leave tell-tell traces (**osteological correlates**) on the bones themselves, which can be used to reconstruct their appearance and function in life. Such traces may take the form of distinctive muscle scars, hollow spaces for the reception of air-filled sacs, vascular structures associated with particular tissue types, and, in some cases, even the complete outline of an entire organ. The latter particularly applies to the shape of the brain, the inner ear, and the organ of balance, whose shapes can be reconstructed and measured using CT scans (sections 3.4.4 and 7.5) (Ekdale and Racicot, 2015; Marino *et al.*, 2003; Spoor *et al.*, 2002).

1.4.2 Evolutionary relationships

Understanding the evolutionary relationships between species helps to clarify their origins, and provides the fundamental framework underlying most paleobiological inquiry. Modern techniques to reconstruct cetacean interrelationships (their **phylogeny**) are also able to determine when two related species first diverged. Together with ongoing refinements in the dating of individual fossils, phylogenies thus can answer such important questions as: What other mammals are whales related to? When did they first evolve? When, and how quickly, did they diversify? And does their evolution follow any particular trends?

By convention, evolutionary relationships are depicted in the form of a **tree**, which may include both living and extinct species. A tree consists of terminal and internal **branches**, all of which connect at **nodes**. Internal branches, and the nodes

joint, help to reconstruct locomotor and feeding abilities (Deméré *et al.*, 2008; Gingerich *et al.*, 1994; Gutstein *et al.*, 2014). Similar insights can be gained from **morphometrics**, which involves the quantification of direct measurements or anatomical **landmarks** (homologous points) based on two- or three-dimensional osteological models (Figure 1.4) (Galatius, 2010; Hampe and Baszio, 2010). This approach has the advantage of suffering less from subjective assessments and individual scoring error than purely descriptive character

Figure 1.5 Illustration of cetacean, mysticete, and odontocete crown and stem groups. Life reconstructions © C. Buell.

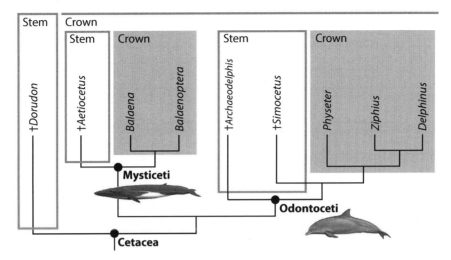

they lead to, are sometimes also interpreted as **hypothetical ancestors**. Related species (i.e., those deriving from a single ancestor) form a **clade**. Both clades and individual species can be referred to as **taxa**. A clade is said to be **monophyletic** if it includes all of its ancestor's descendants, and **paraphyletic** if a particular, usually highly distinctive subgroup of its members has been deliberately excluded. Mammals and birds are examples of monophyletic clades, whereas non-avian dinosaurs (all dinosaurs minus birds) are paraphyletic. Proposed groups that contain members of different clades—say, a group including birds and mammals—are **polyphyletic**, and taxonomically untenable. A related concept is that of the crown and stem group. A **crown group** is a clade defined by all of the extant representatives of a particular taxonomic group (e.g., all extant cetaceans), and it comprises them, their last common ancestor and all those extinct taxa that are descended from the latter. In most cases, a crown group is associated with a **stem group**, which includes all those extinct taxa that fall outside of the crown group, yet are more closely related to it than to any other major clade. Stem groups are often, though not necessarily, paraphyletic (Figure 1.5).

Phylogenetic analyses can be based on two basic types of data: (1) **molecular** sequences, including DNA and, less commonly, proteins; and (2) **morphological** observations. With the exception of extremely young (i.e., Pleistocene) material, fossils do not preserve any usable DNA. Likewise, protein sequences have never been reported from any truly ancient cetacean fossil, although it is possible that some limited information may be preserved under ideal conditions. Reconstructing the evolutionary relationships of fossil taxa must therefore rely solely on morphological data, although molecular sequences still play an important role in the placement of extant species—and thus, by proxy, also that of their close fossil relatives (Wiens, 2009). At the basis of morphological phylogenetics lies anatomical observation (as discussed in this chapter). For the purpose of phylogenetic reconstruction, descriptive morphological data are usually broken down into **discrete characters**, each of which can take two or more **states** (Figure 1.6). For example, a simple character may record the presence (state 0) or absence (state 1) of an external hind limb. The characters are then collated into a **matrix** and analyzed according to cladistic principles.

Cladistics was first proposed by the German entomologist Willi Hennig (Hennig, 1965), who proposed that two species should only be considered as related to each other if they are united by one or more derived characters. In other words, evolutionary relatedness must be demonstrated by the possession of **shared, homologous features** demonstrating an evolutionary change from a primitive state (**plesiomorphy**) to a derived state (**apomorphy**). Each character has therefore a **polarity** (primitive to derived), which is usually reflected in the numbering of states within a cladistic matrix; by convention, 0 denotes the primitive condition. Imposing character polarities naturally raises the question of how the primitive state can be recognized. The most commonly used option is to define an external point of reference, usually in the form of an additional species (**outgroup**), that clearly falls outside the group of interest and therefore is likely to show the primitive state for all of the analyzed characters.

Figure 1.6 provides an example of a simple cladistic analysis. The matrix shown in Figure 1.6a contains five taxa scored for seven characters (Figure 1.6b). The snail represents the outgroup and accordingly shows the primitive state (0) for all characters. All of the other taxa possess a certain number of derived features (synapomorphies; state 1). Whales and humans share most of these derived characters (e.g., possession of hair, constant body temperature, and suckling of young), followed by lizards and, finally, fishes. The matrix therefore suggests a ladder-shaped cladistic hypothesis, or **cladogram**, in which whales and humans form a clade (in this case, Mammalia) to the exclusion of all other taxa in the analysis. Together, mammals are most closely related to lizards, with which they form a clade known as Amniota—animals producing eggs with a protective membrane. Finally, all amniotes share a common ancestor with fishes.

The scenario shown here is the most likely given the available data, but note that there are other possibilities. As shown by character 7 in Figure 1.6, whales and fishes share an aquatic lifestyle, which could be interpreted as evidence for a

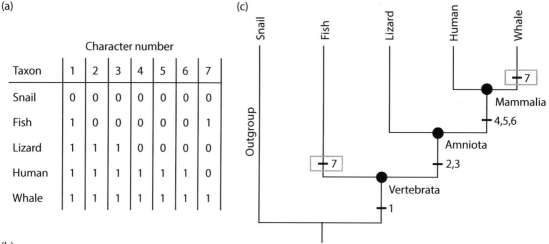

(a)

Character number

Taxon	1	2	3	4	5	6	7
Snail	0	0	0	0	0	0	0
Fish	1	0	0	0	0	0	1
Lizard	1	1	1	0	0	0	0
Human	1	1	1	1	1	1	0
Whale	1	1	1	1	1	1	1

(b)
1) Vertebral column: 0, absent; 1, present
2) Embryo covered by protective membrane: 0, absent; 1, present
3) Four limbs with five fingers each: 0, absent; 1, present
4) Hair: 0, absent; 1, present
5) Constant body temperature: 0, absent; 1, present
6) Suckling of young: 0, absent; 1, present
7) Aquatic lifestyle: 0, absent; 1, present

Figure 1.6 Example of a simple cladistic analysis. (a) Data matrix comprising five taxa (the snail is the outgroup) and seven characters, described in (b). Analysis of the data matrix would result in the cladogram shown in (c). In (c), numbers refer to characters supporting a particular branch. Characters 1–6 are synapomorphies, whereas character 7 is homoplastic and an autapomorphy of fishes and whales, respectively.

close relationship of these two taxa to the exclusion of humans and lizards. If this were true, however, then hair, a constant body temperature and suckling would have had to evolve twice—once in whales, and once in humans. Similarly, a protective egg membrane would have had to arise twice, or else be lost convergently in fishes. More evidence thus speaks for the tree shown in Figure 1.6c, which is more **parsimonious** than the alternative arrangement that allies whales and fishes. In the context of this analysis, being aquatic is thus a **homoplasy** (i.e., a derived feature that is shared but not homologous, having instead arisen via **convergent evolution**). Because an aquatic existence now only characterizes single branches (whales and fishes, respectively), rather than a clade, it is also known as an **autapomorphy**.

This example is a simple demonstration of the principle of **maximum parsimony**, which seeks to minimize the number of transitions between different states. The tree, or trees, with the smallest number of **steps** are considered optimal, and preferred over alternative, less parsimonious arrangements. In computerized form, parsimony analysis has long been one of the most important cladistic tools, and is still widely used to analyze morphological data. Alternative approaches include **maximum likelihood** and **Bayesian** methods, which have largely replaced maximum parsimony in the context of molecular phylogenetics, and are increasingly being adopted by morphologists as well. Unlike parsimony, these approaches include assumptions about how often and how easily changes between certain states can occur. Such models are particularly relevant with regards to molecular data, since it is known that certain mutations are less likely to occur than others. In addition, Bayesian methods offer the advantage of greater control by allowing the inclusion of (well-justified) *a priori* assumptions about tree shape and other analysis parameters.

Recent trends also include the combination of molecular and morphological data into **total evidence** analyses (Deméré *et al.*, 2008; Geisler *et al.*, 2011), and a realization that both data types can be used to estimate the time at which two taxa diverged (section 4.5) (Pyron, 2011; Ronquist *et al.*, 2012). To do so, the total amount of molecular and/or morphological change that occurred along a particular branch is calculated and calibrated against the fossil record, often based on a series of predetermined fossil taxa of known age. This calibration effectively turns the rate of change into a **molecular/morphological clock**, which can either be held constant throughout the tree (**strict clock**) or be allowed to vary across lineages (**relaxed clock**). The latter is often a more likely scenario, since changes in generation times, population sizes, protein functions, species-specific physiological mechanisms, and the strength of natural selection likely conspire to render a universal, strict clock inapplicable (Ayala, 1999).

Once a tree has been constructed, it can be used to reconstruct the combination of morphological character states or molecular sequences that would have been present at each of its internal nodes. **Ancestral state reconstruction** can be carried out within a parsimony, likelihood or Bayesian framework, and is often employed to infer unknown traits for a particular taxon (e.g., soft tissue characters) based on its position in the phylogeny itself—a process also known as **phylogenetic bracketing** (Witmer, 1995). In addition, ancestral state reconstruction can be used to trace the evolution of a particular character over time, or to estimate the morphology of a hypothetical ancestor. Such reconstructions therefore create predictions about particular morphologies that have not yet been found as actual specimens, but are likely to have occurred based on the existing fossil record. One recent example of this approach is the reconstruction of the hypothetical ancestor of all placental mammals, based on a large phylogenetic analysis comprising all major mammalian clades (O'Leary *et al.*, 2013).

1.4.3 Habitat and feeding preferences

The habitat preference of a particular fossil species can often be reconstructed from associated **stratigraphic** and **sedimentological** data. However, such information can be confounded by postmortem transportation of the carcass, and it does not record movement during life. Thus, for example, an archaic whale could well have been at home both in the water and on land, even if its remains are only preserved in marine rocks. Tooth morphology, **wear**, **microwear**, and **tooth marks** can provide data on diet and, by proxy, habitat (section 6.1)

(Fahlke, 2012; Fahlke *et al.*, 2013; Thewissen *et al.*, 2011). However, the study of these features relies on the presence of teeth, which are absent, reduced, or highly simplified in many cetaceans, and thus often fails to distinguish clearly between different habitat and prey types. Other observations related to functional anatomy, such as the ability to rotate the jaw or the estimation of muscle function and maximum bite force via **Finite Element Analysis** (Snively *et al.*, 2015), can offer insights into particular feeding strategies, but generally do not distinguish habitats.

A fourth option is the interpretation of **stable isotope ratios**, particularly those of oxygen and carbon (Clementz *et al.*, 2006; Roe *et al.*, 1998). Oxygen and carbon are both essential components of body tissues, the isotopic composition of which is determined by body and ambient water, as well as an animal's diet. Because of their different physical properties, isotopes vary in the rate at which they take part in environmental and biological processes, such as evaporation, condensation, and tissue formation. Ultimately, this leads to differences in the isotopic compositions of various substances, which can be recorded in the form of stable isotope ratios ($^{18}O/^{16}O$ and $^{13}C/^{12}C$, respectively) and are usually expressed as deviations (δ) from an international standard. Recorded in bone or teeth, such isotopic signals can become "fossilized" along with the remains of the animal itself.

To distinguish marine, freshwater, and terrestrial species (sections 5.1 and 6.1), it is important to consider both the actual value and the variability of their oxygen isotopic signal (Clementz and Koch, 2001; Clementz *et al.*, 2006). ^{16}O isotopes evaporate more easily than ^{18}O, which causes vapor formed over the ocean to be enriched in ^{16}O. As the vapor moves inland, it condenses and falls as rain, which builds up to form freshwater. This process results in a distinct isotopic difference (typically 3‰) between marine and freshwater environments (Roe *et al.*, 1998). In fully aquatic animals, such as modern cetaceans, this isotopic signal ($\delta^{18}O$) is incorporated into body tissues via direct exchange of water through the skin and ingestion of water during feeding (Costa, 2009; Hui, 1981). Because aquatic environments are relatively homogeneous, variations in the isotopic signal tend to be low—with the

exception of some highly variable freshwater systems, such as estuaries. Thus, both freshwater and marine species are characterized by a narrow range of oxygen isotope values (Clementz and Koch, 2001), with freshwater taxa generally scoring lower (Clementz *et al.*, 2006; Thewissen *et al.*, 1996). By contrast, the tissues of terrestrial animals mainly reflect the isotopic composition of dietary and drinking water, the composition of which varies from place to place and over time as a result of evaporation, distance from the sea and differences in elevation. In addition, species-specific physiological processes introduce further variation, which leads to terrestrial species having much more variable $\delta^{18}O$ values than marine animals (Clementz *et al.*, 2006).

Carbon isotope ($\delta^{13}C$) values record the type of primary producer sustaining a particular food web, as well as the trophic level at which an animal feeds. The former is mainly related to the photosynthetic pathway (C3, C4, or CAM) employed by the primary producer and the environmental conditions in which the latter grows. Together, these variables result in a broad range of $\delta^{13}C$ values that distinguish terrestrial from aquatic environments (Thewissen *et al.*, 2011), as well as freshwater and marine offshore habitats (low $\delta^{13}C$) from nearshore habitats (high $\delta^{13}C$) (Clementz *et al.*, 2006). **Trophic fractionation** occurs each time one organism is being fed on by another, and results in a slight enrichment in ^{13}C in the tissues of the consumer (Vander Zanden and Rasmussen, 2001). Cumulatively, this enrichment results in markedly higher $\delta^{13}C$ values in consumers feeding at a high trophic level relative to those feeding at a low one (Figure 1.7) (Clementz *et al.*, 2014). Isotope fractionation also occurs to a different degree in herbivores versus carnivores, and can thus be used to distinguish feeding strategies (Clementz *et al.*, 2006; Thewissen *et al.*, 2011).

In addition to reflecting habitat type and feeding, both oxygen and carbon isotopes correlate negatively with latitude as a result of different temperatures, salinities, and levels of productivity. This spatial variation, which seems to have existed since at least the Eocene, results in greater than expected isotopic variance in migratory species (Clementz and Sewall, 2011; Roe *et al.*, 1998). Because the exact relationship between isotopic composition

Figure 1.7 Graph showing the effect of latitude and trophic level on the stable isotope composition of carbon incorporated in mysticete (white circles) and odontocete (gray circles) teeth and bones. Higher trophic levels reflect longer food chains leading up to the final consumer. The white circle marked by a dot marks an exception: the gray whale, *Eschrichtius robustus*, which differs from other mysticetes in feeding on benthic invertebrates at high latitudes. H Lat., high latitude; L–M Lat., low–mid latitude. Reproduced from Clementz *et al.* (2014), with permission of Elsevier.

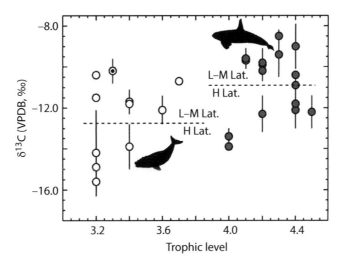

and latitude differs between hemispheres and ocean basins, it can be used to identify consumers foraging in particular geographic areas, as well as to create a map of isotopic composition (isoscape) to track the movements of marine consumers (Clementz *et al.*, 2014; Graham *et al.*, 2010).

1.4.4 Macroevolutionary dynamics

Macroevolution is the study of major, supraspecific evolutionary patterns, such as **adaptive radiations**, evolutionary **trends**, major **turnover events**, and **convergent evolution**, usually over timescales of millions of years (sections 7.1–7.5 and 7.7). To place macroevolutionary events in a temporal context, relevant fossils first need to be dated. Sometimes, this can be done more or less directly by determining the age of fossil-bearing rocks absolutely via **radiometric dating**. Where the latter is impossible—which it often is—rocks are instead dated in a relative fashion by correlating them with units of known age based on their lithology, magnetic profile, chemical composition, or fossil content (**biostratigraphy**). To facilitate comparisons across wide geographical areas, time periods characterized by the occurrence of particular

organisms or magnetic profiles are grouped into biozones and chrons, respectively, and correlated with the global **geological time scale**. The latter groups all of Earth's history into a series of hierarchical units, which, in descending order, comprise eons, eras, periods, epochs, and stages. Cetaceans are only known from the most recent eon (the Phanaerozoic, or time of "visible life") and era (the Caenozoic, or time of "new life"), but span both the Paleogene and Neogene periods, as well as several epochs (Figure 1.8).

Once fossils have been dated, their occurrence can be correlated with other paleontological and paleoenvironmental data to identify potential biotic or physical factors that may have acted as **evolutionary drivers**. In the case of cetaceans, potential candidates may range from **competition** to **key innovations** (e.g., baleen and echolocation), ocean restructuring and changes in climate, sea level and food abundance (Fordyce, 1980; Marx and Uhen, 2010; Pyenson *et al.*, 2014b). To investigate whether any of these phenomena played a role in cetacean evolution, there first needs to be an estimate of past biological diversity (Marx and Uhen, 2010; Slater *et al.*, 2010; Steeman *et al.*, 2009). The

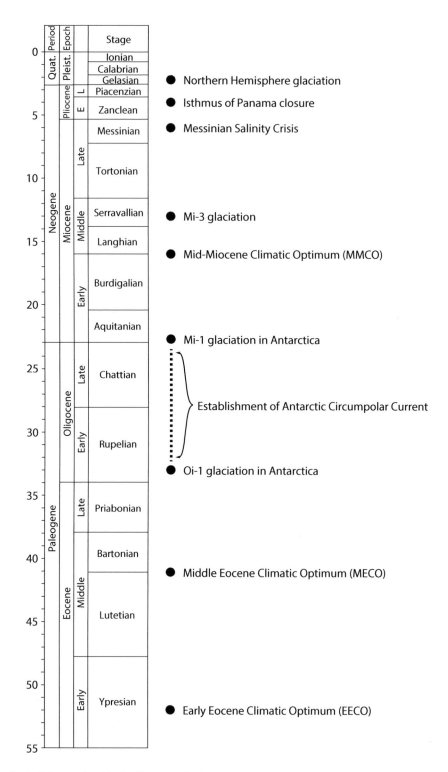

Figure 1.8 Geological time scale with significant earth history events.

latter is a rather inclusive concept comprising aspects of taxonomy, ecology, and morphology, and accordingly can be quantified in a number of ways.

Taxonomic diversity measures the total number of species (or higher ranking taxa) that existed at a particular point in time, as judged from their stratigraphic occurrence (section 7.1). This is the most direct measure of biological diversity, and forms the basis for assessments of lineage **diversification** and **extinction**. However, taxonomic diversity can also be strongly biased, for example, by variable amounts of rock that can be searched for fossils (Smith, 2007; Uhen and Pyenson, 2007). Rather than numbers of taxa, morphological diversity (**disparity**) measures among-species variation in overall body shape (section 7.3) (Foote, 1991). A simple way to think about disparity is to compare an African elephant with an Indian elephant on the one hand, and an elephant with an ant on the other. In both cases, taxonomic diversity is the same (two species), yet it is perfectly obvious even to the non-biologist that the African and Indian elephants look much more similar to each other (i.e., they are less disparate) than either does to the ant. Disparity can be quantified either with regards to overall body shape (Wills *et al.*, 1994) or by focusing on a particular phenotypic trait, such as body or brain size (Lambert *et al.*, 2010; Montgomery *et al.*, 2013; Slater *et al.*, 2010). Finally, **ecological** and **functional disparity** measure variation in life habits, such as diets, feeding styles, or modes of locomotion (Slater *et al.*, 2010).

Although not entirely reliant on it, macroevolutionary analyses greatly benefit from the inclusion of phylogenetic data. Crucially, phylogenies (1) allow the integration of molecular data; (2) provide an alternative way to date lineage divergences (based on molecular/morphological clocks; section 1.4.2), and thus the timing of macroevolutionary events; and (3) provide a framework within which diversity and disparity changes can be analyzed statistically. Phylogeny-dependent analyses include calculating rates of phenotypic and genomic change (Lee *et al.*, 2013), the tempo of lineage diversification, disparification, and extinction (Rabosky, 2014; Thomas and Freckleton, 2012), and the detailed dynamics of evolutionary trends (sections 7.1 and 7.4) (Montgomery *et al.*, 2013).

Recent work has even attempted to estimate past taxonomic diversity from molecular-based phylogenies of extant taxa alone, with potentially promising results (Morlon *et al.*, 2011).

1.4.5 Other methodologies

Beyond the fields of study detailed in this chapter, insights on cetacean evolution have also come from **bone histology**, **pathology**, and **taphonomy**. Thus, increased bone density has been interpreted as ballast enabling early cetaceans to stay underwater (de Buffrénil *et al.*, 1990; Gray *et al.*, 2007; Thewissen *et al.*, 2007); the presence of well-developed columns of spongy bone in the limb bones as providing support for terrestrial locomotion (Madar, 1998); bone fractures in the lower jaw as evidence of benthic feeding (Beatty and Dooley, 2009); bony outgrowths along tooth sockets as a clue to raptorial feeding (Lambert *et al.*, 2014); and localized breakdown of bone as a proxy for diving-related decompression syndrome, commonly known as the "bends" (Beatty and Rothschild, 2008). By contrast, taphonomy generally does not provide insights into cetacean biology itself, but may elucidate causes of past mass strandings, such as toxic algal blooms (Pyenson *et al.*, 2014a).

1.5 Suggested readings

Berta, A., J. L. Sumich, and K. M. Kovacs. 2015. Marine Mammals: Evolutionary Biology, 3rd ed. Academic Press, Burlington, MA.

Perrin, W. F., B. Wursig, and J. G. M. Thewissen. 2009. Encyclopedia of Marine Mammals, 2nd ed. Academic Press, Burlington, MA.

Ridgway, S. H., and R. Harrison. 1985–1989. Handbook of Marine Mammals. Vols. 3–6. Academic Press, Burlington, MA.

References

Aoki, K., M. Amano, M. Yoshioka, K. Mori, D. Tokuda, and N. Miyazaki. 2007. Diel diving behavior of sperm whales off Japan. Marine Ecology Progress Series 349:277–287.

Ayala, F. J. 1999. Molecular clock mirages. BioEssays 21:71–75.

Beatty, B. L., and A. C. Dooley, Jr. 2009. Injuries in a mysticete skeleton from the Miocene of Virginia, with a discussion of buoyancy and the primitive feeding mode in the Chaeomysticeti. Jeffersoniana 20:1–28.

Beatty, B. L., and B. M. Rothschild. 2008. Decompression syndrome and the evolution of deep diving physiology in the Cetacea. Naturwissenschaften 95:793–801.

Brown, M., and P. Corkeron. 1995. Pod characteristics of migrating humpback whales (*Megaptera novaeangliae*) off the East Australian coast. Behaviour 132:163–179.

Brownell, R. L., Jr., and K. Ralls. 1986. Potential for sperm competition in baleen whales. Reports of the International Whaling Commission (Special Issue) 8:97–112.

Buchholtz, E. A. 2007. Modular evolution of the cetacean vertebral column. Evolution & Development 9:278–289.

Carpenter, K. 2006. Biggest of the big: a critical re-evaluation of the mega-sauropod *Amphicoelias fragillimus* Cope, 1878. New Mexico Museum of Natural History and Science Bulletin 36:131–138.

Clementz, M., and P. Koch. 2001. Differentiating aquatic mammal habitat and foraging ecology with stable isotopes in tooth enamel. Oecologia 129:461–472.

Clementz, M. T., R. E. Fordyce, S. Peek, L., and D. L. Fox. 2014. Ancient marine isoscapes and isotopic evidence of bulk-feeding by Oligocene cetaceans. Palaeogeography, Palaeoclimatology, Palaeoecology 400:28–40.

Clementz, M. T., A. Goswami, P. D. Gingerich, and P. L. Koch. 2006. Isotopic records from early whales and sea cows: contrasting patterns of ecological transition. Journal of Vertebrate Paleontology 26:355–370.

Clementz, M. T., and J. O. Sewall. 2011. Latitudinal gradients in greenhouse seawater d^{18}O: evidence from Eocene sirenian tooth enamel. Science 332:445–458.

Committee on Taxonomy. 2014. List of marine mammal species and subspecies. Society of Marine Mammalogy, www. marinemammalscience.org.

Costa, D. P. 2009. Osmoregulation; pp. 801–806 in W. F. Perrin, B. Würsig, and J. G. M. Thewissen (eds.), Encyclopedia of Marine Mammals. Academic Press, Burlington, MA.

Croll, D. A., R. Kudela, and B. R. Tershy. 2006. Ecosystem impact of the decline of large whales in the North Pacific; pp. 202–214 in J. A. Estes, D. P. DeMaster, D. F. Doak, T. M. Williams, and R. L. Brownell (eds.), Whales, Whaling and Ocean Ecosystems. University of California Press, Berkeley.

Dalebout, M. L., C. Scott Baker, D. Steel, K. Thompson, K. M. Robertson, S. J. Chivers, W. F. Perrin, M. Goonatilake, R. Charles Anderson, J. G. Mead, C. W. Potter, L. Thompson, D. Jupiter, and T. K. Yamada. 2014. Resurrection of *Mesoplodon hotaula* Deraniyagala 1963: a new species of beaked whale in the tropical Indo-Pacific. Marine Mammal Science 30:1081–1108.

de Buffrénil, V., A. de Ricqlès, C. E. Ray, and D. P. Domning. 1990. Bone histology of the ribs of the archaeocetes (Mammalia: Cetacea). Journal of Vertebrate Paleontology 10:455–466.

Deméré, T. A., M. R. McGowen, A. Berta, and J. Gatesy. 2008. Morphological and molecular evidence for a stepwise evolutionary transition from teeth to baleen in mysticete whales. Systematic Biology 57:15–37.

Ekdale, E. G., and R. A. Racicot. 2015. Anatomical evidence for low frequency sensitivity in an archaeocete whale: comparison of the inner ear of *Zygorhiza kochii* with that of crown Mysticeti. Journal of Anatomy 226:22–39.

Fahlke, J. M. 2012. Bite marks revisited—evidence for middle-to-late Eocene *Basilosaurus isis* predation on *Dorudon atrox* (both Cetacea, Basilosauridae). Palaeontologia Electronica 15:32A.

Fahlke, J. M., K. A. Bastl, G. M. Semprebon, and P. D. Gingerich. 2013. Paleoecology of archaeocete whales throughout the Eocene: dietary adaptations revealed by microwear analysis. Palaeogeography, Palaeoclimatology, Palaeoecology 386:690–701.

Foote, M. 1991. Morphological and taxonomic diversity in a clade's history: the blastoid record and stochastic simulations. Contributions from the Museum of

Paleontology, University of Michigan 28:101–140.

Fordyce, R. E. 1980. Whale evolution and Oligocene Southern Ocean environments. Palaeogeography, Palaeoclimatology, Palaeoecology 31:319–336.

Foster, E. A., D. W. Franks, S. Mazzi, S. K. Darden, K. C. Balcomb, J. K. B. Ford, and D. P. Croft. 2012. Adaptive prolonged postreproductive life span in killer whales. Science 337:1313.

Galatius, A. 2010. Paedomorphosis in two small species of toothed whales (Odontoceti): how and why? Biological Journal of the Linnean Society 99:278–295.

Geisler, J. H., M. R. McGowen, G. Yang, and J. Gatesy. 2011. A supermatrix analysis of genomic, morphological, and paleontological data from crown Cetacea. BMC Evolutionary Biology 11:1–33.

Geisler, J. H., and J. M. Theodor. 2009. *Hippopotamus* and whale phylogeny. Nature 458:E1–E4.

George, J. C., J. Bada, J. Zeh, L. Scott, S. E. Brown, T. O'Hara, and R. Suydam. 1999. Age and growth estimates of bowhead whales (*Balaena mysticetus*) via aspartic acid racemization. Canadian Journal of Zoology 77:571–580.

Gingerich, P. D., S. M. Raza, M. Arif, M. Anwar, and X. Zhou. 1994. New whale from the Eocene of Pakistan and the origin of cetacean swimming. Nature 368:844–847.

Gingerich, P. D., N. A. Wells, D. E. Russell, and S. M. I. Shah. 1983. Origin of whales in epicontinental remnant seas: new evidence from the early Eocene of Pakistan. Science 220:403–406.

Graham, B., P. Koch, S. Newsome, K. McMahon, and D. Aurioles. 2010. Using isoscapes to trace the movements and foraging behavior of top predators in oceanic ecosystems; pp. 299–318 in J. B. West, G. J. Bowen, T. E. Dawson, and K. P. Tu (eds.), Isoscapes. Springer, Dordrecht.

Gray, N.-M., K. Kainec, S. Madar, L. Tomko, and S. Wolfe. 2007. Sink or swim? Bone density as a mechanism for buoyancy control in early cetaceans. The Anatomical Record 290:638–653.

Gutstein, C. S., M. A. Cozzuol, and N. D. Pyenson. 2014. The antiquity of riverine adaptations in Iniidae (Cetacea, Odontoceti) documented by a humerus from the Late Miocene of the Ituzaingó Formation, Argentina. The Anatomical Record 297:1096–1102.

Hampe, O., and S. Baszio. 2010. Relative warps meet cladistics: a contribution to the phylogenetic relationships of baleen whales based on landmark analyses of mysticete crania. Bulletin of Geosciences 85:199–218.

Hennig, W. 1965. Phylogenetic systematics. Annual Review of Entomology 10:97–116.

Hooker, S. K., and R. W. Baird. 1999. Deep-diving behaviour of the northern bottlenose whale, *Hyperoodon ampullatus* (Cetacea: Ziphiidae). Proceedings of the Royal Society B 266:671–676.

Hui, C. A. 1981. Seawater consumption and water flux in the common dolphin *Delphinus delphis*. Physiological Zoology 54:430–440.

Lambert, O., G. Bianucci, and B. Beatty. 2014. Bony outgrowths on the jaws of an extinct sperm whale support macroraptorial feeding in several stem physeteroids. Naturwissenschaften 101:517–521.

Lambert, O., G. Bianucci, K. Post, C. de Muizon, R. Salas-Gismondi, M. Urbina, and J. Reumer. 2010. The giant bite of a new raptorial sperm whale from the Miocene epoch of Peru. Nature 466:105–108.

Lee, M. S. Y., J. Soubrier, and G. D. Edgecombe. 2013. Rates of phenotypic and genomic evolution during the Cambrian Explosion. Current Biology 23:1889–1895.

Luo, Z.-X., and P. D. Gingerich. 1999. Terrestrial Mesonychia to aquatic Cetacea: transformation of the basicranium and evolution of hearing in whales. University of Michigan Papers on Paleontology 31:1–98.

Madar, S. 1998. Structural adaptations of early archaeocete long bones; pp. 353–378 in J. G. M. Thewissen (ed.), The Emergence of Whales. Plenum Press, New York.

Marino, L. 2009. Brain size evolution; pp. 149–152 in W. F. Perrin, B. Würsig, and J. G. M. Thewissen (eds.), Encyclopedia of Marine Mammals. Academic Press, Burlington, MA.

Marino, L., R. C. Connor, R. E. Fordyce, L. M. Herman, P. R. Hof, L. Lefebvre, D. Lusseau, B. McCowan, E. A. Nimchinsky, A. A. Pack, L. Rendell, J. S. Reidenberg, D. Reiss, M. D. Uhen, E. Van der Gucht, and H. Whitehead. 2007. Cetaceans have complex brains for complex cognition. PLoS Biology 5:e139.

Marino, L., M. D. Uhen, N. D. Pyenson, and B. Frohlich. 2003. Reconstructing cetacean brain evolution using computed tomography. The Anatomical Record Part B: The New Anatomist 272B:107–117.

Marsh, H., and T. Kasuya. 1986. Evidence for reproductive senescence in female cetaceans. Reports of the International Whaling Commission (Special Issue) 8:57–74.

Marx, F. G., and M. D. Uhen. 2010. Climate, critters, and cetaceans: Cenozoic drivers of the evolution of modern whales. Science 327:993–996.

Matsuoka, K., Y. Fujise, and L. A. Pastene. 1996. A sighting of a large school of the pygmy right whale, *Caperea marginata*, in the southeast Indian Ocean. Marine Mammal Science 12:594–597.

May-Collado, L., I. Agnarsson, and D. Wartzok. 2007. Phylogenetic review of tonal sound production in whales in relation to sociality. BMC Evolutionary Biology 7:136.

Mead, J. G., and R. E. Fordyce. 2009. The therian skull: a lexicon with emphasis on the odontocetes. Smithsonian Contributions to Zoology 627:1–248.

Montgomery, S. H., J. H. Geisler, M. R. McGowen, C. Fox, L. Marino, and J. Gatesy. 2013. The evolutionary history of cetacean brain and body size. Evolution 67:3339–3353.

Morlon, H., T. L. Parsons, and J. B. Plotkin. 2011. Reconciling molecular phylogenies with the fossil record. Proceedings of the National Academy of Sciences 108:16327–16332.

Nicol, S., A. Bowie, S. Jarman, D. Lannuzel, K. M. Meiners, and P. Van Der Merwe. 2010. Southern Ocean iron fertilization by baleen whales and Antarctic krill. Fish and Fisheries 11:203–209.

O'Leary, M. A., J. I. Bloch, J. J. Flynn, T. J. Gaudin, A. Giallombardo, N. P. Giannini, S. L. Goldberg, B. P. Kraatz, Z.-X. Luo, J. Meng, X. Ni, M. J. Novacek, F. A. Perini, Z. S. Randall, G. W. Rougier, E. J. Sargis, M. T. Silcox, N. B. Simmons, M. Spaulding, P. M. Velazco, M. Weksler, J. R. Wible, and A. L. Cirranello. 2013. The placental mammal ancestor and the post–K-Pg radiation of placentals. Science 339:662–667.

Pivorunas, A. 1979. The feeding mechanisms of baleen whales. American Scientist 67:432–440.

Pyenson, N. D., C. S. Gutstein, J. F. Parham, J. P. Le Roux, C. C. Chavarría, H. Little, A. Metallo, V. Rossi, A. M. Valenzuela-Toro, J. Velez-Juarbe, C. M. Santelli, D. R. Rogers, M. A. Cozzuol, and M. E. Suárez. 2014a. Repeated mass strandings of Miocene marine mammals from Atacama Region of Chile point to sudden death at sea. Proceedings of the Royal Society B 281:20133316.

Pyenson, N. D., N. P. Kelley, and J. F. Parham. 2014b. Marine tetrapod macroevolution: physical and biological drivers on 250 Ma of invasions and evolution in ocean ecosystems. Palaeogeography, Palaeoclimatology, Palaeoecology 400:1–8.

Pyenson, N. D., and S. N. Sponberg. 2011. Reconstructing body size in extinct crown Cetacea (Neoceti) using allometry, phylogenetic methods and tests from the fossil record. Journal of Mammalian Evolution 18:269–288.

Pyron, R. A. 2011. Divergence time estimation using fossils as terminal taxa and the origins of Lissamphibia. Systematic Biology 60:466–481.

Rabosky, D. L. 2014. Automatic detection of key innovations, rate shifts, and diversity-dependence on phylogenetic trees. PLoS One 9:e89543.

Racicot, R. A., T. A. Deméré, B. L. Beatty, and R. W. Boessenecker. 2014. Unique feeding morphology in a new prognathous extinct porpoise from the Pliocene of California. Current Biology 24:774–779.

Ramp, C., W. Hagen, P. Palsbøll, M. Bérubé, and R. Sears. 2010. Age-related multi-year associations in female humpback whales (*Megaptera novaeangliae*). Behavioral Ecology and Sociobiology 64:1563–1576.

Rendell, L., and H. Whitehead. 2001. Culture in whales and dolphins. Behavioral and Brain Sciences 24:309–324.

Roe, L. J., J. G. M. Thewissen, J. Quade, J. R. O'Neil, S. Bajpai, A. Sahni, and S. T. Hussain. 1998. Isotopic approaches to understanding the terrestrial-to-marine transition of the earliest cetaceans; pp. 399–422 in J. G. M. Thewissen (ed.), The Emergence of Whales. Plenum Press, New York.

Ronquist, F., S. Klopfstein, L. Vilhelmsen, S. Schulmeister, D. L. Murray, and A. P. Rasnitsyn. 2012. A total-evidence approach to dating with fossils, applied to the early radiation of the Hymenoptera. Systematic Biology 61:973–999.

Sasaki, T., M. Nikaido, S. Wada, T. K. Yamada, Y. Cao, M. Hasegawa, and N. Okada. 2006. *Balaenoptera omurai* is a newly discovered baleen whale that represents an ancient evolutionary lineage. Molecular Phylogenetics and Evolution 41:40–52.

Schorr, G. S., E. A. Falcone, D. J. Moretti, and R. D. Andrews. 2014. First long-term behavioral records from Cuvier's beaked whales (*Ziphius cavirostris*) reveal record-breaking dives. PLoS One 9:e92633.

Simpson, G. G. 1945. The principles of classification and a classification of mammals. Bulletin of the American Museum of Natural History 85:1–350.

Slater, G. J., S. A. Price, F. Santini, and M. E. Alfaro. 2010. Diversity versus disparity and the radiation of modern cetaceans. Proceedings of the Royal Society B 277:3097–3104.

Smith, A. B. 2007. Marine diversity through the Phanerozoic: problems and prospects. Journal of the Geological Society 164:731–745.

Smith, C. R., and A. R. Baco. 2003. Ecology of whale falls at the deep-sea floor. Oceanography and Marine Biology: An Annual Review 41:311–354.

Snively, E., J. M. Fahlke, and R. C. Welsh. 2015. Bone-breaking bite force of *Basilosaurus isis* (Mammalia, Cetacea) from the Late Eocene of Egypt estimated by Finite Element Analysis. PLoS One 10:e0118380.

Spoor, F., S. Bajpai, S. T. Hussain, K. Kumar, and J. G. M. Thewissen. 2002. Vestibular evidence for the evolution of aquatic behaviour in early cetaceans. Nature 417:163–166.

Steeman, M. E., M. B. Hebsgaard, R. E. Fordyce, S. Y. W. Ho, D. L. Rabosky, R. Nielsen, C. Rhabek, H. Glenner, M. V. Sørensen, and E. Willerslev. 2009. Radiation of extant cetaceans driven by restructuring of the oceans. Systematic Biology 58:573–585.

Stern, S. J. 2009. Migration and movement patterns; pp. 726–730 in W. F. Perrin, B. Würsig, and J. G. M. Thewissen (eds.), Encyclopedia of Marine Mammals. Academic Press, Burlington, MA.

Stewart, B. S. 2009. Diving behavior; pp. 321–327 in W. F. Perrin, B. Würsig, and J. G. M. Thewissen (eds.), Encyclopedia of Marine Mammals. Academic Press, Burlington, MA.

Thewissen, J. G. M., L. N. Cooper, M. T. Clementz, S. Bajpai, and B. N. Tiwari. 2007. Whales originated form aquatic artiodactyls in the Eocene epoch of India. Nature 450:1190–1195.

Thewissen, J. G. M., L. J. Roe, J. R. O'Neil, S. T. Hussain, A. Sahni, and S. Bajpai. 1996. Evolution of cetacean osmoregulation. Nature 381:379–380.

Thewissen, J. G. M., J. D. Sensor, M. T. Clementz, and S. Bajpai. 2011. Evolution of dental wear and diet during the origin of whales. Paleobiology 37:655–669.

Thomas, G. H., and R. P. Freckleton. 2012. MOTMOT: models of trait macroevolution on trees. Methods in Ecology and Evolution 3:145–151.

Tomilin, A. G. 1957. Mammals of the U.S.S.R. and Adjacent Countries, vol. 9: Cetacea. Akademii Nauk SSSR, Moscow (translated by the Israel Program for Scientific Translations, Jerusalem, 1967).

Trillmich, F. 2009. Sociobiology; pp. 1047–1053 in W. F. Perrin, B. Würsig, and J. G. M. Thewissen (eds.), Encyclopedia of Marine Mammals. Academic Press, Burlington, MA.

Uhen, M. D. 2010. The origin(s) of whales. Annual Review of Earth and Planetary Sciences 38:189–219.

Uhen, M. D., and N. D. Pyenson. 2007. Diversity estimates, biases and historiographic effects: resolving cetacean diversity in the Tertiary. Palaeontologia Electronica 10:10.2.10A.

Vander Zanden, M. J., and J. B. Rasmussen. 2001. Variation in $\delta^{15}N$ and $\delta^{13}C$ trophic fractionation: implications for aquatic food web studies. Limnology and Oceanography 46:2061–2066.

Walsh, B. M., and A. Berta. 2011. Occipital ossification of balaenopteroid mysticetes. The Anatomical Record 294:391–398.

Weinrich, M. T. 1991. Stable social associations among humpback whales (*Megaptera novaeangliae*) in the southern Gulf of Maine. Canadian Journal of Zoology 69:3012–3019.

Werth, A. J. 2000. Feeding in marine mammals; pp. 487–526 in K. Schwenk (ed.), Feeding: Form, Function and Evolution in Tetrapods. Academic Press, San Diego.

Whitehead, H., B. McGill, and B. Worm. 2008. Diversity of deep-water cetaceans in relation to temperature: implications for ocean warming. Ecology Letters 11:1198–1207.

Wiens, J. J. 2004. The role of morphological data in phylogeny reconstruction. Systematic Biology 53:653–661.

Wiens, J. J. 2009. Paleontology, genomics, and combined-data phylogenetics: can molecular data improve phylogeny estimation for fossil taxa? Systematic Biology 58:87–99.

Willis, J. 2014. Whales maintained a high abundance of krill; both are ecosystem engineers in the Southern Ocean. Marine Ecology Progress Series 513:51–69.

Wills, M. A., D. E. G. Briggs, and R. A. Fortey. 1994. Disparity as an evolutionary index; a comparison of Cambrian and Recent arthropods. Paleobiology 20:93–130.

Wing, S. R., L. Jack, O. Shatova, J. J. Leichter, D. Barr, R. D. Frew, and M. Gault-Ringold. 2014. Seabirds and marine mammals redistribute bioavailable iron in the Southern Ocean. Marine Ecology Progress Series 510:1–13.

Witmer, L. M. 1995. The extant phylogenetic bracket and the importance of reconstructing soft tissues in fossils; pp. 19–33 in J. J. Thomason (ed.), Functional Morphology in Vertebrate Paleontology. Cambridge University Press, New York.

2 Cetacean Fossil Record

2.1 A history of exploration

The history of discovery of fossil whales parallels that of vertebrate paleontology in general (Uhen *et al.*, 2013). Over the past 180 years, the number of publications on fossil cetaceans initially fluctuated around a relatively low mean, but then soared from about 1960 onward (Figure 2.1). There is a slight decrease in output toward the present, but this pattern is likely artificial and only reflects the lag between the publication of the latest papers and their entry into relevant databases. Overall, the rate of scientific discovery regarding fossil cetaceans has never been higher than today, and shows no signs of leveling off.

The earliest description of a fossil cetacean also corresponds to one of the very first contributions to paleontology. In 1670, the Italian **Agostino Scilla** (1629–1700) illustrated a fragmentary squalodontid mandible from Malta in his book *La vana speculazione disingannata dal senso* (Figure 2.2) (Scilla, 1670). Following this anecdotal account, more than 150 years passed before the Frenchman **Georges Cuvier** (1769–1832, Figure 2.3), one of the founders of comparative anatomy and paleontology, provided a more detailed account of a variety of fossil mysticetes and odontocetes from Belgium, France, and Italy in his famous *Recherches sur les ossemens fossiles* (Cuvier, 1823). Not long thereafter, **Richard Harlan** (1796–1843) described the first fossil cetacean from North America (Harlan, 1834). Thinking the remains were those of a large aquatic reptile, he named his discovery *Basilosaurus*, or "king lizard." The mistake was spotted by the famous British paleontologist **Richard Owen** (1804–1892), who recognized the specimen as a cetacean and attempted to change its name to *Zeuglodon* (yoke-tooth) *cetoides* (whale). Though well-meant, the new name violated the priority of Harlan's earlier suggestion, and consequently did not stick: after much discussion and revision, Harlan's animal is today known as *Basilosaurus cetoides*. Besides his contribution to this debate, Owen also described a range of fossil cetaceans from the Pliocene and Pleistocene of Great Britain (Owen, 1870).

Not long after the discovery of *Basilosaurus*, German-born **Albert C. Koch** (1804–1867) roamed across much of the United States collecting fossils, antiquities, and other oddities that he displayed in various "museums" and traveling shows in both the United States and Europe. Despite the fact that later in life he went by the appellation "Dr", Albert Koch was an enthusiastic avocational paleontologist—or, perhaps, a fossil showman. Among his more doubtful projects was the assemblage of a gigantic chimaera of bones, collected mostly in Alabama, into the "sea serpent" *Hydrarchos* (aka *Hydrargos*), which probably included elements of the species now known as

Cetacean Paleobiology, First Edition. Felix G. Marx, Olivier Lambert, and Mark D. Uhen.
© 2016 John Wiley & Sons, Ltd. Published 2016 by John Wiley & Sons, Ltd.

Figure 2.1 Plot of the number of publications regarding fossil cetaceans from 1834 to the present. 1834 is the year that the first fossil whale, *Basilosaurus*, was given a scientific name. Note that the number of publications fluctuates around a low average until 1960, but then increases dramatically. The line represents a 3-year running average. This pattern is similar to that for fossil vertebrates in general (Uhen *et al.*, 2013). Data are from the Paleobiology Database (paleobiodb.org). Drawing of *Pakicetus* adapted from Thewissen *et al.* (2001). Drawings of *Dorudon* and *Maiacetus* adapted from Gingerich *et al.* (2009) under a Creative Commons Attribution license.

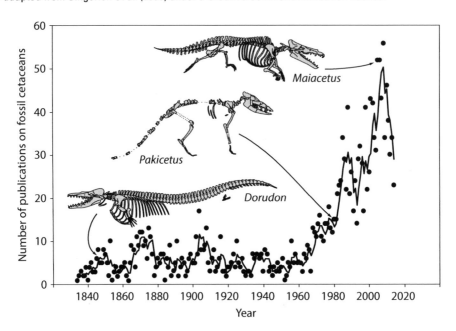

Basilosaurus cetoides, Cynthiacetus maxwelli, and *Zygorhiza kochii* (Figure 2.4) (Koch, 1972). Around the same time, **Robert W. Gibbes** (1809–1866) described the small basilosaurid archaeocete *Dorudon serratus* and, unbeknownst to him at the time, the first deciduous tooth of any cetacean (Gibbes, 1845). Following these early workers, **Joseph Leidy** (1823–1891, Figure 2.5) and **Edward D. Cope** (1840-1897), two of the fathers of vertebrate paleontology in the United States, took up the baton of paleocetology during the second half of the 19th century (Warren, 1998). Between them, they described a series of specimens from the Gulf and East coasts of North America, especially the "Ashley Phosphate Beds" and the Chesapeake Group sediments of Maryland and Virginia.

Meanwhile, in Europe, the German naturalist **Johann Friedrich von Brandt** (1802–1879) had named the first fossil mysticete, *Cetotherium* (Brandt, 1843) and, in the early 1870s, summarized the work on European fossil cetaceans up to that time (Brandt, 1873, 1874). From about 1860

onward, the discovery of vast amounts of Neogene cetacean remains around the city of Antwerp (Belgium) called into action a former pupil of Cuvier, the Belgian zoologist and paleontologist **Pierre-Joseph Van Beneden** (1809–1894; Figure 2.6). After numerous papers on fossil cetaceans from Belgium and other parts of Europe, Van Beneden's efforts culminated in his series *Description des ossements fossiles des environs d'Anvers* (Van Beneden, 1880, 1882) and, in collaboration with the French paleontologist **Paul Gervais** (1816–1879), the exhaustive and beautifully illustrated *Ostéographie des cétacés vivants et fossiles* (Van Beneden and Gervais, 1880). The fossil remains from Antwerp were later reviewed by the Austrian **Othenio Abel** (1875–1946), who also published on a series of fossil cetaceans from across and beyond Europe (Abel, 1905, 1914, 1938).

In Italy, the abundance of the Belgian material was rivalled by a rich, well-preserved record, including several articulated and nearly complete specimens. Early accounts of this material were provided by **Giovanni Capellini** (1833–1922), an enthusiastic supporter of Darwin's theory of evolution and a driving force behind the establishment of the International Geological Congress, **Alessandro Portis** (1853–1931), and **Giorgio Dal Piaz** (1872–1962) (Capellini, 1901, 1904; Dal Piaz, 1916). In Great Britain, the anatomist **William H. Flower** (1831–1899) and the zoologist **John E. Gray** (1800–1875) both worked at the Natural History Museum of London, and between them named the three cetacean suborders: Archaeoceti, named by Flower, in 1883; Mysticeti, by Gray, in 1864; and Odontoceti, by Flower, in 1867.

Besides being a time of abundant fossil discoveries in Europe and North America, the late 19th and early 20th centuries also saw the study of cetacean evolution spread across the globe in earnest. Early work on fossil cetaceans from South America prominently featured studies by the Argentine zoologists **Karl Hermann K. Burmeister** (1807–1892) and **Ángel Cabrera** (1879–1960), as well as the British naturalist **Richard Lydekker** (1849-1915) (Cozzuol, 1996). On the other side of the Pacific, **Frederick McCoy** (1817–1899) and **Thomas H. Huxley** (1825–1895), the latter also known as "Darwin's bulldog," described some of the earliest discovered specimens from Australia and New Zealand, respectively, with later contributions by **James Hector** (1834–1907), **Theodore T. Flynn** (1858–1915), and **William B. Benham** (1860–1950) (Fitzgerald, 2004; Fordyce, 1980). Further north, the first Japanese material was brought to the

attention of the international scientific community by **Hikoshichiro Matsumoto** (1887–1975) (Matsumoto, 1926). Finally, in Africa, the German paleontologists **Ernst Stromer von Reichenbach** (1870–1952) and **Eberhard Fraas** (1862–1915) collected and described some of the first specimens from Egypt, including the first protocetid archaeocete. Around the same time, and also in Egypt, the British geologist **Hugh J. L. Beadnell** (1874–1944) discovered the now iconic locality of Wadi Al-Hitan ("Valley of Whales") (Gingerich, 2010).

The modern era of cetacean paleontology started with the works of **Frederick W. True** (1858–1914), who was Curator of Mammals at the US National Museum of Natural History (USNM), and his successor **Remington Kellogg** (1892–1969; Figure 2.7). Their collection efforts contributed much of the vast holdings of extant and fossil cetaceans at the USNM, and resulted in dozens of papers on fossil and recent marine mammals—including Kellogg's (1936) seminal *A Review of the*

Figure 2.3 Portrait of Georges Cuvier. Line engraving by C. Lorichon, 1826, after N. Jacques. Reproduced under a Creative Commons Attribution license. Courtesy of Wellcome Library, London.

Figure 2.2 Fragmentary squalodontid-like mandible from Malta, as figured by Scilla (1670). This drawing represents the earliest scientific depiction of a fossil cetacean.

Figure 2.4 Poster used by Albert C. Koch to promote his display of the "sea serpent" *Hydrarchos*, which is actually a chimaera of several individuals of several species of archaeocete whales. Digital image from the Library of Congress, Rare Book and Special Collections Division, digital ID rbpe 11903300 http://hdl.loc.gov/loc.rbc/rbpe.11903300.

Archaeoceti, three monographs on the taxonomy of extant cetaceans (True, 1889, 1904, 1910), and a series of papers on a well-preserved cetacean assemblage from the Chesapeake Bay area on the US East Coast (Whitmore, 1975). Late in his career, Kellogg overlapped with the enthusiastic fossil collector **Douglas R. Emlong** (1942–1980), who contributed hundreds of fossil vertebrates from the north-western coast of the United States to the collections of the USNM. Despite his intense collection efforts, Emlong published just a single scientific paper. In it, he named the early toothed mysticete *Aetiocetus cotylalveus*, which proved to be one of the most important discoveries in the study of baleen whale evolution (Emlong, 1966).

2.2 Strengths and weaknesses of the cetacean fossil record

2.2.1 Preservation potential

The deep-time perspective that paleontology contributes to the study of evolution is both its greatest strength and its greatest weakness. Fossils provide a direct window into the process of evolution, rather than just its outcome, and allow the investigation of long-term trends and interactions both between different organisms and between life and the environment as a whole. However, being tied to fossils also means that paleontology is subject to the vagaries of their preservation. Whether a dead organism is turned into a fossil depends on a myriad of factors, such as the preservation potential of the body itself (hard vs soft tissues), the depositional environment, temperature, water chemistry, scavenging, and time before burial.

Cetacean fossils are 5–8 times more abundant in siliciclastic environments than in carbonate ones, even when counting carbonate and mixed

Figure 2.6 Portrait of Pierre-Joseph Van Beneden, holding a sketch of the skeleton of the southern right whale *Eubalaena australis*. In its final form, the drawing would eventually appear in his famous *Ostéographie des cétacés vivants et fossiles*, co-authored with P. Gervais (Van Beneden and Gervais, 1880). Portrait by E. Broerman. Digital image of the KIK-IRPA (Royal Institute for Cultural Heritage), Brussels.

Figure 2.5 Portrait of Joseph Leidy by L. E. Faber. Reproduced under a Creative Commons Attribution license. Courtesy of Wellcome Library, London.

Figure 2.7 Portrait of Remington Kellogg, holding a skull of the extant Amazon river dolphin *Inia*. Courtesy of Smithsonian Institution.

environments together (Uhen, 2015). Cetaceans are both distinctive and large, even for mammals, and thus hard to overlook in deposits in which they occur. Certain parts of their skeleton, such as the bones surrounding the inner and middle ear, are highly mineralized and tend to preserve well even under adverse conditions. Many cetaceans inhabit, or at least frequently approach, marginal marine environments, which affects the chances of their preservation as fossils in various ways. Carcasses deposited in shallow water are prone to refloating, because the ambient water pressure is not enough to counteract the buildup of decompositional gas within the body cavity. Once at the surface, currents will transport the carcass until enough gas has escaped to allow the remains to sink to the seafloor permanently. During this process, some of the skeleton may disarticulate en route, resulting in a string of isolated, often largely non-diagnostic elements (Schäfer, 1972).

After final deposition, preservation in shallow water may be aided by an abundant sediment supply from nearby coastlines or the deposition of mineralized microorganisms in areas of high productivity, both of which are capable of rapidly burying any remains. Where such a supply is absent, the chances and quality of fossilization depend on the characteristics of the depositional setting: in low-energy environments, such as deep or restricted shallow waters, the durable nature, and large weight of cetacean bones may still cause them to be preserved in articulation (Peters *et al.*, 2009). By contrast, deposition in high-energy environments often leads to partial or complete disarticulation, as well as the fracturing and reworking of individual elements; if there is a protracted period of low sediment input, a **bone bed** may form (Pyenson *et al.*, 2009). In either case, prolonged exposure on the sea floor may lead to intense **scavenging** and **bioerosion** (colonization and degradation of the bone by boring organisms) (section 6.2.2). Alongside **bioturbation** (reworking of sediments by burrowers), these processes can severely compromise the articulation and, in the case of bioerosion, structural integrity of the bones.

2.2.2 Biases affecting fossil recovery

Once fossilized, a specimen may still be damaged or even destroyed by tectonic forces, either directly by being crushed or through exposure and subsequent erosion. Frustratingly, the factors determining preservation constantly vary—not only geographically but also over time. Changes in sea level; climate; and the shape, arrangement, and topography of the tectonic plates all may raise or lower the chances of fossil preservation on a grand scale. For example, higher sea levels may inundate vast tracts of low-lying continental areas, thus creating wide, shallow seas and long coastlines. The latter, in turn, provide more opportunities for rocks and fossils to form (Raup, 1972; Smith, 2007). Conversely, a drop in sea level not only causes shallow seas to retreat, but also exposes marine rocks already formed to large-scale erosion, thus creating a temporal gap in fossil preservation. Such drops likely affected the cetacean record (Fordyce, 2003), for example, during the Early Oligocene and possibly again during the Early Miocene—two of the least known periods in cetacean evolution (section 7.1).

Assuming that preservation of a particular fossil has occurred, the chances that it will ever be recovered still vary both globally and regionally. This is because the likelihood of a fossil being found heavily depends on the accessibility of fossil-bearing rocks, as well as the means and willingness to study them (Smith, 2007). From the standpoint of physical geography, dense vegetation cover may obscure suitable outcrops, while particular climates may hasten the erosion of exposed material. From a human perspective, greater population densities may lead to more chance discoveries of fossils, or the collecting habits of particularly prolific workers might bias the record toward particular areas or time intervals (Uhen and Pyenson, 2007). In addition, affluent areas tend to sustain more researchers and interested laymen, thus resulting in a relatively greater local research effort. This is clearly evident in the global distribution of the fossil cetacean material discovered so far, which is heavily concentrated in Europe, Japan, and the United States, as well as, to a somewhat lesser degree, Australia and New Zealand (Figure 2.8).

Last, but not least, a recovered fossil is also subject to varying taxonomic practices. The beginning of paleocetology, like that of paleontology in general, saw many specimens described and named that today would usually be considered non-diagnostic (e.g., Steeman, 2010). On top of these changing norms, taxonomic practices vary even today between workers with rather different ideas as to what constitutes a species or a genus (Smith, 2007). Most species of extinct cetaceans are based only on the holotype, or else a very small number of specimens. From a biological point of view, this situation is highly unrealistic, and likely reflects (1) a genuine scarcity of identifiable material, (2) a tendency to over-split the fossil record into too many species, and (3) a lack of studies formally referring fragmentary or subsequently discovered specimens to existing taxa.

With the fossil record being subject to so many **geological, geographical**, and **human biases**, one of the biggest challenges facing paleontologists is to disentangle their effects from genuine biological signals. This is not a trivial task, especially since the very same processes that can bias the record (**biases hypothesis**) can also influence organisms in a direct, biological way. For example, higher sea levels not only produce more rock, but also create a large number of semi-connected shallow-water habitats. Such variable environments in turn may promote speciation, thus conflating heightened

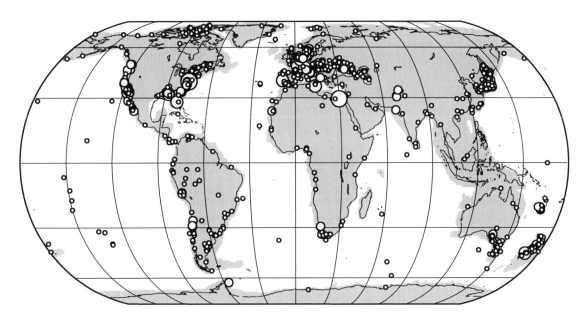

Figure 2.8 Global cetacean collections, with the size of the circles being proportional to the number of collections from each area. Note the obvious bias toward Europe, North America, Japan, New Zealand, and Australia. Data are from the Paleobiology Database (paleobiodb.org). Plotting software by J. Alroy.

preservation potential with a real increase in biological diversity (**common cause hypothesis**) (Peters, 2005; Smith, 2007). In spite of these problems, there are three lines of evidence which indicate that our understanding of the cetacean fossil record has substantially improved over time.

First, increased sampling of the oldest known cetaceans (archaeocetes) has led to well-constrained estimates for the time of origin of the group as a whole. Figure 2.9 illustrates how the "best estimate" for this event has evolved over historical time. Successive discoveries of ever-older specimens have pushed back the supposed origin of cetaceans further and further into the past, from circa 37 Ma in 1840 to around 53 Ma today. However, the more material is added to the record (i.e., the more **sampling density** increases) the more confident we can be that the oldest known

fossil is indeed close to the actual time of origin of the group (Gingerich and Uhen, 1998). The current estimate, based on *Himalayacetus* from northern India, is associated with an almost negligible confidence interval, and corresponds well with recent molecular clock estimates of the split between cetaceans and their proposed closest living relatives, the hippopotamids (Gatesy, 2009; Zhou *et al.*, 2011).

Second, better knowledge of the cetacean fossil record has made it increasingly hard to discover previously unknown genera. Since 1820, the number of fossil cetacean genera has grown steadily by about 20 genera per decade. At the same time, however, the number of occurrences has grown at a much higher rate, approximating an exponential growth curve (Figure 2.10). This shows that, at least statistically speaking, researchers

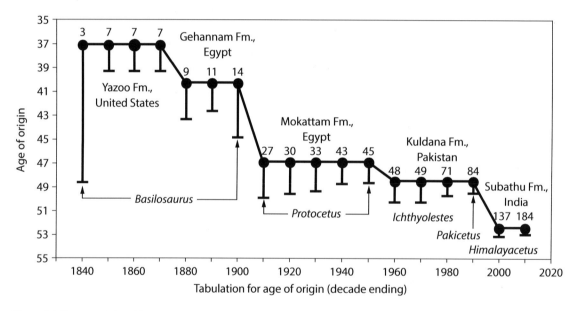

Figure 2.9 The age of the oldest cetacean fossil over historical time, along with a measure of how much the oldest fossil age might underestimate the actual origin of whales. Circles represent the oldest fossil whales known at a particular time in the historical past, with their ages based on modern interpretations of the stratigraphic record. Their names and the formations from which they originated are shown on either side of the graph. Jumps in the age line indicate discoveries of older fossil whales. The vertical black bars represent a 95% confidence interval on the time of origin of Cetacea at each decade in the past (Gingerich and Uhen, 1998). Numbers above the circles indicate the number of archaeocete fossil horizons used in each calculation. Note the large confidence interval in the first decade (1840), and its tendency to shrink toward the next jump in the age line as the number of fossil horizons increases. 1900 is an exception caused by the discovery of a much younger fossil than the previously youngest archaeocete. Also note the almost negligible confidence interval on the present decade. Adapted from Uhen (2010).

need to collect a considerable number of specimens today in order to uncover something genuinely unknown, whereas in the past the discovery rate of new genera was much higher. It is important to note that this effect may also reflect changed taxonomic practices, with modern paleontologists generally being more cautious about the creation of new taxa than their 19th-century counterparts. As is so often the case, the truth most likely lies somewhere in between.

Last, a series of studies directly compared the diversity (number of genera) of fossil cetaceans through time with proxy measures of rock abundance or exposure, such as the number of formations or outcrop area (Marx, 2009; Marx and Uhen, 2010; Uhen and Pyenson, 2007). To achieve this, they mostly relied on a comprehensive archive of marine mammal fossil data stored in the **Paleobiology Database** (www.paleobiodb.org). None of these analyses found strong evidence for a correlation between rock abundance and diversity, as would have had to exist if the cetacean record were indeed primarily driven by varying preservation potential. Nevertheless, they did

point out particular intervals (e.g., the Early Oligocene and the Early Miocene) in which cetaceans are clearly undersampled, and diversity counts thus unreliable (Marx and Fordyce, 2015; Uhen and Pyenson, 2007).

Other potential biasing factors, such as population size and collecting habits, also seem to have little effect, with the single exception of research effort. The latter is measured in terms of the number of publications on fossil cetaceans, and clearly correlated with diversity counts (Uhen and Pyenson, 2007). There is no easy way of determining whether the research interests of particular scientists influence which specimens are collected and described, or whether paleontologists simply study what is available in the first place. It does seem, however, that certain areas and collections (e.g., the Oligocene material from Oregon, United States, collected by D. R. Emlong) have received less attention than their scope and quality would warrant—possibly because of the high initial investment of having to prepare difficult material, or the availability of abundant and more easily accessed material from other localities.

2.2.3 Outlook

Several issues affecting the reliability of the cetacean fossil record persist and will need to be addressed as further specimens are discovered and described. For example, there is a good chance that wider geographical coverage and the eventual exploitation of "underappreciated" collections will have a considerable impact on our current understanding of cetacean diversity—especially as regards such undersampled periods as the Early Oligocene and the Early Miocene (Marx and Fordyce, 2015; Uhen and Pyenson, 2007). Previous models testing the effects of geological biases on past cetacean diversity may have been too simplistic in using outcrop area or formation number as a sampling proxy (Dunhill *et al.*, 2014), and the question remains whether, and to what degree, the bias and common cause hypotheses apply to the cetacean record. Taxonomic practice still largely depends on the individual researcher, and much potentially referable material remains undescribed, keeping the number of specimens per taxon artificially low. It is partly because of

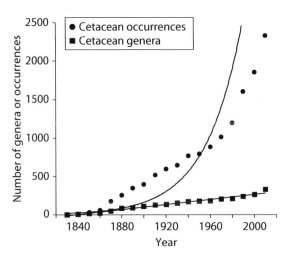

Figure 2.10 Cetacean fossil record growth curves. Plot of the number of occurrences and the number of genera known exclusively as fossils over historical time. Note that the number of genera has increased linearly over historical time (R^2=0.972), while the number of occurrences has grown at a much higher rate, approximating exponential growth (R^2=0.774).

this that no one has yet attempted to correct the cetacean record for potential sampling biases by using (rather data-hungry) subsampling techniques, which have featured prominently in the analysis of much larger, global datasets of animal diversity (e.g., Alroy, 2010).

Nevertheless, on the whole, the outlook for paleocetology is positive: research output is booming (Figure 2.1), the cetacean record is becoming demonstrably better known (Figures 2.9 and 2.10), associated stratigraphic and age data are becoming more refined (e.g., Marx and Fordyce, 2015), and new taxa are constantly being described, including from areas that have historically been undersampled. Of particular note here are highly fossiliferous localities located along the western coast of South America (Peru and Chile), Australasia, and parts of Africa outside Egypt (section 2.3) (Brand *et al.*, 2004; Fordyce, 2006; Gingerich, 2010; Pyenson *et al.*, 2014). These new developments are on track to fill the major geographical gaps that still plague the cetacean record (Figure 2.8) and raise hopes for a relatively comprehensive, global sample of fossil cetaceans within the next few decades. Comparisons of

modern cetacean stranding data with surveys of live animals at sea show that death assemblages can faithfully record the diversity and even relative abundances of a particular region. The same, with some caveats, may be true for temporally well-constrained fossil accumulations, thus opening a window into the structure of ancient cetacean communities (Pyenson, 2010, 2011).

2.3 Major fossil localities

This section presents an overview of some of the major localities, grouped by ocean basin, that have yielded either abundant or particularly important material (Figure 2.11). The list is by necessity incomplete, but should provide a flavor of which areas have shaped our current understanding of cetacean evolution the most. The stratigraphic ranges for all localities are shown in Figure 2.12.

2.3.1 Tethys

Three areas of Indo-Pakistan have produced early whales in great abundance. These are **Khyber Pakhtunkhwa** (formerly known as the North-West

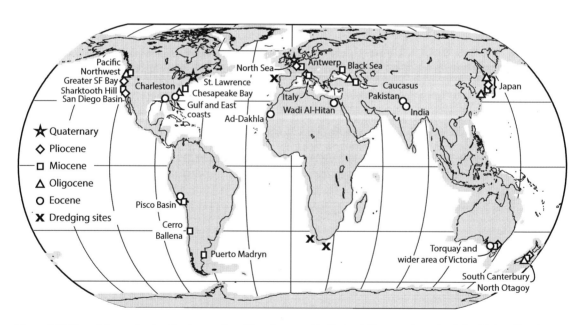

Figure 2.11 Map of the major fossil cetacean localities discussed in the text. Data are from the Paleobiology Database (paleobiodb.org). Plotting software by J. Alroy.

Figure 2.12 Ages of the major fossil localities discussed in the text and shown in Figure 2.11.

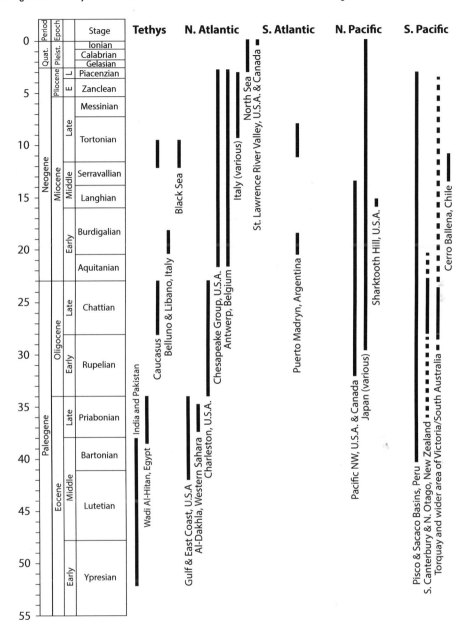

Frontier Province) and the **Sulaiman Mountains** of Pakistan, and the **Kutch** area of India, all of which expose claystones, siltstones, and limestones that were deposited in the **Tethys**—the shallow sea that originally separated India from Asia proper. The oldest rocks in the region that contain fossil whales are Early Eocene (Ypresian) in age, and belong to the Kuldana and Kohat formations in Pakistan, and the Subathu and Panandhro formations in India. These are followed by younger Lutetian and Bartonian layers forming part of the Kohat, Habib Rahi, Domanda, and Drazinda formations in Pakistan, as well as the Harudi Formation in India (Gingerich, 2003; Gingerich et al., 2001b). Between them, these units have produced not only all of the oldest and most basal known cetaceans, the pakicetids, but also members of all of the other archaeocete families, including ambulocetids, remingtonocetids, several protocetids, and the basilosaurids *Basilosaurus* and *Basiloterus* (sections 5.1 and 5.4.1) (Gingerich et al., 1997, 2001a, 2001b, 2009; Williams, 1998). With the exception of a single undescribed remingtonocetid from Egypt, all pakicetids, ambulocetids, and remingtonocetids are known exclusively from Indo-Pakistan.

Further west within the province of the ancient Tethys lies **Wadi Al-Hitan** ("Valley of Whales"), an area within the Fayum Province of Egypt that for many decades has been famous for producing abundant remains of Eocene archaeocetes (Gingerich, 1992; King et al., 2014; Peters et al., 2009). The cetacean-bearing rocks at this locality mostly belong to the latest Bartonian–Priabonian Gehannam, Birket Qarun, and Qasr el-Sagha formations (Underwood et al., 2011). In addition, a single remingtonocetid has been recovered from the older, late Lutetian–early Bartonian Midawara Formation that is exposed nearby (Bebej et al., 2012). Wadi Al-Hitan is especially famous for its rich assemblage of basilosaurids, including *Ancalecetus*, *Basilosaurus*, *Dorudon* (section 5.4.1), *Masracetus*, *Saghacetus*, and *Stromerius* (Gingerich, 2010). A protocetid was recently reported from the lower Gehannam Formation, making this the first clear co-occurrence of protocetids and basilosaurids, and the first record of a protocetid from the Late Eocene (Gingerich et al., 2014).

Despite nearly two centuries of paleocetological research in Europe, the Oligocene of both the Tethys and the eastern North Atlantic is rather poorly known. Some of the most significant specimens come from the Late Oligocene of the **Caucasus** region and include mostly archaic odontocetes, such as mirocetids, possible eurhinodelphinids, and the oldest known potential physeteroid, *Ferecetotherium* (Mchedlidze, 1964, 1970; Sanders and Geisler, 2015). Unfortunately, most of these specimens are poorly preserved and remain obscure because of language barriers, access issues, and the lack of high-quality illustrations. In the absence of a good Oligocene record, one of the best sites to trace the early evolution of Tethyan neocetes is the area surrounding the north-eastern Italian town of **Belluno** and the neighboring village of **Libano**, which exposes the Early Miocene (likely Burdigalian) Libano Sandstone. This unit represents a molasse deposited under warm climatic conditions at the foot of the Alps, and has produced a large collection of well-preserved archaic odontocetes, including potential dalpiazinids, eoplatanistids, eurhinodelphinids, physeteroids, squalodelphinids, and squalodontids (Bianucci and Landini, 2002; Pilleri, 1985). Surprisingly, no remains of any other kind of marine mammal have ever been found here.

Continued northward movements of India and Africa finally led to the closure of the Tethys by the Early–Middle Miocene, leaving behind only the Mediterranean and a vast inland sea covering large parts of Central and Eastern Europe—the **Paratethys**. The final stage of the "Tethyan" cetacean fauna that survived in the Paratethys is preserved in Late Miocene (Tortonian) deposits located in the Caucasus and around the northern and eastern shores of the **Black Sea**—especially in the area around Maykop, on the Taman and Crimean peninsulas and near Odessa. The fossiliferous strata mostly belong to the Blinovo, Chersonian, and Krasnooktyabr'sk formations, which have yielded a variety of seemingly endemic cetotheriids, including *Cetotherium* itself (Brandt, 1873; Gol'din and Startsev, 2014; Gol'din et al., 2014; Tarasenko, 2014; Tarasenko and Lopatin, 2012).

2.3.2 North Atlantic

The cetacean history of the eastern North Atlantic starts with a Late Eocene basilosaurid assemblage from the Samlat Formation exposed near **Ad-Dakhla** in the Western Sahara (Plate 14a), which is very similar in composition to that from Wadi Al-Hitan (section 2.3.1) (Zouhri et al., 2014). In the western North Atlantic, all of the archaeocetes so far described come from Eocene deposits found along the **East** and **Gulf coasts** of the United States (Kellogg, 1936; Uhen, 2013). The East Coast sediments, exposed on the US Coastal Plain stretching from New Jersey to Florida, comprise mostly carbonates and glauconitic sands belonging to (1) the Lutetian and Bartonian Shark River, Piney Point, Castle Hayne, Blue Bluff, and Santee formations, all of which have produced either rare protocetids and/or specimens of the basal basilosaurid *Basilotritus*; and (2) the Priabonian Harleyville, Tupelo Bay, Twiggs Clay, Clinchfield, and Sandersville formations, which have yielded only basilosaurids (Uhen, 2013; Williams, 1998). On the Gulf Coast, marine shelf deposits are found from Texas to Florida and along the Mississippi River Valley up to Arkansas and Tennessee. Protocetids are known from the late Lutetian and Bartonian Yegua, Cook Mountain, and Lisbon formations, whereas the Priabonian White Bluff, Ocala, Crystal River, and Yazoo formations have produced an abundance of basilosaurids. This is particularly true for the Pachuta Member of the Yazoo Formation, which in the past was informally known as the "*Zeuglodon* bed" (Uhen, 2013).

The earliest evolution of neocetes in the North Atlantic is captured on the East Coast of the United States near **Charleston**, South Carolina, by Oligocene rocks belonging to the Ashley, Chandler Bridge, and Tiger Leap formations. These units have produced well-preserved toothed mysticetes and eomysticetids (Barnes and Sanders, 1996; Sanders and Barnes, 2002a,b), as well as a diverse odontocete assemblage including agorophiids, ashleycetids, patriocetids, and xenorophids, such as *Cotylocara* (sections 5.2 and 5.4) (Allen, 1921; Dubrovo and Sanders, 2000; Geisler *et al.*, 2014; Sanders and Geisler, 2015). Following the specimens from Charleston, the sandstones, claystones, siltstones, and diatomites of the wider **Chesapeake Bay** tell the story of North Atlantic cetaceans from the Early Miocene (Aquitanian) to the Pliocene (Piacenzian). In ascending order, these deposits belong to the Calvert (Plate 13b), Choptank, St Mary's, Eastover, and Yorktown formations, with the latter also being prominently exposed further south at the **Lee Creek phosphate mine** in Aurora, North Carolina (Browning *et al.*, 2009; Whitmore and Kaltenbach, 2008). The composition of the fossil assemblage changes through time: whereas the Calvert and Choptank formations are dominated by relatively archaic forms, such as stem balaenopteroids, eurhinodelphinids, kentriodontids, pomatodelphinine platanistids, squalodontids, physeteroids, and rare ziphiids (e.g.,

Kellogg, 1924, 1955, 1965a,b), the upper portion of the sequence (especially the Yorktown Formation) has produced mostly modern-looking taxa, including balaenids, balaenopterids, cetotheriids, delphinids, eschrichtiids, kogiids, monodontids, physeterids, pontoporiids, and ziphiids (Whitmore and Barnes, 2008; Whitmore and Kaltenbach, 2008).

On the European side of the North Atlantic, excavations near the Belgian city of **Antwerp**, carried out mostly during the 19th century, have revealed the richest cetacean assemblage anywhere on the continent. The sediments exposed here range from the Early Miocene (late Aquitanian) to the Pliocene (Piacenzian) and, in ascending order, comprise the Berchem, Diest, Kattendijk, and Lillo formations (De Schepper *et al.*, 2009; Louwye *et al.*, 2010). They consist of glauconitic sands with clayey intercalations and a few basal gravel layers, indicative of discontinuous sedimentation and repeated incursions of oceanic waters into a relatively confined area. Erosion and reworking of sediments during these incursions partly explain why the vast majority of the local fossils are highly fragmentary and dissociated. Random collecting habits by 19th-century construction workers likely also contributed to this situation, and account for the lack of precise stratigraphic data for much of the material. Nevertheless, the fossil assemblage is extremely rich and has inspired scientific studies for nearly 200 years—starting with Cuvier (1823) and leading to a series of well-known publications by Van Beneden and, later, Abel (e.g., Abel, 1905, 1938; Van Beneden, 1880, 1882). Today, the list of taxa known to occur at Antwerp includes balaenids, balaenopterids, and cetotheriids (all mysticetes), as well as eurhinodelphinids, kentriodontids, monodontids, phocoenids, physeteroids, platanistids, pontoporiids, squalodontids, and ziphiids (Bisconti *et al.*, 2013; Lambert, 2006, 2008a,b; Lambert and Gigase, 2007; Steeman, 2010).

The appearance and early evolution of many modern taxa can be traced at not only Antwerp, but also a series of fossiliferous Late Miocene and Pliocene localities scattered across **Italy**—for example, the Zanclean–Piacenzian Sabbie d'Asti Formation exposed near Turin, which has produced well-preserved balaenids, balaenopterids, eschrichtiids, and delphinids (Bianucci, 2013; Bisconti, 2000, 2008). By contrast, the final stage of cetacean evolution, the establishment of the modern fauna, is surprisingly difficult to trace, because many of the most recent (Pleistocene) marine deposits were inundated when sea levels rose at the end of the last glacial period. In Europe, Pleistocene (including Gelasian) sediments that formed in the area of the **North Sea** can be found in Belgium, the Netherlands, and the United Kingdom, as well as on the current sea floor itself. Early Pleistocene deposits from this area have yielded a diverse cetacean assemblage including rorquals (*Balaenoptera*) and the right whale *Eubalaena*, in addition to various extant and extinct delphinids (*Globicephala*, *Hemisyntrachelus*, *Orcinus*, *Platalearostrum*, ?*Stenella*, and *Tursiops*), the beluga *Delphinapterus*, the beaked whale *Mesoplodon*, and the giant sperm whale *Physeter* (Post and Bosselaers, 2005; Post and Kompanje, 2010).

Late Pleistocene faunas from the North Sea appear to be more restricted, and include only *Delphinapterus*, rare balaenopterids, and representatives of the (now extinct) North Atlantic population of the gray whale *Eschrichtius* (Post, 2005). This is roughly in accord with the fauna recovered from the **St Lawrence River Valley** in eastern Canada, which was briefly inundated by the Atlantic Ocean around 15 000 years ago, thus forming the **Champlain Sea**. The cetacean assemblage from this area is entirely modern-looking and dominated (ca 80%) by the remains of the living beluga (*Delphinapterus*), with balaenopterids (*Balaenoptera* and *Megaptera*), balaenids (*Balaena*), and phocoenids (*Phocoena*) making up the remainder (Harington, 1977; Harington *et al.*, 2006).

2.3.3 South Atlantic

The history of cetaceans in the South Atlantic is extremely poorly known, and largely based on specimens from the Early Miocene (Burdigalian) Gaiman and the Late Miocene (Tortonian) Puerto Madryn formations, both exposed near **Puerto Madryn** in the province of Chubut, Patagonia, Argentina. To date, this site has yielded balaenids, stem balaenopteroids and a possible fossil relative of the pygmy right whale *Caperea*, as well as eurhinodelphinids, kentriodontids, the problematic *Prosqualodon*, squalodelphinids, squalodontids,

stem physeteroids, and ziphiids (Buono and Cozzuol, 2013; Buono *et al.*, 2014; Cabrera, 1926; Cozzuol, 1996; Lydekker, 1894). Some of the fossils from Chubut represent the oldest members of their particular lineages, such as the archaic balaenid *Morenocetus*, and have thus been crucial in establishing at what times certain groups of modern cetaceans first diverged from each other (section 4.5).

2.3.4 North Pacific

Cetacean evolution in the western North Pacific is well documented by numerous localities in **Japan**. Overall, the country boasts close to 100 localities ranging from the Early Oligocene all the way to the Pleistocene. Unlike in many other parts of the world, however, these collections are scattered across the country and not concentrated in one particular area (Oishi and Hasegawa, 1995; Uhen, 2015). Especially noteworthy among the many localities are **Ashoro** (Hokkaido; Morawan Formation) and **northern Kyushu** (Ashiya Group), both of which have produced a wealth of Rupelian and Chattian material including archaic toothed mysticetes, exquisitely preserved eomysticetids, and agorophiids (Barnes *et al.*, 1995; Oishi and Hasegawa, 1995; Okazaki, 2012). In addition, Middle Miocene deposits belonging to the Bessho Formation exposed in **Nagano** Prefecture (central Honshu) have produced some of the oldest specimens potentially referable to Balaenopteridae and Delphinidae, as well as physeteroids and ziphiids (Kohno *et al.*, 2007). Finally, the Early Pliocene (Zanclean) Tatsunokuchi Formation has yielded abundant and well-preserved material of the seemingly globally distributed cetotheriid *Herpetocetus*, mainly from **Iwate** Prefecture (north-eastern Honshu) (Bouetel and de Muizon, 2006; Kimura and Hasegawa, 2004; Oishi and Hasegawa, 1995).

In the eastern North Pacific, the period spanning the latest Eocene–Early Miocene (Burdigalian) is represented in the states forming the **Pacific Northwest** (Oregon and Washington in the United States and British Columbia in Canada) by the Alsea, Astoria, Clallam, Lincoln Creek, Makah, Nye Mudstone, Pysht, Sooke, and Yaquina formations (Prothero, 2001). These units, at least some of which were deposited in relatively deep water (Goedert *et al.*, 1995), have yielded several kinds of (mostly undescribed) toothed and baleen-bearing mysticetes, including the type species and genus of Aetiocetidae, *Aetiocetus cotylalveus* (sections 5.2.1 and 5.4.2) (Barnes *et al.*, 1995; Emlong, 1966; Goedert *et al.*, 2007). In addition, they have been the source of several archaic odontocetes, including *Simocetus* (section 5.4.3), the ziphiid-like *Squaloziphius*, and the reportedly oldest and most archaic odontocetes discovered to date (Barnes *et al.*, 2001; de Muizon, 1991; Fordyce, 2002). Its age and duration give the assemblage from the Pacific Northwest a special significance, as it provides one of very few windows into the otherwise extremely poorly known Rupelian and Aquitanian stages.

South of Washington and Oregon, an abundance of fossils has been collected from the Middle Miocene (Langhian) Round Mountain Silt exposed at **Sharktooth Hill** in Kern County, California, United States (Plate 13a)—so named because the hill looks like a shark tooth in profile. Hanna (1925: p. 72) considered the Sharktooth Hill bone bed to be "the most valuable deposit of marine vertebrates thus far discovered in western North America." The fossiliferous deposits form part of the Sharktooth Hill bone bed, which accumulated over roughly 700 000 years and concentrates the mostly dissociated remains of a variety of sharks, bony fishes, turtles, some land mammals, pinnipeds, and, of course, cetaceans—but no large invertebrates or microfossils (Pyenson *et al.*, 2009). The cetacean assemblage includes stem balaenids and balaenopteroids, as well as allodelphinids, kentriodontids, physeterids, and pontoporiids (Barnes *et al.*, 2015; Kellogg, 1931).

The Pliocene history of eastern North Pacific cetaceans is well captured by the Purisima and San Diego formations, exposed in the **Greater San Francisco Bay** area and within the **San Diego Basin** (both in California, United States), respectively. Both have yielded similar cetacean assemblages, including balaenids, balaenopterids, and herpetocetine cetotheriids, as well as albireonids (Purisima only), delphinids, lipotids, monodontids, phocoenids, and physeteroids (Boessenecker, 2013a). As elsewhere, however, the final episode of cetacean evolution during the Pleistocene is difficult to trace, owing to the scarcity of Pleistocene deposits. A single herpetocetid cranium, the youngest described to date, has been recovered from the Pleistocene Falor Formation exposed in northern California (Boessenecker, 2013b). In addition, the top of the San Diego Formation reaches the Gelasian, and thus may be broadly comparable with Early Pleistocene sediments from Europe (section 2.3.2). Otherwise, Pleistocene data from the US West Coast are sparse (Boessenecker, 2013a).

2.3.5 South Pacific

One of the best records of fossil cetaceans from the Southern Hemisphere comes from a series of Middle Eocene–Early Miocene rocks exposed in the **South Canterbury** and **North Otago** regions of the South Island of New Zealand (Plate 14b). The core of the fossiliferous deposits is formed by the Late Oligocene Kokoamu Greensand and the Otekaike Limestone, found mainly along the valleys of the Waitaki River and its Hakataramea and Maerewhenua tributaries. Other important units within the wider area include the Middle Eocene Waihao Greensand, the Early Oligocene (Rupelian) Ototara Limestone, and the Early Miocene (Aquitanian) Mount Harris Formation (Fordyce, 2006, 2008). Together, these formations have yielded a diverse cetacean assemblage, including: basilosaurid and kekenodontid archaeocetes (Clementz *et al.*, 2014; Köhler and Fordyce, 1997); archaic toothed mysticetes, including a possible llanocetid and a mammalodontid (Fordyce, 2003; Fordyce and Marx, 2011); baleen-bearing mysticetes, including a range of eomysticetids and stem balaenopteroids (Boessenecker and Fordyce, 2015a,b,c; Steeman, 2007; Tsai and Fordyce, 2015); and a variety of odontocetes, including potential dalpiazinids, squalodontids, stem delphinoids, and waipatiids, such as the iconic *Waipatia* (section 5.4.3) (Clementz *et al.*, 2014; Fordyce, 1994; Tanaka and Fordyce, 2014, 2015).

Much of the New Zealand material has not yet been formally described, and even more awaits final preparation and announcement. Beyond the central South Island, cetacean fossils ranging from the Oligocene all the way to the Pleistocene are known from other, as yet little explored parts of New Zealand (e.g., Aguirre-Fernández and Fordyce, 2014; Bearlin, 1987), which thus promises to remain productive for a long time to come. Across the Tasman

Sea, well-preserved Late Oligocene specimens have also been found in the vicinity of **Torquay**, in Victoria, Australia. Here, the Jan Juc Marl has yielded most of the known material of mammalodontids, as well as *Prosqualodon* and a possible waipatiid (Fitzgerald, 2004, 2006). Other sites located throughout **Victoria**, easternmost South Australia, and Tasmania have produced further mammalodontids, a possible aetiocetid, balaenids, balaenopterids, stem balaenopteroids, and a possible fossil relative of *Caperea*, as well as delphinids, eurhinodelphinids, physeterids, *Prosqualodon*, possible squalodontids, and ziphiids. These specimens range in age from the Early Oligocene (Rupelian) all the way to the Pliocene (Fitzgerald, 2004).

On the eastern side of the South Pacific, the **Pisco** and **Sacaco** basins extend for more than 250 km along the arid southern coast of Peru, where they expose a series of highly fossiliferous, Middle Eocene–Pliocene marine sediments. Both areas are known as fossil **Lagerstätten** because of the exceptional quality of fossil preservation, including fully articulated skeletons and, uniquely, frequent examples of fossilized baleen (Figure 2.13; Plates 12a and 15b) (Esperante *et al.*, 2008). The stratigraphic framework of the fossil-bearing layers is not yet fully understood, but preliminarily they comprise the Eocene (?late Lutetian–early Priabonian) Paracas Formation, the Eocene–Oligocene (Priabonian–Rupelian) Otuma Formation, the Oligocene–Miocene (Chattian–Aquitanian or Burdigalian) Chilcatay Formation, and the Miocene–Pliocene (Serravallian–Zanclean) Pisco Formation (de Muizon and DeVries, 1985; DeVries, 1998; Ehret *et al.*, 2012; Uhen *et al.*, 2011).

Between them, these units have yielded an astonishing variety of fossil taxa, including: basilosaurid archaeocetes (Martínez Cáceres and de Muizon, 2011; Uhen *et al.*, 2011); archaic toothed mysticetes (Martínez Cáceres *et al.*, 2011); various baleen-bearing mysticetes, including the only certainly known fossil relative of *Caperea* (Bisconti, 2012), cetotheriids (Bouetel and de Muizon, 2006) and numerous balaenopterids (e.g., Pilleri, 1989); and a bewildering variety of odontocetes, including delphinids, eurhinodelphinids, kentriodontids, kogiids, monodontids, odobenocetopsids, phocoenids, pontoporiids, squalodelphinids, stem physeteroids, and ziphiids (e.g., Bianucci *et al.*,

Figure 2.13 Fossil skeleton of a large balaenopteroid mysticete, exposed at Cerro la Bruja, Pisco Formation, Pisco Basin, Peru. Courtesy of G. Bianucci.

2010; de Muizon, 1984, 1988, 1993; Lambert *et al.*, 2010, 2013). Of particular note are the walrus-like delphinoid *Odobenocetops* and the gigantic "killer sperm whale" *Livyatan* (sections 5.4.3 and 7.7).

Despite this remarkable diversity, work up to now has only scratched the surface: much of the fossil assemblage from the Pisco Formation, and even more so from the older Paleogene and Early Miocene deposits, remains to be described. The same holds true for a second South American locality rivalling the Peruvian deposits in the quality of preservation: **Cerro Ballena**, in the Atacama Region of Chile. Here, the Late Miocene Bahía Inglesa Formation has yielded numerous articulated balaenopterid skeletons (Plate 12b), in addition to *Odobenocetops*, another delphinoid and a physeteroid. The fossils exposed at this site occur in distinct, densely packed horizons, which may have formed as the result of mass mortality events caused by harmful algal blooms (Pyenson *et al.*, 2014).

2.3.6 Dredge sites: South Africa and Iberia

Commercial and scientific deep-sea dredging has brought to light cetacean fossils since at least the mid-19th century. In general, however, this technique only leads to the recovery of isolated elements which, except for ear bones, are rarely diagnostic (Murray and Renard, 1891). Nevertheless, in some instances enough of a skull is preserved to determine the genus or even the species. This is particularly true for the waters off South Africa and the Atlantic coast of the Iberian Peninsula, both of which have yielded rich samples of well-preserved fossil ziphiid remains—together representing 14 new species—from depths sometimes exceeding 1000 m (Bianucci *et al.*, 2007, 2013). Although they lack stratigraphic context, the specimens from these sites provide unique data on animals living in remote, deep oceanic regions, which are only rarely represented by rock outcrops on land.

2.4   Suggested Readings

de Muizon, C. 1988. Les vertébrés fossiles de la Formation Pisco (Pérou). Troisième partie: Les Odontocètes (Cetacea, Mammalia) du Miocène. Travaux de l'Institut Français d'Etudes Andines 42:1–244.

Fordyce, R. E. 2006. A southern perspective on cetacean evolution and zoogeography; pp. 755–782 in J. R. Merrick, M. Archer, G. M. Hickey and M. S. Y. Lee (eds.), Evolution and Biogeography of Australasian Vertebrates. Auscipub, Oatlands, Australia.

Fordyce, R. E. 2009. Cetacean fossil record; pp. 201–207 in J. G. M. Thewissen, W. F. Perrin, and B. Würsig (eds.), Encyclopedia of Marine Mammals, 2nd ed. Elsevier, Burlington, MA.

Gingerich, P. D. 2008. Early evolution of whales: a century of research in Egypt; pp. 107–124 in J. G. Fleagle and C. C. Gilbert (eds.), Elwyn Simons: A Search for Origins. Springer, New York.

Gottfried, M. D., Bohaska, D. J., and Whitmore, F. C., Jr. 1994. Miocene cetaceans of the Chesapeake Group. Proceedings of the San Diego Society of Natural History 29:228–238.

References

Abel, O. 1905. Les Odontocètes du Boldérien (Miocène supérieur) d'Anvers. Mémoires du Musée royal D'Histoire Naturelle de Belgique 3:1–155.

Abel, O. 1914. Die Vorfahren der Bartenwale. Denkschriften der Kaiserlichen Akademie der Wissenschaften 90:155–224.

Abel, O. 1938. Vorlaeufige Mitteilungen über die Revision der fossilen Mystacoceten aus dem Tertiaer Belgiens. Bulletin du Musée royal d'Histoire naturelle de Belgique 14:1–34.

Aguirre-Fernández, G., and R. E. Fordyce. 2014. *Papahu taitapu*, gen. et sp. nov., an early Miocene stem odontocete (Cetacea) from New Zealand. Journal of Vertebrate Paleontology 34:195–210.

Allen, G. M. 1921. A new fossil cetacean. Bulletin of the Museum of Comparative Zoology 65:1–13.

Alroy, J. 2010. The shifting balance of diversity among major marine animal groups. Science 329:1191–1194.

Barnes, L. G., J. L. Goedert, and H. Furusawa. 2001. The earliest known echolocating toothed whales (Mammalia; Odontoceti): preliminary observations of fossils from Washington State. Mesa Southwest Museum Bulletin 8:91–100.

Barnes, L. G., M. Kimura, H. Furusawa, and H. Sawamura. 1995. Classification and distribution of Oligocene Aetiocetidae (Mammalia; Cetacea; Mysticeti) from western North America and Japan. The Island Arc 3:392–431.

Barnes, L. G., and A. E. Sanders. 1996. The transition from archaeocetes to mysticetes: Late Oligocene toothed mysticetes from near Charleston, South Carolina. Paleontological Society Special Publication 8:24.

Barnes, L. G., L. Tohill, and S. Tohill. 2015. A new pontoporiid dolphin from the Sharktooth Hill Bonebed, central California; the oldest known member of its family. PaleoBios 32 (Suppl.):4.

Bearlin, R. K. 1987. The morphology and systematics of Neogene Mysticeti from Australia and New Zealand. PhD Thesis, University of Otago.

Bebej, R. M., I. S. Zalmout, A. A. Abed El-Aziz, M. S. M. Antar, and P. D. Gingerich. 2012. First evidence of Remingtonocetidae (Mammalia, Cetacea) outside Indo-Pakistan: new genus from the early Middle Eocene of Egypt. Journal of Vertebrate Paleontology: Program and Abstracts, 62.

Bianucci, G. 2013. *Septidelphis morii*, n. gen. et sp., from the Pliocene of Italy: new evidence of the explosive radiation of true dolphins (Odontoceti, Delphinidae). Journal of Vertebrate Paleontology 33:722–740.

Bianucci, G., O. Lambert, and K. Post. 2007. A high diversity in fossil beaked whales (Mammalia, Odontoceti, Ziphiidae) recovered by trawling from the sea floor off South Africa. Geodiversitas 29:561–618.

Bianucci, G., O. Lambert, and K. Post. 2010. High concentration of long-snouted beaked whales (genus *Messapicetus*) from the Miocene of Peru. Palaeontology 53:1077–1098.

Bianucci, G., and W. Landini. 2002. Change in diversity, ecological significance and biogeographical relationships of the Mediterranean Miocene toothed whale fauna. Geobios 35 (Suppl. 1):19–28.

Bianucci, G., I. Miján, O. Lambert, K. Post, and O. Mateus. 2013. Bizarre fossil beaked whales (Odontoceti, Ziphiidae) fished from the Atlantic Ocean floor off the Iberian Peninsula. Geodiversitas 35:105–153.

Bisconti, M. 2000. New description, character analysis and preliminary phyletic assessment of two Balaenidae skulls from the Italian Pliocene. Palaeontographia Italica 87:37–66.

Bisconti, M. 2008. Morphology and phylogenetic relationships of a new eschrichtiid genus (Cetacea: Mysticeti) from the early Pliocene of northern Italy. Zoological Journal of the Linnean Society 153:161–186.

Bisconti, M. 2012. Comparative osteology and phylogenetic relationships of *Miocaperea pulchra*, the first fossil pygmy right whale genus and species (Cetacea, Mysticeti, Neobalaenidae). Zoological Journal of the Linnean Society 166:876–911.

Bisconti, M., O. Lambert, and M. Bosselaers. 2013. Taxonomic revision of *Isocetus depauwi* (Mammalia, Cetacea, Mysticeti) and the phylogenetic relationships of archaic "cetothere" mysticetes. Palaeontology 56:95–127.

Boessenecker, R. W. 2013a. A new marine vertebrate assemblage from the Late Neogene Purisima Formation in Central California, part II: pinnipeds and cetaceans. Geodiversitas 35:815–940.

Boessenecker, R. W. 2013b. Pleistocene survival of an archaic dwarf baleen whale (Mysticeti: Cetotheriidae). Naturwissenschaften 100:365–371.

Boessenecker, R. W., and R. E. Fordyce. 2015a. A new eomysticetid (Mammalia: Cetacea) from the Late Oligocene of New Zealand and a re-evaluation of "*Mauicetus*" *waitakiensis*. Papers in Palaeontology 1:107–140.

Boessenecker, R. W., and R. E. Fordyce. 2015b. Anatomy, feeding ecology, and ontogeny of a transitional baleen whale: a new genus and species of Eomysticetidae (Mammalia: Cetacea) from the Oligocene of New Zealand. PeerJ 3:e1129.

Boessenecker, R. W., and R. E. Fordyce. 2015c. A new genus and species of eomysticetid (Cetacea: Mysticeti) and a reinterpretation of "*Mauicetus*" *lophocephalus* Marples, 1956: transitional baleen whales from the upper Oligocene of New Zealand. Zoological Journal of the Linnean Society 175:607–660.

Bouetel, V., and C. de Muizon. 2006. The anatomy and relationships of *Piscobalaena nana* (Cetacea, Mysticeti), a Cetotheriidae s.s. from the early Pliocene of Peru. Geodiversitas 28:319–396.

Brand, L. R., R. Esperante, A. V. Chadwick, O. P. Porras, and M. Alomía. 2004. Fossil whale preservation implies high diatom accumulation rate in the Miocene–Pliocene Pisco Formation of Peru. Geology 32:165–168.

Brandt, J. F. 1843. De cetotherio, novo balaenarum familiae genre in Rossia Meridionali ante aliquot annos effoso. Bulletin de l'Académie impériale des Sciences de Saint Pétersbourg 1:145–148.

Brandt, J. F. 1873. Untersuchungen über die fossilen und subfossilen Cetaceen Europa's. Mémoires de l'Académie impériale des sciences de St.-Pétersbourg 20:1–372.

Brandt, J. F. 1874. Ergänzungen zu den fossilen Cetaceen Europa's. Mémoires de l'Académie impériale des sciences de St.-Pétersbourg 21:1–54.

Browning, J. V., K. G. Miller, P. P. McLaughlin, L. E. Edwards, A. A. Kulpecz, D. S. Powars, B. S. Wade, M. D. Feigenson, and J. D. Wright. 2009. Integrated sequence stratigraphy of the postimpact sediments from the Eyreville core holes, Chesapeake Bay impact structure inner basin. Geological Society of America Special Paper 458:775–810.

Buono, M. R., and M. A. Cozzuol. 2013. A new beaked whale (Cetacea, Odontoceti) from the Late Miocene of Patagonia, Argentina. Journal of Vertebrate Paleontology 33:986–997.

Buono, M. R., M. T. Dozo, F. G. Marx, and R. E. Fordyce. 2014. A Late Miocene potential neobalaenine mandible from Argentina sheds light on the origins of the living pygmy right whale. Acta Palaeontologica Polonica 59:787–793.

Cabrera, Á. 1926. Cetáceos fósiles del Museo de La Plata. Revista del Museo de La Plata 29:363–411.

Capellini, G. 1901. Balenottera miocenica del Monte Titano, Repubblica di S. Marino. Memorie della Reale Accademia delle Scienze dell'Istituto di Bologna ser. 5, 9:237–260.

Capellini, G. 1904. Balene fossili toscane. II: *Balaena montalionis*. Memorie della Reale Accademia delle Scienze dell'Istituto di Bologna ser. 6, 1:47–55.

Clementz, M. T., R. E. Fordyce, S. Peek, and D. L. Fox. 2014. Ancient marine isoscapes and isotopic evidence of bulk-feeding by Oligocene cetaceans. Palaeogeography, Palaeoclimatology, Palaeoecology 400:28–40.

Cozzuol, M. A. 1996. The record of the aquatic mammals in southern South America. Münchner Geowissenschaftliche Abhandlungen, Reihe A 30:321–342.

Cuvier, G. 1823. Recherches sur les ossements fossiles, tome 5 (1ère partie). G. Dufour et E. D'Ocagne, Paris.

Dal Piaz, G. 1916. Gli Odontoceti del Miocene Bellunese. Parte II: *Squalodon*. Memorie dell'Istituto geologico della Reale Università di Padova 4:3–94.

de Muizon, C. 1984. Les vertébrés de la Formation Pisco (Pérou). Deuxième partie: Les Odontocètes (Cetacea, Mammalia) du Pliocène inférieur de Sud-Sacaco. Travaux de l'Institut Français d'Etudes Andines 27:1–188.

de Muizon, C. 1988. Les vertébrés fossiles de la Formation Pisco (Pérou). Troisième partie: Les Odontocètes (Cetacea, Mammalia) du Miocène. Travaux de l'Institut Français d'Etudes Andines 42:1–244.

de Muizon, C. 1991. A new Ziphiidae (Cetacea) from the Early Miocene of Washington State (USA) and phylogenetic analysis of the major groups of odontocetes. Bulletin du Musée National d'Histoire Naturelle (Paris) 12:279–326.

de Muizon, C. 1993. Walrus-like feeding adaptation in a new cetacean from the Pliocene of Peru. Nature 365:745–748.

de Muizon, C., and T. J. DeVries. 1985. Geology and paleontology of the Cenozoic marine deposits in the Sacaco area (Peru). Geologische Rundschau 74:547–563.

De Schepper, S., M. J. Head, and S. Louwye. 2009. Pliocene dinoflagellate cyst stratigraphy, palaeoecology and sequence stratigraphy of the Tunnel-Canal Dock, Belgium. Geological Magazine 146:92–112.

DeVries, T. J. 1998. Oligocene deposition and Cenozoic sequence boundaries in the Pisco Basin (Peru). Journal of South American Earth Sciences 11:217–231.

Dubrovo, I. A., and A. E. Sanders. 2000. A new species of *Patriocetus* (Mammalia, Cetacea) from the Late Oligocene of Kazakhstan. Journal of Vertebrate Paleontology 20:577–590.

Dunhill, A. M., B. Hannisdal, and M. J. Benton. 2014. Disentangling rock record bias and common-cause from redundancy in the British fossil record. Nature Communications 5:4818.

Ehret, D. J., B. J. Macfadden, D. S. Jones, T. J. Devries, D. A. Foster, and R. Salas-Gismondi. 2012. Origin of the white shark *Carcharodon* (Lamniformes: Lamnidae) based on recalibration of the Upper Neogene Pisco Formation of Peru. Palaeontology 55:1139–1153.

Emlong, D. R. 1966. A new archaic cetacean from the Oligocene of northwest Oregon. Bulletin of the Oregon University Museum of Natural History 3:1–51.

Esperante, R., L. Brand, K. E. Nick, O. Poma, and M. Urbina. 2008. Exceptional occurrence of fossil baleen in shallow marine sediments of the Neogene Pisco Formation, Southern Peru. Palaeogeography Palaeoclimatology Palaeoecology 257:344–360.

Fitzgerald, E. M. G. 2004. A review of the Tertiary fossil Cetacea (Mammalia) localities in Australia. Memoirs of Museum Victoria 61:183–208.

Fitzgerald, E. M. G. 2006. A bizarre new toothed mysticete (Cetacea) from Australia and the early evolution of baleen whales. Proceedings of the Royal Society B 273:2955–2963.

Fordyce, R. E. 1980. The fossil Cetacea of New Zealand (A catalogue of described genera and species with an annotated literature guide and reference list). New Zealand Geological Survey Report 90:1–60.

Fordyce, R. E. 1994. *Waipatia maerewhenua*, new genus and new species (Waipatiidae, new family), an archaic Late Oligocene dolphin (Cetacea: Odontoceti: Platanistoidea) from New Zealand. Proceedings of the San Diego Society of Natural History 29:147–176.

Fordyce, R. E. 2002. *Simocetus rayi* (Odontoceti: Simocetidae, new family): a bizarre new archaic Oligocene dolphin from the eastern North Pacific. Smithsonian Contributions to Paleobiology 93:185–222.

Fordyce, R. E. 2003. Cetacean evolution and Eocene-Oligocene oceans revisited; pp. 154–170 in D. R. Prothero, L. C. Ivany, and E. A. Nesbitt (eds.), From Greenhouse to Icehouse: The Marine Eocene-Oligocene Transition. Columbia University Press, New York.

Fordyce, R. E. 2006. A southern perspective on cetacean evolution and zoogeography; pp. 755–782 in J. R. Merrick, M. Archer, G. M. Hickey, and M. S. Y. Lee (eds.), Evolution and Biogeography of Australasian Vertebrates. Auscipub, Oatlands, Australia.

Fordyce, R. E. 2008. Fossil mammals; pp. 415–428 in M. J. Winterbourn, G. A. Knox, C. J. Burrows, and I. Marsden (eds.), Natural History of Canterbury. University of Canterbury Press, Christchurch.

Fordyce, R. E., and F. G. Marx. 2011. Toothed mysticetes and ecological structuring of Oligocene whales and dolphins from New Zealand. Geological Survey of Western Australia Annual Record 2011/9:33.

Gatesy, J. 2009. Whales and even-toed ungulates (Cetartiodactyla); pp. 511–515 in S. B. Hedges, and S. Kumar (eds.), The TimeTree of Life. Oxford University Press, New York.

Geisler, J. H., M. W. Colbert, and J. L. Carew. 2014. A new fossil species supports an early origin for toothed whale echolocation. Nature 508:383–386.

Gibbes, R. W. 1845. Description of the teeth of a new fossil animal found in the Green Sand of South Carolina. Proceedings of the Academy of Natural Sciences of Philadelphia 2:254–256.

Gingerich, P. D. 1992. Marine mammals (Cetacea and Sirenia) from the Eocene of Gebel Mokattam and Fayum, Egypt: stratigraphy, age and paleoenvironments. University of Michigan Papers on Paleontology 30:1–84.

Gingerich, P. D. 2003. Stratigraphic and micropaleontological constraints on the Middle Eocene age of the mammal-bearing Kuldana Formation of Pakistan. Journal of Vertebrate Paleontology 23:643–651.

Gingerich, P. D. 2010. Cetacea; pp. 873–899 in L. Werdelin, and W. J. Sanders (eds.), Cenozoic Mammals of Africa. University of California Press, Berkeley.

Gingerich, P. D., M. S. M. Antar, and I. Zalmout. 2014. Skeleton of new protocetid (Cetacea, Archaeoceti) from the lower Gehannam Formation of Wadi al Hitan in Egypt: survival of a protocetid into the Priabonian Late Eocene. Journal of Vertebrate Paleontology: Program and Abstracts, 138.

Gingerich, P. D., M. Arif, M. A. Bhatti, M. Anwar, and W. J. Sanders. 1997. *Basilosaurus drazindai* and *Basiloterus hussaini*, new Archaeoceti (Mammalia, Cetacea) from the Middle Eocene Drazinda Formation, with a revised interpretation of ages of whale-bearing strata in the Kirthar Group of the Sulaiman Range, Punjab (Pakistan). Contributions from the Museum of Paleontology, University of Michigan 30:55–81.

Gingerich, P. D., M. u. Haq, I. S. Zalmout, I. H. Khan, and M. S. Malakani. 2001a. Origin of whales from early artiodactyls: hands and feet of Eocene Protocetidae from Pakistan. Science 293:2239–2242.

Gingerich, P. D., and M. D. Uhen. 1998. Likelihood estimation of the time of origin of Cetacea and the time of divergence of Cetacea and Artiodactyla. Palaeontologia Electronica 1:1.2.5A.

Gingerich, P. D., M. Ul-Haq, I. H. Khan, and I. S. Zalmout. 2001b. Eocene stratigraphy and archaeocete whales (Mammalia, Cetacea) of Drug Lahar in the eastern Sulaiman Range, Balochistan (Pakistan). Contributions from the Museum of Paleontology, University of Michigan 30:269–319.

Gingerich, P. D., M. Ul-Haq, W. von Koenigswald, W. J. Sanders, B. H. Smith, and I. S. Zalmout. 2009. New protocetid whale from the Middle Eocene of Pakistan: birth on land, precocial development, and sexual dimorphism. PLoS One 4:1–20.

Goedert, J. L., L. G. Barnes, and H. Furusawa. 2007. The diversity and stratigraphic distribution of cetaceans in early Cenozoic strata of Washington State, U.S.A. Geological Society of Australia Abstracts 85:44.

Goedert, J. L., R. L. Squires, and L. G. Barnes. 1995. Paleoecology of whale-fall habitats from deep-water Oligocene rocks, Olympic Peninsula, Washington state. Palaeogeography, Palaeoclimatology, Palaeoecology 118:151–158.

Gol'din, P., and D. Startsev. 2014. *Brandtocetus*, a new genus of baleen whales (Cetacea, Cetotheriidae) from the Late Miocene of Crimea, Ukraine. Journal of Vertebrate Paleontology 34:419–433.

Gol'din, P., D. Startsev, and T. Krakhmalnaya. 2014. The anatomy of the Late Miocene baleen whale *Cetotherium riabinini* from Ukraine. Acta Palaeontologica Polonica 59:795–814.

Hanna, G. D. 1925. Miocene marine vertebrates in Kern County, California. Science 61:71–72.

Harington, C. R. 1977. Marine mammals in the Champlain Sea and the Great Lakes. Annals of the New York Academy of Sciences 288:508–537.

Harington, C. R., S. Lebel, M. Paiement, and A. d. Vernal. 2006. Félix: a Late Pleistocene white whale (*Delphinapterus leucas*) skeleton from Champlain Sea deposits at Saint-Félix-de-Valois, Québec. Géographie physique et Quaternaire 60:183–198.

Harlan, R. 1834. Notice of fossil bones found in the Tertiary formation of the state of Louisiana. Transactions of the American Philosophical Society Philadelphia 4:397–403.

Kellogg, R. 1924. A fossil porpoise from the Calvert Formation of Maryland. Proceedings of the United States National Museum 63:1–39.

Kellogg, R. 1931. Pelagic mammals of the Temblor Formation of the Kern River region, California. Proceedings of the California Academy of Sciences 19:217–397.

Kellogg, R. 1936. A review of the Archaeoceti. Carnegie Institution of Washington Publication 482:1–366.

Kellogg, R. 1955. Three Miocene porpoises from the Calvert Cliffs, Maryland. Proceedings of the United States National Museum 105:101–154.

Kellogg, R. 1965a. Fossil marine mammals from the Miocene Calvert Formation of Maryland and Virginia, part 1: a new whalebone whale from the Miocene Calvert Formation. United States National Museum Bulletin 247:1–45.

Kellogg, R. 1965b. The Miocene Calvert sperm whale *Orycterocetus*. United States National Museum Bulletin 247:47–63.

Kimura, T., and Y. Hasegawa. 2004. An outline of the Miocene cetotheres of Japan. Bulletin of the Gunma Museum of Natural History 8:79–88.

King, C., C. J. Underwood, and E. Steurbaut. 2014. Eocene stratigraphy of the Wadi Al-Hitan World Heritage Site and adjacent areas (Fayum, Egypt). Stratigraphy 11:185–234.

Koch, A. C. 1972. Journey Through a Part of the United States of North America in the Years 1844 to 1846. Southern Illinois University Press, Carbondale.

Köhler, R., and R. E. Fordyce. 1997. An archaeocete whale (Cetacea: Archaeoceti) from the Eocene Waihao Greensand, New Zealand. Journal of Vertebrate Paleontology 17:574–583.

Kohno, N., H. Koike, and K. Narita. 2007. Outline of fossil marine mammals from the Middle Miocene Bessho and Aoki Formations, Nagano Prefecture, Japan. Research Report of the Shinshushinmachi Fossil Museum 10:1–45.

Lambert, O. 2006. First record of a platanistid (Cetacea, Odontoceti) in the North Sea Basin: a review of *Cyrtodelphis* Abel, 1899 from the Miocene of Belgium. Oryctos 6:69–79.

Lambert, O. 2008a. A new porpoise (Cetacea, Odontoceti, Phocoenidae) from the Pliocene of the North Sea. Journal of Vertebrate Paleontology 28:863–872.

Lambert, O. 2008b. Sperm whales from the Miocene of the North Sea: a re-appraisal. Bulletin de l'Institut Royal des Sciences Naturelles de Belgique: Sciences de la Terre 78:277–216.

Lambert, O., G. Bianucci, K. Post, C. de Muizon, R. Salas-Gismondi, M. Urbina, and J. Reumer. 2010. The giant bite of a new raptorial sperm whale from the Miocene epoch of Peru. Nature 466:105–108.

Lambert, O., C. de Muizon, and G. Bianucci. 2013. The most basal beaked whale *Ninoziphius platyrostris* Muizon, 1983: clues on the evolutionary history of the family Ziphiidae (Cetacea: Odontoceti). Zoological Journal of the Linnean Society 167:569–598.

Lambert, O., and P. Gigase. 2007. A monodontid cetacean from the Early Pliocene of the North Sea. Bulletin de l'Institut Royal des Sciences Naturelles de Belgique: Sciences de la Terre 77:197–210.

Louwye, S., R. Marquet, M. Bosselaers, and O. Lambert. 2010. Stratigraphy of an Early–Middle Miocene sequence near Antwerp in northern Belgium (southern North Sea basin). Geologica Belgica 13:269–284.

Lydekker, R. 1894. Cetacean skulls from Patagonia. Annales del Museo de la Plata 2:1–13.

Martínez Cáceres, M., and C. de Muizon. 2011. A new basilosaurid (Cetacea, Pelagiceti) from the Late Eocene to Early Oligocene Otuma Formation of Peru. Comptes Rendus Palevol 10:517–526.

Martínez Cáceres, M., C. de Muizon, O. Lambert, G. Bianucci, R. Salas-Gismondi, and M. Urbina Schmidt. 2011. A toothed mysticete from the Middle Eocene to Lower Oligocene of the Pisco Basin, Peru: new data on the origin and feeding evolution of Mysticeti. Sixth Triennial Conference on Secondary Adaptation of Tetrapods to Life in Water:56–57.

Marx, F. G. 2009. Marine mammals through time: when less is more in studying palaeodiversity. Proceedings of the Royal Society B 276:887–892.

Marx, F. G., and R. E. Fordyce. 2015. Baleen boom and bust: a synthesis of mysticete phylogeny, diversity and disparity. Royal Society Open Science 2:140434.

Marx, F. G., and M. D. Uhen. 2010. Climate, critters, and cetaceans: Cenozoic drivers of the evolution of modern whales. Science 327:993–996.

Matsumoto, H. 1926. On some fossil cetaceans of Japan. Science Reports of the Tohoku Imperial University 10:17–27.

Mchedlidze, G. A. 1964. Fossil Cetacea of the Caucasus. Metsniereba, Tbilisi.

Mchedlidze, G. A. 1970. Nekotorye obshchie cherty istorii kitoobraznykh (Some General Features of the History of Cetaceans). Metsniereba, Tbilisi.

Murray, J., and A. F. Renard. 1891. Report on Deep-Sea Deposits Based on the Specimens Collected during the Voyage of HMS Challenger in the Years 1872 to 1876. Neill and Company, Edinburgh.

Oishi, M., and Y. Hasegawa. 1995. A list of fossil cetaceans in Japan. The Island Arc 3:493–505.

Okazaki, Y. 2012. A new mysticete from the upper Oligocene Ashiya Group, Kyushu, Japan and its significance to mysticete evolution. Bulletin of the Kitakyushu Museum of Natural History and Human History, Series A (Natural History) 10:129–152.

Owen, R. 1870. Monograph on the British Fossil Cetacea from the Red Crag. Palaeontographical Society, London.

Peters, S. E. 2005. Geologic constraints on the macroevolutionary history of marine animals. Proceedings of the National Academy of

Sciences of the United States of America 102:12326–12331.

Peters, S. E., M. S. M. Antar, I. S. Zalmout, and P. D. Gingerich. 2009. Sequence stratigraphic control on preservation of Late Eocene whales and other vertebrates at Wadi Al-Hitan, Egypt. PALAIOS 24:290–302.

Pilleri, G. 1985. The Miocene Cetacea of the Belluno sandstones (Eastern Southern Alps). Memorie di Scienze Geologiche 37:1–250.

Pilleri, G. 1989. *Balaenoptera siberi*, ein neuer spätmiozäner Bartenwal aus der Pisco-Formation Perus; pp. 63–84 in G. Pilleri (ed.), Beiträge zur Paläontologie der Cetaceen Perus. Hirnanatomisches Institut Ostermundigen, Bern.

Post, K. 2005. A Weichselian marine mammal assemblage from the southern North Sea. Deinsea 11:21–27.

Post, K., and M. Bosselaers. 2005. Late Pliocene occurrence of *Hemisyntrachelus* (Odontoceti, Delphinidae) in the southern North Sea. Deinsea 11:29–45.

Post, K., and E. J. O. Kompanje. 2010. A new dolphin (Cetacea, Delphinidae) from the Plio-Pleistocene of the North Sea. Deinsea 14:1–12.

Prothero, D. R. 2001. Chronostratigraphic calibration of the Pacific coast Cenozoic: a summary. Pacific Section SEPM Book 91:377–394.

Pyenson, N. D. 2010. Carcasses on the coastline: measuring the ecological fidelity of the cetacean stranding record in the eastern North Pacific Ocean. Paleobiology 36:453–480.

Pyenson, N. D. 2011. The high fidelity of the cetacean stranding record: insights into measuring diversity by integrating taphonomy and macroecology. Proceedings of the Royal Society B 278:3608–3616.

Pyenson, N. D., C. S. Gutstein, J. F. Parham, J. P. Le Roux, C. C. Chavarría, H. Little, A. Metallo, V. Rossi, A. M. Valenzuela-Toro, J. Velez-Juarbe, C. M. Santelli, D. R. Rogers, M. A. Cozzuol, and M. E. Suárez. 2014. Repeated mass strandings of Miocene marine mammals from Atacama Region of Chile point to sudden death at sea. Proceedings of the Royal Society B 281:20133316.

Pyenson, N. D., R. B. Irmis, J. H. Lipps, L. G. Barnes, E. D. Mitchell, and S. A. McLeod. 2009. Origin of a widespread marine bonebed deposited during the middle Miocene Climatic Optimum. Geology 37:519–522.

Raup, D. M. 1972. Taxonomic diversity during the Phanerozoic. Science 177:1065–1071.

Sanders, A. E., and L. G. Barnes. 2002a. Paleontology of the Late Oligocene Ashley and Chandler Bridge Formations of South Carolina, 2: *Micromysticetus rothauseni*, a primitive cetotheriid mysticete (Mammalia, Cetacea). Smithsonian Contributions to Paleobiology 93:271–293.

Sanders, A. E., and L. G. Barnes. 2002b. Paleontology of the Late Oligocene Ashley and Chandler Bridge Formations of South Carolina, 3: Eomysticetidae, a new family of primitive mysticetes (Mammalia: Cetacea). Smithsonian Contributions to Paleobiology 93:313–356.

Sanders, A. E., and J. H. Geisler. 2015. A new basal odontocete from the upper Rupelian of South Carolina, U.S.A., with contributions to the systematics of *Xenorophus* and *Mirocetus* (Mammalia, Cetacea). Journal of Vertebrate Paleontology 35:e890107.

Schäfer, W. 1972. Ecology and Palaeoecology of Marine Environments. University of Chicago Press, Chicago.

Scilla, A. 1670. La vana speculazione disingannata dal senso. Lettera responsiva circa i corpi marini che petrificati si truovano in varii luoghi terrestri. Appresso Andrea Colicchia, Napoli.

Smith, A. B. 2007. Marine diversity through the Phanerozoic: problems and prospects. Journal of the Geological Society 164:731–745.

Steeman, M. E. 2007. Cladistic analysis and a revised classification of fossil and recent mysticetes. Zoological Journal of the Linnean Society 150:875–894.

Steeman, M. E. 2010. The extinct baleen whale fauna from the Miocene-Pliocene of Belgium and the diagnostic cetacean ear bones. Journal of Systematic Palaeontology 8:63–80.

Tanaka, Y., and R. E. Fordyce. 2014. Fossil dolphin *Otekaikea marplesi* (Latest Oligocene, New Zealand) expands the morphological and taxonomic diversity of Oligocene cetaceans. PLoS One 9:e107972.

Tanaka, Y., and R. E. Fordyce. 2015. Historically significant late Oligocene dolphin *Microcetus hectori* Benham 1935: a new species of *Waipatia* (Platanistoidea). Journal of the Royal Society of New Zealand: 1–16.

Tarasenko, K. K. 2014. New genera of baleen whales (Cetacea, Mammalia) from the Miocene of the northern Caucasus and Ciscaucasia: 3. *Zygiocetus* gen. nov. (Middle Sarmatian, Adygea). Paleontological Journal 48:551–562.

Tarasenko, K. K., and A. V. Lopatin. 2012. New baleen whale genera (Cetacea, Mammalia) from the Miocene of the northern Caucasus and Ciscaucasia: 1. *Kurdalagonus* gen. nov. from the middle-late Sarmatian of Adygea. Paleontological Journal 46:531–542.

Thewissen, J. G. M., E. M. Williams, L. J. Roe, and S. T. Hussain. 2001. Skeletons of terrestrial cetaceans and the relationship of whales to artiodactyls. Nature 413:277–281.

True, F. W. 1889. Contributions to the natural history of the cetaceans: a review of the family Delphinidae. Bulletin of the United States National Museum 36:1–191.

True, F. W. 1904. The whalebone whales of the western North Atlantic compared with those occurring in European waters, with some observations on the species of the North Pacific. Smithsonian Contributions to Knowledge 23:1–332.

True, F. W. 1910. An account of the beaked whales of the family Ziphiidae in the collection of the United States National Museum, with remarks on some specimens in other American museums. Bulletin of the United States National Museum 73:1–89.

Tsai, C.-H., and R. E. Fordyce. 2015. The earliest gulp-feeding mysticete (Cetacea: Mysticeti) from the Oligocene of New Zealand. Journal of Mammalian Evolution 22:535–560.

Uhen, M. D. 2010. The origin(s) of whales. Annual Review of Earth and Planetary Sciences 38:189–219.

Uhen, M. D. 2013. A review of North American Basilosauridae. Alabama Museum of Natural History Bulletin 31:1–45.

Uhen, M. D. 2015. Cetacea: Paleobiology Database Data Archive 9.

Uhen, M. D., A. D. Barnosky, B. Bills, J. Blois, M. T. Carrano, M. A. Carrasco, G. M. Erickson, J. T. Eronen, M. Fortelius, R. W. Graham, E. C. Grimm, M. A. O'Leary, A. Mast, W. H. Piel, P. D. Polly, and L. K. Säilä. 2013. From card catalogs to computers: databases in vertebrate paleontology. Journal of Vertebrate Paleontology 33:13–28.

Uhen, M. D., and N. D. Pyenson. 2007. Diversity estimates, biases and historiographic effects: resolving cetacean diversity in the Tertiary. Palaeontologia Electronica 10:10.2.10A.

Uhen, M. D., N. D. Pyenson, T. J. Devries, M. Urbina, and P. R. Renne. 2011. New Middle Eocene whales from the Pisco Basin of Peru. Journal of Paleontology 85:955–969.

Underwood, C. J., D. J. Ward, C. King, S. M. Antar, I. S. Zalmout, and P. D. Gingerich. 2011. Shark and ray faunas in the Middle and Late Eocene of the Fayum Area, Egypt. Proceedings of the Geologists' Association 122:47–66.

Van Beneden, P.-J. 1880. Description des ossements fossiles des environs d'Anvers. Deuxième partie. Cétacés. Genres *Balaenula*, *Balaena* et *Balaenotus*. Annales du Musée Royal d'Histoire Naturelle de Belgique 4:1–82.

Van Beneden, P.-J. 1882. Description des ossements fossiles des environs d'Anvers. Troisième partie. Cétacés. Genres *Megaptera*, *Balaenoptera*, *Burtinopsis* et *Erpetocetus*. Annales du Musée Royal d'Histoire Naturelle de Belgique 7:1–87.

Van Beneden, P.-J., and P. Gervais. 1880. Ostéographie des cétacés vivants et fossiles. Bertrand, Paris.

Warren, L. 1998. Joseph Leidy: The Last Man Who Knew Everything. Yale University Press, New Haven.

Whitmore, F. C., Jr. 1975. Remington Kellogg, 1892–1969. Biographical Memoirs of the National Academy of Sciences 46:159–189.

Whitmore, F. C., Jr., and L. G. Barnes. 2008. The Herpetocetinae, a new subfamily of extinct baleen whales (Mammalia, Cetacea, Cetotheriidae). Virginia Museum of Natural History Special Publication 14:141–180.

Whitmore, F. C., Jr., and J. A. Kaltenbach. 2008. Neogene Cetacea of the Lee Creek Phosphate Mine, North Carolina. Virginia Museum of Natural History Special Publication 14:181–269.

Williams, E. M. 1998. Synopsis of the earliest cetaceans; pp. 1–28 in J. G. M. Thewissen (ed.), The Emergence of Whales. Plenum Press, New York.

Zhou, X., S. Xu, Y. Yang, K. Zhou, and G. Yang. 2011. Phylogenomic analyses and improved resolution of Cetartiodactyla. Molecular Phylogenetics and Evolution 61:255–264.

Zouhri, S., P. D. Gingerich, N. Elboudali, S. Sebti, A. Noubhani, M. Rahali, and S. Meslouh. 2014. New marine mammal faunas (Cetacea and Sirenia) and sea level change in the Samlat Formation, Upper Eocene, near Ad-Dakhla in southwestern Morocco. Comptes Rendus Palevol 13:599–610.

3 Morphology

3.1 Overview

At first sight, the skeleton of cetaceans may seem rather different from that of land mammals, yet, in its fundamentals, it is no more than a variation on a common theme. All mammals, and indeed all tetrapods (four-legged animals), have a skeleton that is divided into two main components: the **axial skeleton**, comprising the skull, vertebral column, and rib cage; and the **appendicular skeleton**, which includes the shoulder girdle, the pelvis, and both the fore- and hind limbs (Figure 3.1). Modern whales and dolphins stand apart from both other mammals and most of their archaeocete ancestors in having a greatly altered facial region with posteriorly displaced nostrils, markedly different ear bones, a shortened neck, a largely immobilized forelimb, and virtually no trace of the pelvis and hind limb (Figure 3.2). Nevertheless, they still retain all of the bones and basic arrangements typical of mammals, such as the presence of a secondary palate, three middle ear ossicles, and a lower jaw that consists of just a single element and articulates with the skull via a bone known as the squamosal (Kemp, 2004).

The skull is by far the most informative part of any cetacean, both because of its inherent complexity and because most of the remaining bones have become comparatively simplified (Figure 3.2). As in all mammals, the skull comprises the **cranium**, the **rostrum** (snout), the lower jaw or **mandible**, and, arguably, the **hyoid apparatus**, which supports the muscles of the tongue and the throat. The cranium in turn is largely equivalent to the **braincase**, which is flanked on either side by a large opening known as the **temporal fossa** (Figures 3.1 and 3.3). The ventral portion of the braincase mainly consists of the ear bones, the internal openings of the narial passages, various openings (foramina) for cranial nerves and blood vessels, and—in modern cetaceans—a complex network of air spaces that isolate the ear bones from the remainder of the skull. Together, these structures form the base of the skull, or **basicranium**. Throughout the skeleton, individual bones articulate through mobile **joints**, or are firmly connected via **sutures**. In addition, sutures also exist between separate ossification centers of otherwise single bony elements, such as the shaft (**diaphysis**) and articular ends (**epiphyses**) of a limb bone, or the plate-like epiphyses and the main body of a vertebra. At birth, the sutures separating individual bones and epiphyses are open, but they later close as the animal matures (section 6.7).

Cetacean Paleobiology, First Edition. Felix G. Marx, Olivier Lambert, and Mark D. Uhen.
© 2016 John Wiley & Sons, Ltd. Published 2016 by John Wiley & Sons, Ltd.

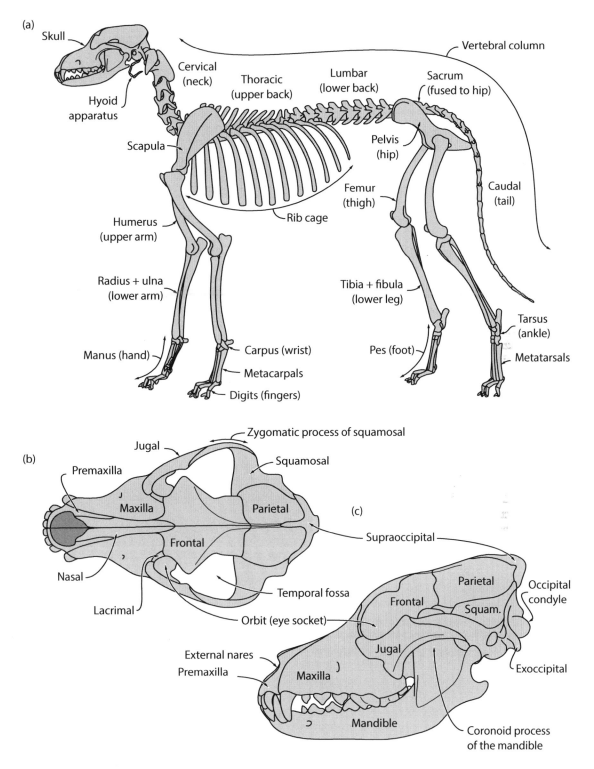

Figure 3.1 Skeleton of a typical land mammal, the dog, with (a) the complete skeleton in lateral view and the skull in (b) dorsal and (c) lateral views. Adapted from Ellenberger *et al.* (1911).

Figure 3.2 Generalized overview of the major types of cetacean skeletons, including (a) the protocetid archaeocete *Maiacetus*, (b) the oceanic dolphin *Delphinus*, and (c) the rorqual *Balaenoptera*. Drawing of *Maiacetus* adapted from Gingerich *et al.* (2009) under a Creative Commons Attribution license.

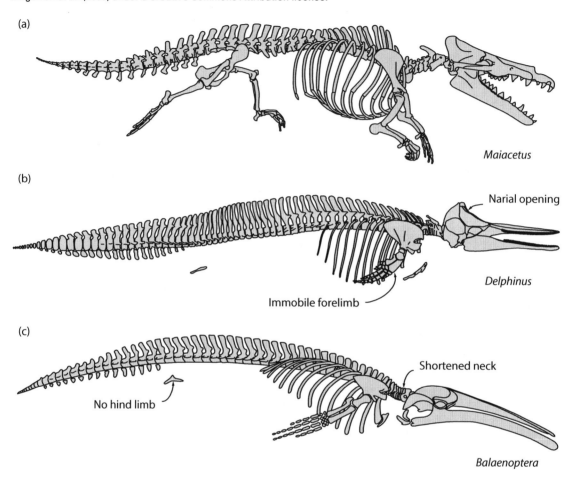

(a)

Maiacetus

(b)

Narial opening

Delphinus

Immobile forelimb

(c)

Shortened neck

No hind limb

Balaenoptera

3.2 The skull

3.2.1 Rostrum and central facial region

With few exceptions, the rostrum (snout) of cetaceans tends to be elongated, thus forming the "beak" of dolphins and beaked whales, the gavial-like jaws of the Ganges river dolphin (*Platanista*), and the enormous palate of rorquals. As in all mammals, the rostrum is mainly composed of five bones (Figures 3.1, 3.3, and 3.4): (1) the **maxilla**, which contributes nearly all of the rostral and palatal surfaces and, in archaeocetes, most odontocetes and archaic (baleen) whales, bears most of the teeth (section 3.2.7); in all other mysticetes,

the maxilla is toothless and instead bears a series of nutrient foramina and sulci that carry blood vessels supplying the baleen plates (section 3.4.2); (2) the **premaxilla**, which bears the anteriormost teeth (except in mysticetes and some odontocetes) and forms the ventral portion of the external narial opening, or external naris; (3) the **nasal**, which roofs or posteriorly delimits the external naris; (4) the **palatine**, a largely flat bone that forms the posterior portion of the bony palate; and (5) the **vomer**, an elongate, unpaired bone which runs along the center of the rostrum and divides the **choanae** (internal narial openings) into left and right passages.

Figure 3.3 Cranium of the minke whale *Balaenoptera acutorostrata* in dorsal view. Photograph shows specimen M42450 of the National Museum of Nature and Science (NMNS), Tsukuba, Japan.

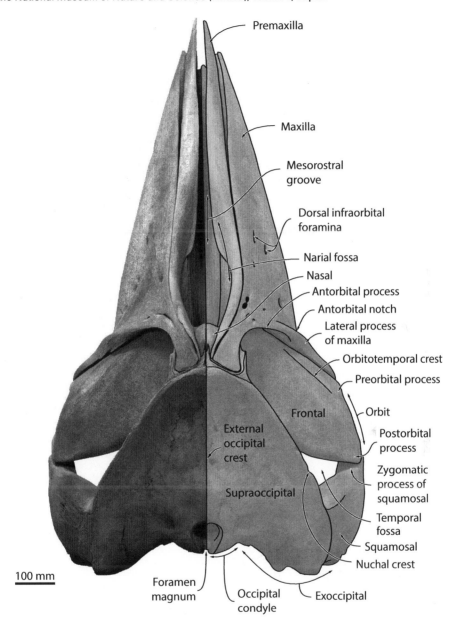

For the most part, the vomer is externally concealed by the premaxilla, maxilla, and palatine, but is exposed dorsally as the floor of the **mesorostral groove**, ventrally within a small window or along the midline of the rostrum between the maxillae, and posteriorly as part of the basicranium. Within the nasal cavity, the vomer dorsally contacts the **ethmoid** bone, which in most mammals forms the **bony nasal septum** dividing the left and right narial passages. In addition, the ethmoid gives rise to the **ethmoturbinates**, an accumulation of delicate bony scrolls carrying the tissue responsible for

Figure 3.4 Cranium of the minke whale *Balaenoptera acutorostrata* (NMNS M42450) in ventral view.

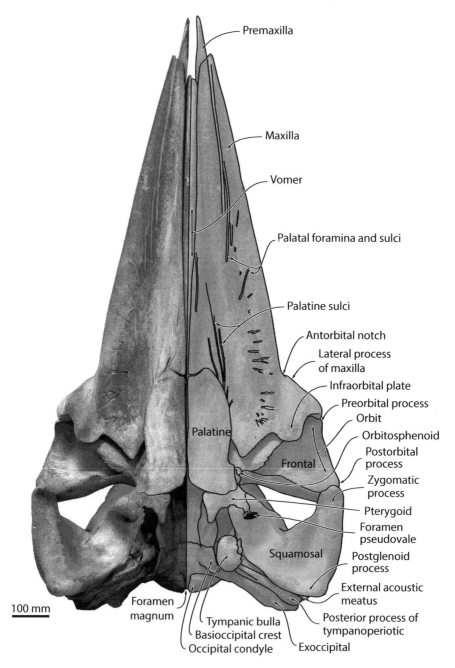

detecting odors (the olfactory epithelium) (Berta *et al.*, 2014). In cetaceans, the ethmoid is reduced, and the bony nasal septum is entirely formed by the vomer (Mead and Fordyce, 2009). Similarly, the ethmoturbinates are effectively absent in all modern odontocetes, but they are retained and seemingly still functioning in mysticetes (sections 3.4.5 and 5.1.3). The presence of the **mesethmoid**, which in most mammals contributes to the nasal septum, is disputed (Ichishima, 2011; Rommel, 1990). Posteriorly, the **cribriform plate** of the ethmoid forms the anterior border of the braincase and is generally pierced by multiple foramina for the passage of nerves connecting the olfactory epithelium with the brain (sections 3.4.4 and 3.4.5) (Thewissen *et al.*, 2011). Anterior to the ethmoid proper, the mesorostral groove is filled by the **mesorostral cartilage**, which has been interpreted as an unossified anterior extension of either the mesethmoid (Mead, 1975; Schulte, 1917) or the presphenoid (Rommel, 1990).

Compared to other mammals, the cetacean rostrum is relatively flattened. The palate tends to be flat or broadly convex, except in baleen-bearing mysticetes (Chaeomysticeti), in which the maxilla and vomer bend downward along the midline of the rostrum to form a distinct **palatal keel**. In neocetes (mysticetes plus odontocetes), the posterolateral portion of the maxilla bears a **lateral process**, which is divided from the lateral edge of the rostrum by the **antorbital notch** (Figure 3.3). Laterally (in odontocetes) or dorsally (in mysticetes), this notch is further defined by the ridge-like **antorbital process**. In life, the antorbital notch carries the facial nerve, which innervates the enlarged maxillonasolabialis muscles surrounding the blowhole and, in odontocetes, various structures related to sound production (sections 3.4.3, 3.4.4, and 5.2.2) (Huggenberger *et al.*, 2009; Mead, 1975). Posteroventrally, the maxilla terminates in front of the orbit (eye socket) in all odontocetes save one (*Mirocetus*), but partially underlaps it in archaeocetes and mysticetes (Figure 3.5). In archaeocetes, some archaic mysticetes and *Mirocetus*, this **infraorbital process** of the maxilla bears the posteriormost tooth (Fitzgerald, 2010; Sanders and Geisler, 2015). In all other mysticetes, it is expanded into a broad, toothless **infraorbital plate**.

The dorsal surface of the maxilla bears a varying number of **dorsal infraorbital foramina**, which represent the distal outlets of the **infraorbital canal** (Figures 3.3 and 3.6). The latter originates near the orbit at the **ventral infraorbital foramen**, and represents a major passageway for the infraorbital nerve and blood vessels (section 3.4.4) (Rommel, 1990). Close to its origin near the orbit,

the infraorbital canal gives rise to the **alveolar canal** for the superior alveolar artery, which runs anteriorly through the rostrum and supplies the teeth or—in mysticetes—baleen via alveolar and/ or palatal foramina (section 3.4.2) (Ekdale *et al.*, 2015). In very young individuals, the alveolar canal is ventrally open and forms a distinct alveolar sulcus. In odontocetes, the infraorbital canal branches out into the premaxilla and forms the **premaxillary foramen** (Figure 3.7). From the premaxillary foramen, three distinct **sulci** run anteromedially, posteromedially, and posterolaterally. Between the posteromedial and posterolateral sulci and the mesorostral groove, the premaxilla of odontocetes bears a variably developed **premaxillary sac fossa** for one of a series of air sacs involved in sound production (section 3.4.3) (Schenkkan, 1973). Anterior to this fossa, the posteromedial and anteromedial sulci define the **prenarial triangle,** which marks the origin of the nasal plug muscle (section 3.4.1) (Fordyce, 1994; Mead and Fordyce, 2009).

Most of the rostral bones of neocetes are affected by **telescoping** (Miller, 1923), which is the stacking of several bones caused mainly by the backward displacement of the external narial opening toward the top of the skull (section 5.1.3). Thus, the premaxilla, maxilla, and nasal have come to overlie the central portion of the forehead while, at the same time, the anterior border of the nasal has itself moved posteriorly to uncover much of the narial passage. Besides the migration of the nares, which presumably facilitates breathing while being submerged (Heyning and Mead, 1990), telescoping was likely also driven by the adoption of filter feeding in mysticetes, where it led to the development of the expanded infraorbital plate; and by the evolution of echolocation in odontocetes, in which the enlargement of the maxillonasolabialis muscles caused the posterior portion (**ascending process**) of the maxilla to become exceedingly large, covering most of the forehead up to the nuchal crest (see below and sections 3.4 and 5.2.2) (Geisler *et al.*, 2014). In addition to telescoping, many odontocetes show strong cranial **asymmetry**, especially in the area surrounding the external nares (Mead, 1975). This asymmetry is expressed in both the relative size and position of bilateral skull elements

Figure 3.5 Skull of the minke whale *Balaenoptera acutorostrata* (NMNS M42450) in lateral view: (a) photograph and (b) line drawing.

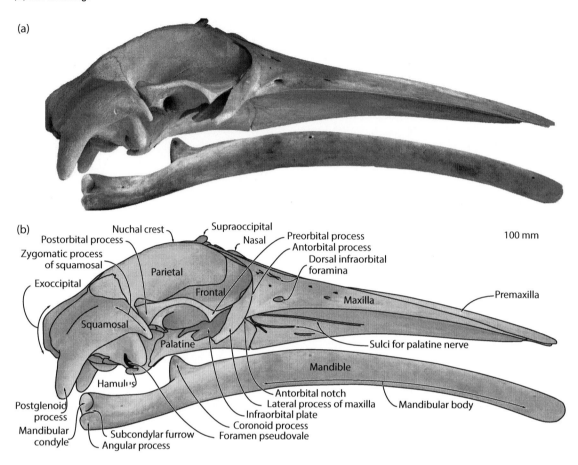

(a)

(b)

Nuchal crest
Postorbital process
Zygomatic process of squamosal
Exoccipital
Parietal
Frontal
Squamosal
Palatine
Hamulus
Postglenoid process
Mandibular condyle
Subcondylar furrow
Angular process
Supraoccipital
Nasal
Preorbital process
Antorbital process
Dorsal infraorbital foramina
Maxilla
Premaxilla
Sulci for palatine nerve
Mandible
Antorbital notch
Lateral process of maxilla
Infraorbital plate
Coronoid process
Foramen pseudovale
Mandibular body
100 mm

(especially the premaxillae, maxillae, and nasals), and generally results in a left-sided skew of the affected bones and sutures (Figure 3.7) (Heyning, 1989). To some degree, asymmetry is also developed in archaeocetes, in which it is generally right-sided and mostly results in a curved and axially twisted rostrum (Fahlke *et al.*, 2011).

3.2.2 Forehead, skull vertex, and posterior cranium

As in other mammals, the forehead of cetaceans is formed by the **frontal** bone, but has become markedly reduced because of the effects of facial telescoping and, especially in mysticetes, the concurrent forward projection of bones forming the back of the skull (see below). The frontal forms the upper rim of the **orbit** (eye socket), which in the earliest whales lies at the level of the forehead and relatively close to the midline of the skull (Nummela *et al.*, 2006). In more crownward taxa, the development of the **supraorbital process** of the frontal gradually displaces the orbit from its original position—first horizontally and eventually ventrally, with the orbit being located far below the dorsalmost portion of the frontal in all living cetaceans.

The supraorbital process roofs the **optic canal** and is flanked by the **preorbital process** (the anterior border of the orbit) anteriorly and the **postorbital process** (posterior border of the orbit) posteriorly. In archaeocetes, odontocetes, and archaic mysticetes, a variably developed **orbitotemporal crest** runs along, or close to, the posterior rim of the supraorbital process, where it marks

Figure 3.6 Cranium of the common dolphin *Delphinus delphis* in (a) dorsal, (b) ventral, and (c) lateral views.

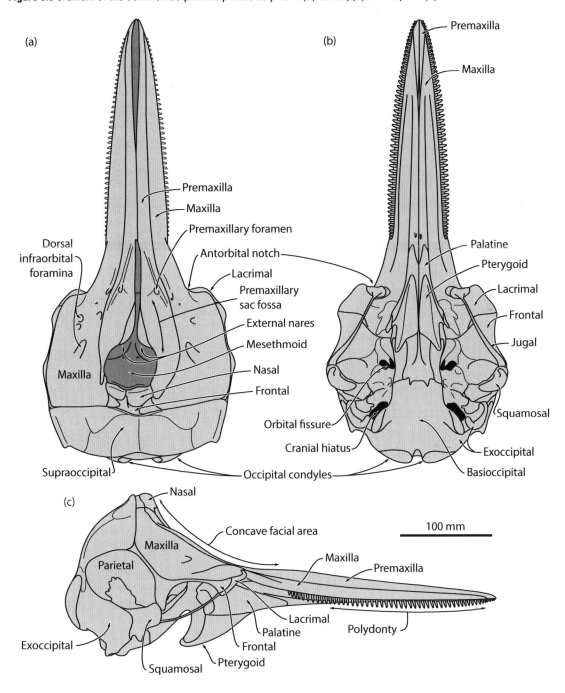

the anterior extent of the temporal muscle (section 3.4.1) (El Adli and Deméré, 2015; Schulte, 1916). In crown mysticetes, this crest is usually located further forward on the dorsal surface of the supraorbital, or otherwise is indistinct. The lower rim of the orbit is formed by the **jugal**,

Figure 3.7 Detailed view of the narial region of the common dolphin *Delphinus delphis* in dorsal view.

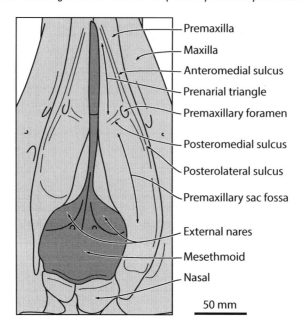

- Premaxilla
- Maxilla
- Anteromedial sulcus
- Prenarial triangle
- Premaxillary foramen
- Posteromedial sulcus
- Posterolateral sulcus
- Premaxillary sac fossa
- External nares
- Mesethmoid
- Nasal

50 mm

which in cetaceans is generally gracile and, in many odontocetes, reduced to a thin rod. At the anterior rim of the orbit, wedged in between the supraorbital process of the frontal and the maxilla, there is a variably developed **lacrimal** bone, which in many odontocetes fuses with the jugal. In archaeocetes, the lacrimal bears the lacrimal canal and associated channels, whereas it is relatively featureless in neocetes (Kellogg, 1936; Uhen, 2004).

The **vertex**, or top, of the skull is located just behind the forehead, and is formed by the posteriormost portions of rostral bones (premaxilla, maxilla, and nasal), the frontal, and the **parietal**. In beaked whales and some delphinids, the central portion of the anterior vertex is transversely compressed between the maxillae, and stands out from the facial surface of the skull as a prominent, elevated structure sometimes called the **synvertex** (Fordyce *et al.*, 2002; Moore, 1968). The parietal is broadly exposed on the skull in most mammals and archaic cetaceans (Figures 3.1 and 3.8), but strongly compressed anteroposteriorly in crown mysticetes and odontocetes as a result of facial telescoping (Figures 3.3 and 3.6). Crown mysticetes and, to a lesser degree, odontocetes are also affected by a second type of telescoping, namely, the

anterior displacement of the parietal and the **supraoccipital shield** (Miller, 1923). This forward projection of the posterior skull bones has led to the parietal overriding much of the vertex exposure of the frontal and, in some cases, the posteromedial corner of the supraorbital process. Concurrently, the parietal is itself increasingly covered by the anterior portion of the supraoccipital (Figure 3.3). In the most extreme cases (e.g., *Balaenoptera siberi* and *Caperea marginata*), both the frontal and parietal are thus entirely excluded from the skull vertex, with the supraoccipital virtually contacting the rostral bones. Note that the supraoccipital shield is a compound bone consisting of the **supraoccipital** proper and the **interparietal**. Some cetaceans, including certain rorquals and cetotheriids, seem to preserve a distinct interparietal, but its distribution across Cetacea as a whole is still poorly known (Hampe *et al.*, 2015; Mead and Fordyce, 2009; Ridewood, 1923).

Posteriorly, the supraoccipital widens to form much of the back of the skull and continues to the dorsal edge of the **foramen magnum** (i.e., the opening through which the spinal cord connects with the brain) (Figure 3.3). In some species, the supraoccipital bears a variably developed **external**

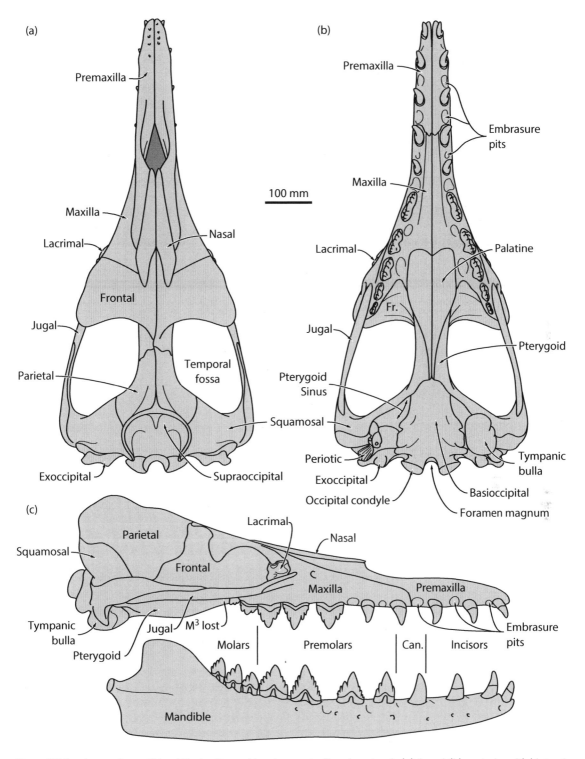

Figure 3.8 Cranium and mandible of the basilosaurid archaeocete *Dorudon atrox* in (a) dorsal, (b) ventral, and (c) lateral views. can., canine; fr., frontal. Adapted from Uhen (2004).

occipital crest. The lateral borders of the foramen magnum are formed by the **occipital condyles**, which articulate with the vertebral column posteriorly. The occipital condyle is part of the **exoccipital**, which also forms the posterolateral portion of the skull and fuses with the supraoccipital during ontogeny (Walsh and Berta, 2011). Ventrolateral to the occipital condyle, the exoccipital completely surrounds the **jugular notch** for the passage of the internal jugular vein, the internal carotid artery, and several cranial nerves (section 3.4.4) (Beauregard, 1894; Fraser and Purves, 1960). Immediately lateral to the jugular notch, the exoccipital descends ventrally to form the **paroccipital process**. Ventral and anterior to the supraoccipital and exoccipital, the lateral skull wall is mostly formed by the parietal and the more posteriorly located **squamosal**. Where these two bones meet the overlying supraoccipital, they contribute to the formation of the **nuchal crest**. In mysticetes, the telescoping of the supraoccipital may cause this crest to move outward and overhang the lateral skull wall, thus obscuring the parietal and the medial portion of the squamosal when viewed from above (Figure 3.3).

3.2.3 Temporal fossa and basicranium

The squamosal is morphologically complex and contributes to the lateral skull wall, the basicranium, and the **temporomandibular joint**, which articulates the cranium with the lower jaw. Laterally, the squamosal includes a well-developed, anteriorly directed **zygomatic process**, which contacts the jugal and, in most neocetes, closely approaches the postorbital process of the frontal (Figure 3.5). Together with the parietal and the posterior margin of the supraorbital process of the frontal, the squamosal surrounds the often large, dorsally open **temporal fossa** for the attachment of the temporal muscle (section 3.4.1) (El Adli and Deméré, 2015). However, in many crown odontocetes, the extreme telescoping of the rostral bones is accompanied by a concurrent backward shift of the supraorbital process (generally itself covered by the maxilla), which thus comes to roof the temporal fossa in dorsal view (Figure 3.6).

In archaeocetes, mysticetes, and archaic odontocetes, the dorsal border of the temporal fossa, or **temporal crest**, is formed jointly by the orbitotemporal crest, the nuchal crest, and the **supramastoid crest**, which represents a continuation of the nuchal crest on to the dorsal surface of the zygomatic process. Modern odontocetes show a similar arrangement, except that the roofing of the temporal fossa by the frontal causes the supramastoid crest to assume a more ventral position relative to the orbitotemporal and nuchal crests. In archaeocetes and archaic mysticetes, the temporal fossa is usually longer anteroposteriorly than it is wide transversely, with the portion of the frontals and parietals separating them forming a relatively narrow **intertemporal constriction** (Figure 3.8) (e.g., Fitzgerald, 2010; Kellogg, 1936; Sanders and Geisler, 2015). By contrast, the temporal fossa of modern mysticetes and most odontocetes is anteroposteriorly compressed and the intertemporal region is transversely wide, likely as a result of telescoping. Inside the temporal fossa, the squamosal of some mysticetes is bifurcated, with the two portions being separated by a deep **squamosal cleft** (Ridewood, 1923).

Ventrally, the squamosal bears the **glenoid fossa** (or mandibular fossa) and the **postglenoid process** for the articulation with the lower jaw (Figure 3.9). The structure of the cetacean temporomandibular joint has rarely been described in detail, but it appears to be **synovial** (surrounded by a capsule and containing a lubricating fluid) in right whales (Eschricht and Reinhardt, 1866), partially synovial in the gray whale (El Adli and Deméré, 2015), and non-synovial (i.e., fibrous and without a fluid-filled joint cavity) in rorquals and at least some dolphins and porpoises (Lambertsen et al., 1995; McDonald et al., 2015). In archaeocetes, most odontocetes, and archaic mysticetes, the postglenoid process is dorsoventrally short and does not extend below the level of the paroccipital process. By contrast, the process is extremely well developed and elongated in crown mysticetes (Figure 3.5). Anteromedial to the postglenoid process is the **falciform process** of the squamosal; in odontocetes, the latter is separated from the postglenoid by the **tympanosquamosal recess**, which receives an air-filled sinus emanating from the middle ear (section 3.4.3) (Fraser and Purves, 1960). In mysticetes, some archaeocetes, and some odontocetes, the falciform process contributes to a large, external opening for the

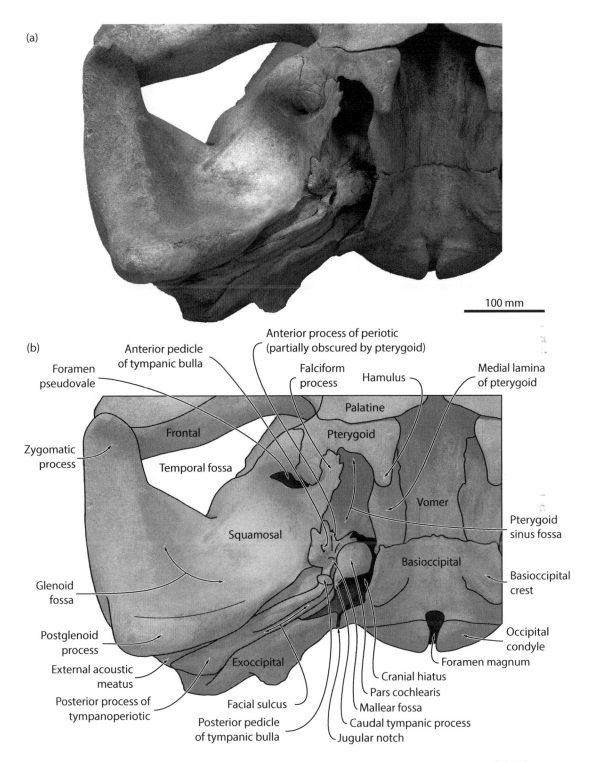

Figure 3.9 Right half of basicranium (including periotic) of the minke whale *Balaenoptera acutorostrata* (NMNS M42450) in ventral view: (a) photograph and (b) line drawing.

mandibular nerve commonly termed the **foramen pseudovale** (Figure 3.9) (Fordyce, 1994; Ridewood, 1923). Medial to the postglenoid process, the squamosal contributes to the rim of a large opening into the braincase, the **cranial hiatus**, which presumably arose as a result of the (partial) isolation of the ear bones from the rest of the skull (sections 3.4.3 and 5.1.2) (Fraser and Purves, 1960). Anatomically, the cranial hiatus represents the confluent paths of several cranial nerves and associated blood vessels (section 3.4.4) (Mead and Fordyce, 2009). Directly behind the postglenoid process, the **external auditory meatus** (ear canal; nonfunctional in modern cetaceans) runs from the ear bones to the outer surface of the skull.

The base of the braincase is formed by a succession of three unpaired bones, including (from anterior to posterior): (1) the **presphenoid**, which carries the paired orbitosphenoids; (2) the **basisphenoid**, which laterally connects to the paired alisphenoids; and (3) the **basioccipital**. Ventrally, the presphenoid, the basisphenoid, and the anteriormost portion of the basioccipital are concealed by the posterior extension of the vomer. Most of the basioccipital is visible externally and forms the lower rim of the foramen magnum, as well as the medial border of the cranial hiatus in the shape of the paired **basioccipital crests** (Figure 3.9). The **orbitosphenoid** is exposed near the base of the supraorbital process of the frontal and forms the anterior portion of the optic canal, which carries the optic nerve. The **alisphenoid** is a relatively large element, variably exposed in and/or just below the temporal fossa in archaeocetes, odontocetes and most mysticetes (e.g., Fraser and Purves, 1960; Muller, 1954). The alisphenoid encloses the **foramen ovale,** which transmits the mandibular nerve (section 3.4.4). Where present, the foramen pseudovale (see above) lies external to the foramen ovale (Fraser and Purves, 1960).

The **pterygoid** bone occupies a central position between the basicranium and the hard palate. Medially, the pterygoid forms the lateral margin of the choanae, and is continuous with the basioccipital crest further posteriorly. Ventrally, the pterygoid develops a hook-shaped process known as the **pterygoid hamulus**. Internally, the pterygoid is excavated and dominated by the large **pterygoid sinus fossa**, which houses a major component of the accessory air sinus system (Figure 3.9) (section 3.4.3). The development of this fossa markedly differs between archaeocetes and mysticetes on the one hand and odontocetes on the other, in line with the much more extensive development of the accessory sinus system in the latter (Fraser and Purves, 1960).

In archaeocetes and mysticetes, the pterygoid sinus is almost entirely enclosed by the pterygoid and delimited by its **lateral, medial, and dorsal laminae**, as well as the alisphenoid. In mysticetes, the pterygoid forms a continuous lateral lamina, which contacts the falciform process of the pterygoid posteriorly and often contributes to the formation of the foramen pseudovale (Figure 3.9). Likewise, in all but the most archaic mysticetes, the dorsal lamina extends posteriorly to roof most or all of the pterygoid sinus fossa, thus covering the alisphenoid in ventral view (Fitzgerald, 2010; Fraser and Purves, 1960). In some mysticetes, the pterygoid furthermore extends upward on to the lateral skull wall, where it may largely or entirely exclude the alisphenoid from the temporal fossa (Muller, 1954). At the same time, much of the ventral surface of the mysticete pterygoid is itself overridden by a posterior extension of the palatine. This arrangement is developed most extremely in right whales, in which the palatine broadly covers most of the pterygoid up to the hamulus (Eschricht and Reinhardt, 1866). In contrast to mysticetes, extensions of the pterygoid sinus reach further anteriorly on to the palatine, or dorsolaterally into the orbit, in many crown odontocetes (section 3.4.3). As a result, the dorsal and lateral laminae of the pterygoid are usually less well developed or absent, leaving the alisphenoid broadly exposed within the pterygoid sinus fossa (Fraser and Purves, 1960). Likely for the same reason, the pterygoid of odontocetes has a tendency to extend forward on to the palatine, partially or entirely covering the latter in ventral view.

3.2.4 Periotic

The **periotic** is located in the basicranium, between the squamosal, exoccipital, basioccipital, and pterygoid. It houses the inner ear (sections 3.4.5) and, along with the tympanic bulla and the three auditory ossicles (malleus, incus and stapes), is one of the five cetacean ear bones. Owing to its high degree of complexity, the ear region provides one of the richest sources of morphological data for classification purposes and phylogenetic analyses (e.g., Ekdale *et al.*, 2011; Geisler and Luo, 1996; Luo and Marsh, 1996; O'Leary, 2010). Cetaceans differ from other mammals in that their ear bones are partially or entirely isolated from the rest of the cranium via the formation of the cranial hiatus and the development of an extensive air sinus system (sections 3.4.3 and 5.1.2) (Fraser and Purves, 1960; Oelschläger, 1986a). In archaeocetes and mysticetes, the outer portion of the periotic is still firmly anchored within the squamosal (Figures 3.8 and 3.9). By contrast, this connection is more tenuous in many crown odontocetes (especially delphinidans), and the ear bones are hence more easily lost during fossilization or specimen preparation (Luo and Gingerich, 1999).

The periotic consists of four main parts: the anterior process, the body of the periotic, the pars cochlearis, and the posterior process (Figure 3.10). As its name suggests, the **anterior process** is the

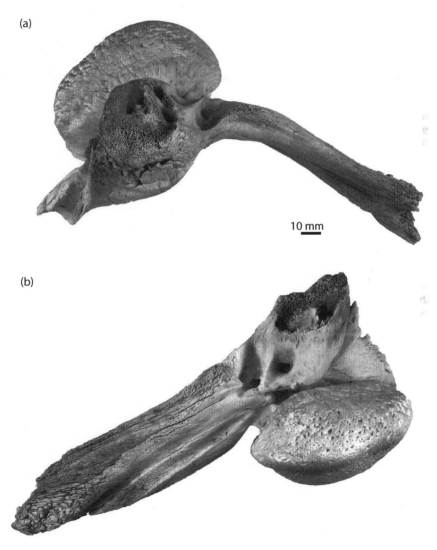

(a)

10 mm

(b)

Figure 3.10 Isolated left periotic and tympanic bulla of the minke whale *Balaenoptera acutorostrata* (NMNS M42450) in (a) dorsal and (b) posteromedial views: (a, b) photographs and (c, d) line drawings.

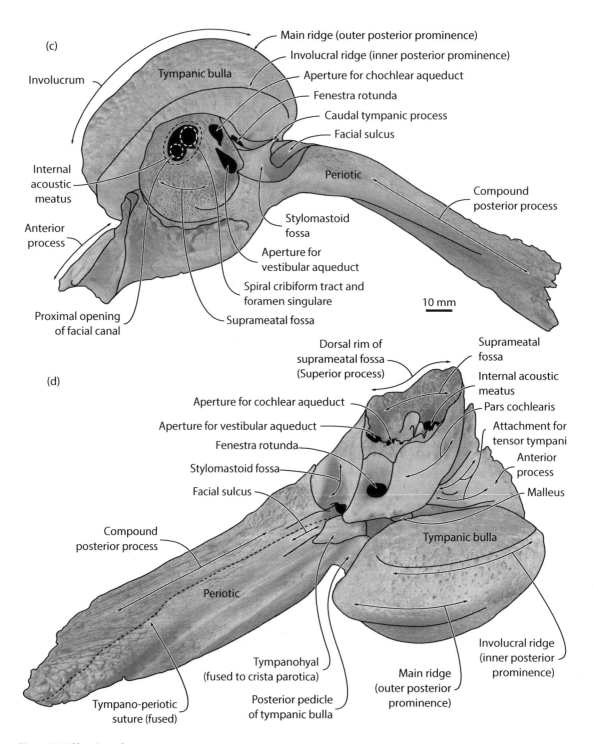

(c)

Main ridge (outer posterior prominence)

Involucral ridge (inner posterior prominence)

Aperture for chochlear aqueduct

Tympanic bulla

Fenestra rotunda

Involucrum

Caudal tympanic process

Facial sulcus

Periotic

Compound posterior process

Internal acoustic meatus

Stylomastoid fossa

Anterior process

Aperture for vestibular aqueduct

Spiral cribiform tract and foramen singulare

10 mm

Proximal opening of facial canal

Suprameatal fossa

(d)

Dorsal rim of suprameatal fossa (Superior process)

Suprameatal fossa

Internal acoustic meatus

Aperture for cochlear aqueduct

Pars cochlearis

Aperture for vestibular aqueduct

Attachment for tensor tympani

Fenestra rotunda

Anterior process

Stylomastoid fossa

Malleus

Facial sulcus

Compound posterior process

Tympanic bulla

Periotic

Tympanohyal (fused to crista parotica)

Involucral ridge (inner posterior prominence)

Tympano-periotic suture (fused)

Posterior pedicle of tympanic bulla

Main ridge (outer posterior prominence)

Figure 3.10 (Continued)

anteriormost portion of the periotic, and it is usually squared or triangular in medial view. With the periotic *in situ* in the skull, the anterior process extends toward the pterygoid sinus fossa; in some taxa (e.g., rorquals), its anterior portion is partially covered by the dorsal or lateral lamina of the pterygoid (Figure 3.9). Laterally, the anterior process gives rise to the **lateral tuberosity** (Figure 3.11) and furthermore bears the **anteroexternal sulcus**, which may carry the meningeal artery (Fordyce, 1994). On its medial side, the anterior process may develop a distinct set of sulci or a crest for the attachment of the **tensor tympani**, which acts upon the malleus and, thus, the tympanic membrane (sections 3.2.6 and 3.4.5) (e.g., Geisler and Luo, 1996; Tsai and Fordyce, 2015). Ventrally, the anterior process articulates with the tympanic bulla via a depression known as the **fovea epitubaria**, which receives the **accessory ossicle** of the bulla (section 3.2.5). In some odontocetes, the anterior process additionally carries a separate **anterior bullar facet** anterior to the fovea epitubaria for articulation with the anterodorsal portion of the bulla (Fordyce, 1994). In archaeocetes, most odontocetes, and archaic mysticetes, the contact between the accessory ossicle and the fovea epitubaria is loose, except for a small area of fusion near the posterior portion of the ossicle (Figures 3.12 and 3.13). By contrast, the accessory ossicle entirely fuses to the anterior process in crown mysticetes and some odontocetes (e.g., sperm whales). In mysticetes, the ossicle is furthermore generally located at the end of—or partially transformed into—an elongate stalk connecting it to the tympanic bulla, and thus more commonly known as the **anterior pedicle** (Mead and Fordyce, 2009; Yamato and Pyenson, 2015).

Near its junction with the body of the periotic, medial to the lateral tuberosity and posterior to the anterior pedicle of the tympanic bulla, the anterior process bears a circular fossa for the head of the malleus (**mallear fossa**), which is relatively well defined in all cetaceans except crown mysticetes (Ekdale *et al.*, 2011). Just posterior to the mallear fossa, two foramina open side by side (Figures 3.12 and 3.13): the **distal opening of the facial canal**, transmitting the facial nerve; and the **fenestra ovalis**, through which the most proximal of the auditory ossicles, the stapes, makes contact

with the vestibule of the inner ear (section 3.4.5). Immediately posterior to the fenestra ovalis lies the **fossa for the stapedius muscle**, which, true to its name, serves to stabilize the stapes. The facial nerve passes from the distal opening of the facial canal into the **facial sulcus**, which runs posteriorly on to the posterior process of the periotic (Figure 3.10) (section 3.4.4). Somewhat ventral to the facial sulcus, at the base of the posterior process of the periotic, is the **fossa incudis**, which receives the crus breve of the incus (section 3.2.6).

The globular **pars cochlearis** is located medial to the body of the periotic and ascends from the fenestra ovalis dorsomedially toward the cranial cavity. The pars cochlearis houses the inner ear (including the organ of balance), which communicates with the brain and the surrounding area via a variety of foramina (section 3.4.5) (Geisler and Luo, 1996; Luo and Eastman, 1995). On the dorsal surface of the pars cochlearis, these foramina include the **internal acoustic meatus** anteriorly and the **apertures for the cochlear and vestibular aqueducts** posteriorly (Figure 3.10). The internal acoustic meatus is itself divided into two distinct openings by the **transverse crest**, with the more anterior foramen representing the **proximal opening of the facial canal** and the more posterior the **spiral cribriform tract** for the passage of the cochlear nerve (section 3.4.4). The spiral cribriform tract may be partially or entirely confluent with the **foramen singulare**, which transmits the vestibular nerve; however, in some species, the two openings are separated by a septum whose height may rival or even exceed that of the transverse crest. In archaeocetes and archaic neocetes, the area dorsal to the internal acoustic meatus is developed into a broad **suprameatal fossa**, which is laterally bounded by a tall crest termed the **superior process of the periotic** (Figure 3.13) (Kellogg, 1936). In crown odontocetes and most crown mysticetes, the superior process is mostly (but not always) indistinct and only preserved as a **dorsal crest** (Boessenecker and Fordyce, 2014; Fordyce, 1994). Nevertheless, it is still often possible to discern a more or less well-developed, and occasionally hypertrophied, suprameatal fossa.

Anterior to the internal acoustic meatus, a usually small foramen located at or near the junction

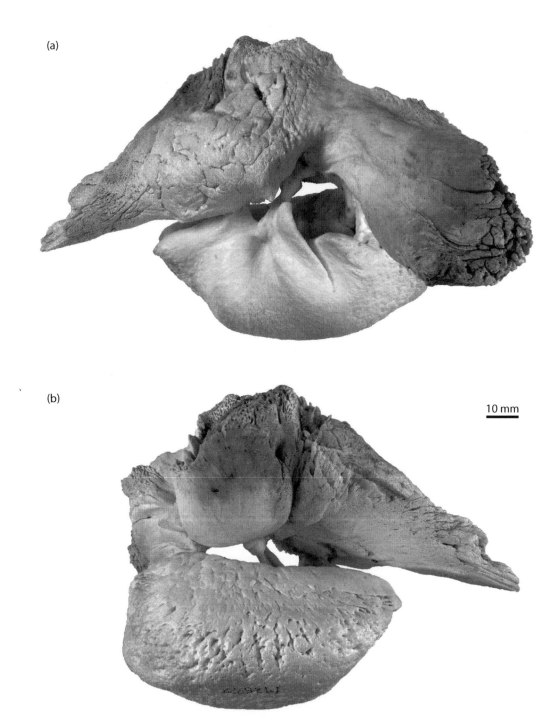

(a)

(b)

10 mm

Figure 3.11 Isolated left periotic and tympanic bulla of the minke whale *Balaenoptera acutorostrata* (NMNS M42450) in (a) lateral and (b) medial views: (a, b) photographs and (c, d) line drawings.

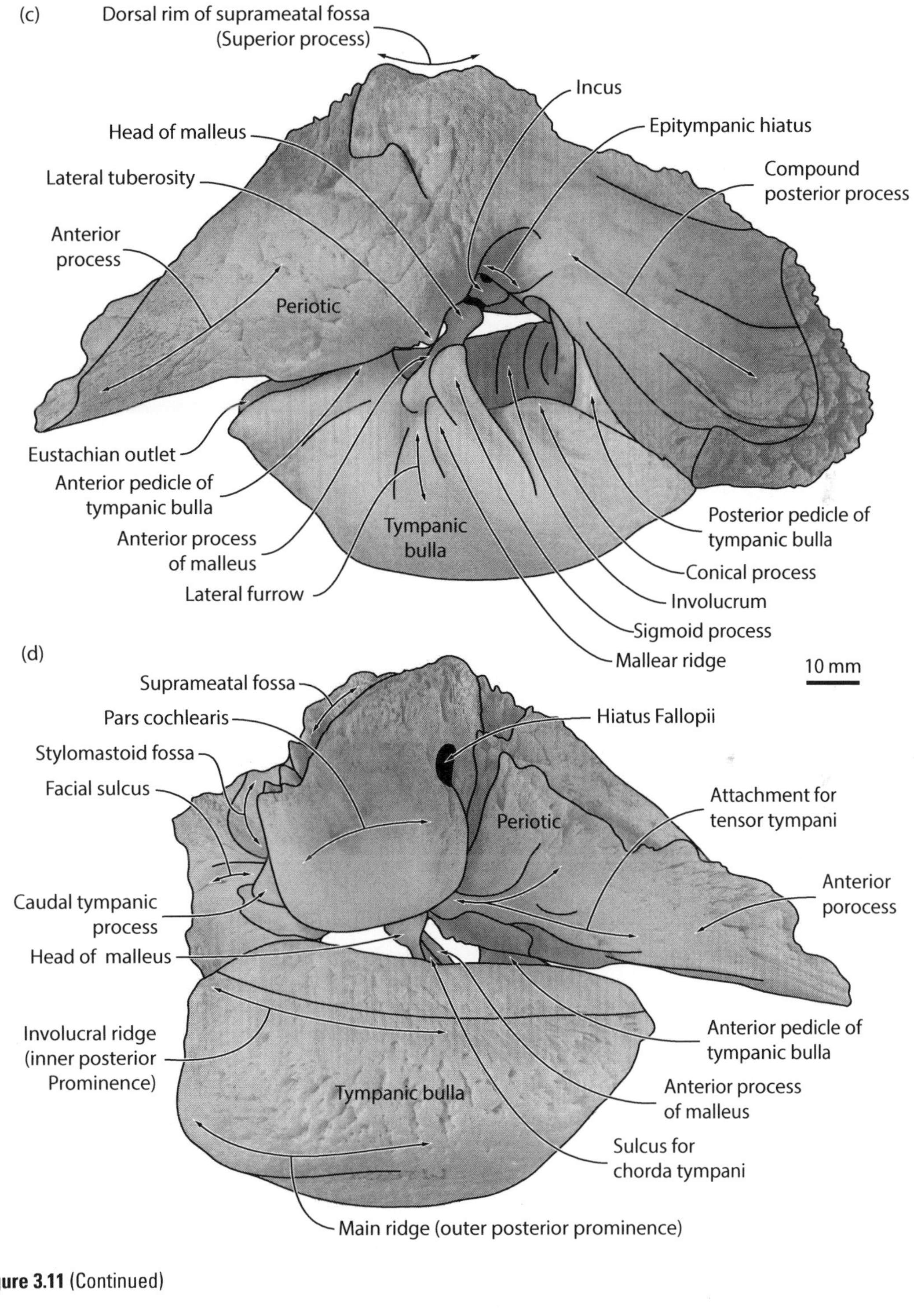

(c)

Dorsal rim of supramental fossa
(Superior process)

Incus

Epitympanic hiatus

Compound
posterior process

Head of malleus

Lateral tuberosity

Anterior
process

Periotic

Eustachian outlet

Anterior pedicle of
tympanic bulla

Anterior process
of malleus

Lateral furrow

Tympanic
bulla

Posterior pedicle of
tympanic bulla

Conical process

Involucrum

Sigmoid process

Mallear ridge

10 mm

(d)

Supramental fossa

Pars cochlearis

Stylomastoid fossa

Facial sulcus

Hiatus Fallopii

Attachment for
tensor tympani

Periotic

Anterior
porocess

Caudal tympanic
process

Head of malleus

Involucral ridge
(inner posterior
Prominence)

Tympanic bulla

Anterior pedicle of
tympanic bulla

Anterior process
of malleus

Sulcus for
chorda tympani

Main ridge (outer posterior prominence)

Figure 3.11 (Continued)

Figure 3.12 Left ear bones of the common dolphin *Delphinus delphis*. Tympanic bulla in (a) lateral and (b) medial views, and periotic in (c) ventral and (d) dorsal views. inner post. prom., inner posterior prominence.

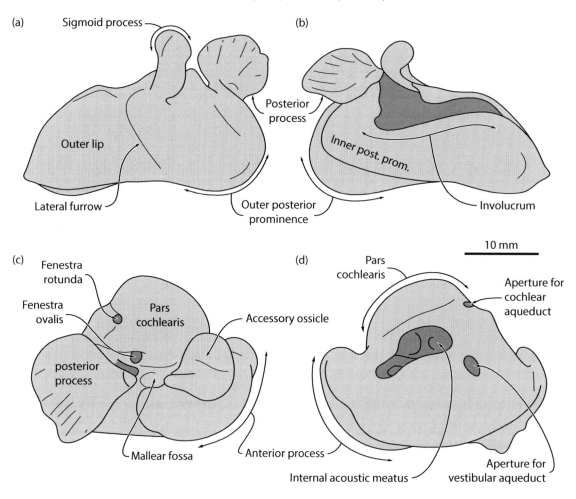

of the pars cochlearis with the anterior process marks the opening for the major petrosal nerve, or **hiatus Fallopii** (Figure 3.11). In some species, the latter may be anteriorly confluent with the internal opening of the facial canal, or else indistinct (Geisler and Luo, 1996; Mead and Fordyce, 2009). A further opening, the **fenestra rotunda**, is located on the posterior or posteroventral face of the pars cochlearis, and sometimes is linked to the aperture for the cochlear aqueduct by a fissure or open sulcus. Laterally, the fenestra rotunda is separated from the fossa for the stapedius muscle by the plate-like **caudal tympanic process**. Lateral to this process, the posterior face of the pars cochlearis

is excavated by the **stylomastoid fossa** (Figure 3.10), which likely houses a portion of the accessory sinus system (section 3.4.3) (Geisler and Luo, 1996).

As its name suggests, the **posterior process of the periotic** forms the hindmost portion of this bone. The posterior process projects posteriorly and/or laterally away from the pars cochlearis, and carries the distal portion of the facial sulcus (Figures 3.12 and 3.13). In archaeocetes, but not neocetes, it is exposed on the lateral skull wall. In some taxa, the stylomastoid fossa extends from the posterior face of the pars cochlearis on to the base of the posterior process. Ventrally, the posterior

Figure 3.13 Right ear bones of the basilosaurid archaeocete *Zygorhiza kochii*. Tympanic bulla in (a) lateral and (b) posteromedial views, and periotic in (c) ventral and (d) dorsal views.

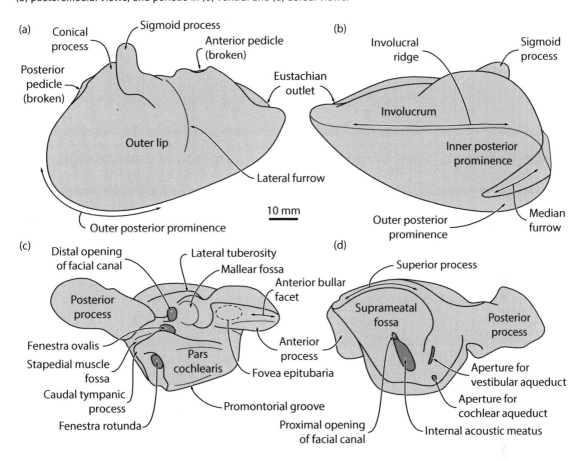

process bears a well-developed surface for the articulation with the posterior process of the tympanic bulla, the **posterior bullar facet**. In crown mysticetes, the two posterior processes are fused (see below).

3.2.5 Tympanic bulla

The **tympanic bulla** is a distinct bean- or box-shaped element enclosing the (fundus of the) **tympanic cavity**, and is located directly below the periotic (Figures 3.10 and 3.11). Except for that of extant mysticetes (Ekdale *et al.*, 2011), the tympanic bulla of most cetaceans is longitudinally divided into a lateral and a medial lobe, also known as the **outer and inner posterior prominences** (Kasuya, 1973; Okazaki, 2012). The latter are separated from each other by the

interprominential notch and its anterior continuation, the **median furrow** (Figure 3.13). Viewed from above, the medial portion of the tympanic bulla—the **involucrum**—is thickened, rounded, and seemingly rolled inward, thus giving this bone its characteristic shape. In some species, the involucrum bears a series of transverse creases on its dorsal surface, and can be distinguished into a thickened posterior portion and a thinner anterior one. Medially, the involucrum may give rise to a low transverse ridge separating the fundus of the tympanic cavity into anterior and posterior portions. Anteriorly, the fundus opens into the **Eustachian outlet** for the passage of the Eustachian tube, which connects the tympanic cavity with the nasopharynx (Figures 3.11 and 3.13) (Purves, 1966).

In contrast to the involucrum, the lateral portion of the tympanic bulla, or **outer lip**, is transversely thin and bears a series of distinct processes. From anterior to posterior, these include: (1) the **accessory ossicle/anterior pedicle**, which articulates with the anterior process of the periotic; (2) the **mallear ridge**, which marks the point of fusion of the anterior process of the malleus (section 3.2.6), and is separated from the anterior pedicle by the sulcus for the chorda tympani; (3) the **sigmoid process**, a vertically oriented, tongue-shaped structure that dorsally approximates or contacts, but never fuses with, the squamosal or periotic; (4) the **conical process**, a small projection located immediately behind the sigmoid process; and (5) a part of the base of the **posterior process of the tympanic bulla** (Figure 3.11). Except in crown mysticetes, the latter is split into an **outer and an inner posterior pedicle**, arising from the posteriormost portions of the outer lip and the dorsal surface of the involucrum, respectively. The opening between the two pedicles is known as the **elliptical foramen** and transmits the posterior sinus (section 3.4.3) (Fraser and Purves, 1960).

The sigmoid process, conical process, and (outer) posterior pedicle together form the **tympanic ring**, or annulus, for the attachment of the tympanic membrane (section 3.4.5). In most cetaceans except crown mysticetes, the sigmoid process is ventrally limited by a clearly defined ventral border, and may override the anterior portion of the conical process in lateral view (Figures 3.12 and 3.13) (Geisler and Sanders, 2003). Anteriorly, the sigmoid process and the mallear ridge are separated from the anterior pedicle by the **lateral furrow**. The posterior process of the tympanic bulla is usually distinctly larger than that of the periotic, and largely covers the latter in ventral view. In crown mysticetes, the two processes fuse to form the **compound posterior process** (Figure 3.10); in archaeocetes, odontocetes, and archaic mysticetes, the processes remain unfused but may be firmly sutured. In some taxa (e.g., cetotheriids, physeteroids, and ziphiids), the distal portion of the posterior process of the bulla is markedly expanded and variably exposed on the lateral skull wall (Fleischer, 1975; Kasuya, 1973; Whitmore and Barnes, 2008).

3.2.6 Auditory ossicles

The auditory ossicles are located in the middle ear, and they comprise a chain of small bones (the **malleus**, **incus**, and **stapes**) stretching from the outer lip of the tympanic bulla to the fenestra ovalis on the ventral side of the periotic (Figure 3.14). The **malleus** is the most distal of the three; it projects from its location adjacent to the sigmoid process into the tympanic cavity, where it articulates with the incus. In nonmammalian terrestrial vertebrates, the homologues of the malleus and incus are known as the quadrate and articular, respectively, and form the articulation of the lower jaw with the skull (Reichert, 1837). In archaic mammals, a secondary jaw articulation arose between the dentary, which originally only formed the anterior portion of the mandible, and the squamosal. Concurrently, the quadrate and articular bones, now free from bearing the jaw, became specialized to function in hearing only and moved from the outer surface of the skull into the middle ear (Kemp, 2004).

The malleus is attached to the outer lip of the tympanic bulla via its elongate **anterior process**, which runs along most of the medial border of the sigmoid process and fuses with the outer lip at the mallear ridge (Figure 3.11). As is also the case in humans, the origins of the anterior process lie not with the malleus itself, but with a separate embryonic element, the **goniale** (Yamato and Pyenson, 2015). The sulcus for the chorda tympani runs from the outer lip on to and along the anterior process toward the globular **head of the malleus**. The latter carries two rounded facets for articulation with the incus and is confluent with the robustly built **tubercule**, which itself is divided into the **manubrium** and the **muscular process**. The manubrium represents the site of attachment of the tympanic membrane (via the tympanic ligament), whereas the muscular process provides a point of insertion for the tensor tympani muscle (Purves, 1966).

The **incus** forms the center of the ossicular chain and connects the malleus with the stapes. The main part of the bone is formed by the body of the incus, which articulates with the head of the malleus and bears two projections: the **crus breve**, which articulates with the fossa incudis of the periotic; and the **crus longum**, which articulates with the stapes via the **lenticular process**. The **stapes**

Figure 3.14 Left auditory ossicles (malleus, incus, and stapes) of the minke whale *Balaenoptera acutorostrata* (NMNS M25927) in posterolateral view: (a) photograph and (b) line drawing. anterior ped., anterior pedicle; lateral tub., lateral tuberosity; mal., malleus.

(a)

(b)

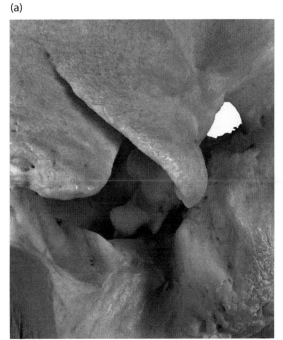

is the most proximal of the auditory ossicles. It connects the ossicular chain to the vestibule of the inner ear via its **footplate**, or base, which is lodged inside the fenestra ovalis of the periotic (section 3.4.5). Above the base, the stapes is elongate and perforated by the variably developed **stapedial foramen**, separating the bone into **anterior and posterior crura**. Distally, the latter pass into the **neck of the stapes**, which carries the insertion of the stapedius muscle, and finally the **head of the stapes**, which articulates with the lenticular process of the incus.

3.2.7 Dentition

Like those of all mammals, cetacean teeth consist of an external **crown** and a variable number of **roots** that anchor the tooth in the jaw. Internally, the core of the entire tooth is made up of **dentine** surrounding an internal **pulp cavity**. Externally, the dentine underlying the crown is usually covered by hard **enamel**, whereas the roots are covered by

cementum. Archaeocetes have a differentiated (**heterodont**) dentition, including, from front to back, **incisors**, **canines**, **premolars**, and **molars** (Figure 3.8) (Kellogg, 1936). In addition, the earliest whales retain the ancestral distinction between an anterior shearing end (**trigonid**) and a posterior crushing basin (**talonid**) in the crowns of the lower molars. However, the talonid is markedly reduced compared to that of other mammals, and entirely disappears in later archaeocetes and neocetes (Thewissen and Bajpai, 2001).

Except for basilosaurids, which lack the third upper molar, archaeocetes have a dental formula of 3.1.4.3/3.1.4.3 (i.e., three incisors, one canine, four premolars, and three molars in each **quadrant** of both the upper and the lower jaws). Basilosaurids and some archaic neocetes have accessory denticles on the mesial and distal margins of their premolars and molars, and reduce the number of upper cheek tooth roots from three to two (Fitzgerald, 2006; Uhen, 2004). With few exceptions, later odontocetes simplify their dentitions to the point where all of their teeth become uniformly conical and single-rooted—a condition known as **homodonty**. The extant odontocete tooth count is usually also far beyond that of terrestrial mammals and thus considered to be **polydont** (Armfield *et al.*, 2013).

3.2.8 Mandible

As is typical of mammals, the mandible of cetaceans is constructed from a single bone (the **dentary**) and divided into the horizontal, tooth-bearing **mandibular body** and the vertically oriented **ramus** (Figure 3.5). Like the rostrum, the cetacean mandible tends to be elongated. In archaeocetes and odontocetes, the left and right mandibles are usually sutured or fused anteriorly at the **symphysis**, and they are externally concave when viewed from above. By contrast, the connection between the mandibles is ligamentous only, and the mandibles are straight or convex in all but the most archaic mysticetes (Fitzgerald, 2012). Inside the body, the **mandibular canal** runs along the entire length of the tooth row and carries the mandibular nerve and associated blood vessels. Along nearly the entire length of the body, the mandibular canal gives rise to a series of medial and lateral openings known as **gingival** and **mental foramina**, respectively; posteriorly, it opens into the **mandibular foramen** (Figure 3.15). In most archaeocetes, archaic mysticetes, and odontocetes, the latter is rather large and occupies nearly the entire posteromedial face of the mandible, thus forming a **mandibular fossa**. In odontocetes in particular, the lateral wall of the mandibular fossa is exceedingly thin and commonly known as the **pan bone** (Barroso *et al.*, 2012). By contrast, the mandibular foramen is markedly smaller and its lateral wall thickened in all living mysticetes.

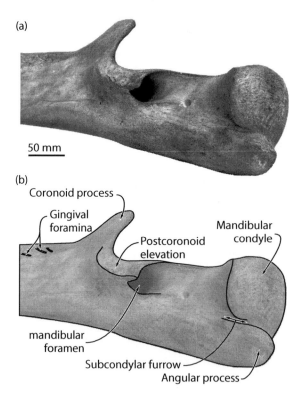

Figure 3.15 Posterior portion of the right mandible of the minke whale *Balaenoptera acutorostrata* (NMNS M42450) in posteromedial view: (a) photograph and (b) line drawing.

Dorsal to the mandibular foramen, the upper portion of the ramus is formed by the **coronoid process**, which serves as attachment site for the temporal muscle (section 3.4.1). The coronoid process tends to be large and plate-like in archaeocetes and archaic neocetes (Kellogg, 1936; Uhen, 2008b), but low and less distinct in modern odontocetes. In mysticetes, the development of the coronoid process is variable, ranging from finger-like to triangular or even absent (Tsai and Fordyce, 2015). Where it is present, it is usually bent outward (Figure 3.15). Ventral to the mandibular foramen, the lower part of the ramus is formed by the **angular process**. Except in crown mysticetes, the latter is medially excavated and confluent with the mandibular fossa. In some sperm whales, the angular process is greatly reduced (Bianucci and Landini, 2006), whereas in some mysticetes it is robust and elongated posteriorly (Bouetel and de Muizon, 2006). Posterior to the ramus, the **mandibular condyle** connects the mandible with the squamosal via the temporomandibular joint. In archaeocetes, odontocetes, and archaic mysticetes, the condyle is relative small and sometimes offset from the ramus by a short **mandibular neck**. In crown mysticetes, the condyle and mandibular neck tend to be enlarged, massive, and closely apposed to the angular process. The latter is usually massive, and it is separated from the condyle by the **subcondylar furrow**.

3.2.9 Hyoid apparatus

The **hyoid apparatus** comprises a chain of often loosely interconnected elements surrounding the throat. Functionally, the hyoid bones provide an attachment site for muscles involved in swallowing and supporting the larynx, and, possibly, they may play a role in locomotion (Reidenberg and Laitman, 2011). In cetaceans, the hyoid apparatus comprises three main ossified elements: (1) the single, unpaired **basihyal**, which forms the center of the hyoid chain; (2) the **thyrohyal**, which anteriorly articulates with the basihyal; in adult crown mysticetes and odontocetes, these two elements are generally fused (Omura, 1964; Reidenberg and Laitman, 1994); and (3) the **stylohyal**, a rod-shaped element that articulates with the anterior portion of the basihyal and connects the hyoid apparatus with the base of the skull (Figure 3.16). Additionally, a variable developed, ossified **tympanohyal** may be present and fused to the posterior process of the periotic (Flower, 1885; Oelschläger, 1986b). The tympanohyal is the most proximal of the hyoid elements and, in other mammals, forms the main connection between the hyoid apparatus (via the stylohyal) and the

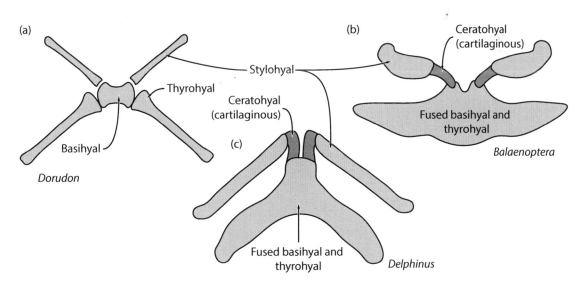

Figure 3.16 Hyoid apparatus of (a) the basilosaurid archaeocete *Dorudon atrox*, (b) the minke whale *Balaenoptera acutorostrata*, and (c) *Delphinus delphis*, all in ventral view. Drawing of *Dorudon* adapted from Uhen (2004).

periotic. However, in cetaceans the stylohyal additionally has a direct, ligamentous attachment to the paroccipital process, with the tympanohyal being mostly cartilaginous (Flower, 1885). Other hyoid elements typically present in mammals (the ceratohyal and epihyal) are not ossified in cetaceans, and they are only present as cartilage forming the connection between the basihyal and the stylohyal.

3.3 The postcranial skeleton

3.3.1 Vertebral column and rib cage

The **vertebral column** is the main support structure of the body to which all other parts of the skeleton (skull, rib cage, and limbs) attach, and forms a protective sheath around the spinal cord. A typical vertebra comprises a main body (**centrum**), a **neural arch** housing the spinal cord, a **neural spine**, and **transverse processes** for the

attachment of the epaxial musculature, and pairs of anterior (**prezygapophysis**) and posterior (**postzygapophysis**) articular processes connecting the individual vertebrae (Flower, 1885). In addition, most vertebrae also bear a **metapophysis**—a robust, anterolaterally pointing process usually situated lateral to the prezygapophysis. Together, the neural arches of successive vertebrae form the **neural canal**.

As in all mammals, the vertebral column of cetaceans is divided into five distinct sections (Figure 3.17): the neck (**cervical**), chest (**thoracic**), lower back (**lumbar**), hip (**sacral**), and tail (**caudal**). Individual vertebrae within these regions are generally labeled C, T, L, S, and Ca, respectively, and numbered according to their relative position (e.g., C7, T5, and Ca1). Like the vast majority of mammals, cetaceans have seven cervical vertebrae (Narita, 2005; Todd, 1922). By contrast, the composition of the thoracic, lumbar, and caudal regions varies considerably between species,

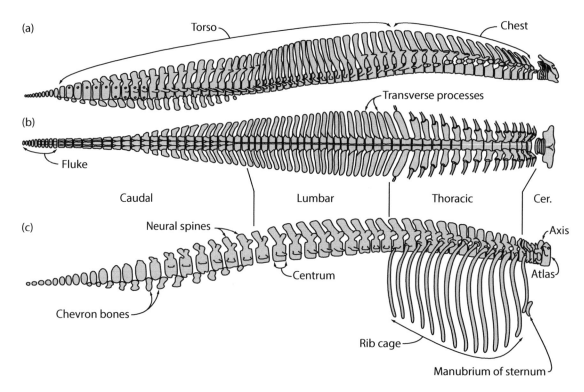

Figure 3.17 Vertebral column of the common dolphin *Delphinus delphis* in (a) lateral and (b) ventral views, and of the humpback whale *Megaptera novaeangliae* in (c) lateral view. cer., cervical. Adapted from Van Beneden and Gervais (1880).

although thoracics usually number more than 10 and rarely more than 15–16. Important exceptions to this rule include the extinct basilosaurids (e.g., *Dorudon* and *Basilosaurus*), Dall's porpoise (*Phocoenoides dalli*), and the pygmy right whale (*Caperea marginata*), in which the thoracics number up to 18 (Buchholtz, 2011; Jefferson, 1988; Uhen, 2004). Lumbar and caudal vertebrae seem to be less constrained in their numbers, with lumbars in particular ranging from as few as 1–2 in *Caperea* to as many as 33 in *Lissodelphis* (Buchholtz, 2011; Buchholtz and Schur, 2004; Jefferson, 1988). In general, however, cetaceans have more thoracic and lumbar vertebrae than most terrestrial mammals, including artiodactyls (Buchholtz, 2007; Narita, 2005).

The **cervical series** is headed by the **atlas** (C1) and **axis** (C2) (Figure 3.17). The atlas is a bowl-shaped bone which accommodates the occipital condyles of the skull, and allows the latter to move up and down as well as sideways. Unlike all of the other vertebrae, the atlas does not possess a massive centrum; instead, its main body forms a ring which is able to rotate around a spine-like projection (**odontoid process**) on the anterior face of the axis. Posterior to the axis, the centra of C3–C7 tend to be round- or square-shaped in anterior view, and bear horizontally oriented zygapophyses that allow twisting of the column and side-to-side motions. However, in most cetaceans the mobility of the neck is severely restricted, with the cervical vertebrae being anteroposteriorly compressed or even fused to one another (e.g., in right whales, sperm whales, and oceanic dolphins). With the exception of the atlas, the cervical vertebrae typically carry two transverse processes on each side (the upper **diapophysis** and the lower **parapophysis**), which together enclose a distinct **vertebrarterial** or **transverse foramen** for the passage of blood vessels. Together, the transverse foramina of successive cervical vertebrae form the **vertebrarterial canal**. On C7, the lower transverse process is often reduced or completely absent.

Posterior to the neck, the **thoracic vertebrae** support the chest, with each vertebra being connected to its respective **rib** (Figure 3.17). In anterior view, thoracic vertebrae usually have a heart-shaped (cordiform) or oval centrum, and generally only a single (upper) transverse process.

Toward the posterior portion of the thorax, the neural spines become gradually taller, and the transverse processes transversely longer and more robust. The orientation of the zygapophyses also changes, from horizontal in the anterior thorax to nearly vertical further posteriorly. This reorientation facilitates dorsoventral motion of the posterior column, which is typical of mammalian locomotory styles. The ribs attach to the vertebrae in two ways: the more anterior ribs are often double-headed, with a distinct head (**capitulum**) and tubercle (**tuberculum**); where this is the case, the capitulum articulates with a facet located at the junction between the centra of two adjacent vertebral centra, whereas the tubercle contacts the transverse process. By contrast, the posterior ribs have often lost the capitulum, and hence only articulate (sometimes rather loosely) with the transverse process of their respective vertebra (Rommel, 1990). In some cases, the posteriormost pair of articulated ribs may be followed by unattached "floating" ribs (Slijper, 1936), thus blurring the boundary between the thoracic and lumbar regions of the vertebral column.

The first pair of ribs is always the stoutest, and ventrally articulates with the **manubrium**, or anteriormost element, of the **sternum** (Figure 3.17). In archaeocetes, odontocetes, and archaic mysticetes, the manubrium is followed by a series of **mesosternal** segments, which in some species may fuse into a single mass. Like the vertebrae, all of the sternal elements (sternebrae) are unpaired and aligned along the sagittal plane. However, developmentally each sternebra does arise from two separate ossification centers, and in some taxa (e.g., the sperm whale, *Physeter*) the gap between these two does not close during ontogeny, thus leaving behind a median fontanelle (Flower, 1885; Rommel, 1990). The **xiphisternum**, the posteriormost sternebra commonly found in mammals, is present in archaeocetes, but it is absent in extant odontocetes (Flower, 1885; Uhen, 2004).

Modern mysticetes have reduced the sternum even further and only retain a single sternebra formed entirely by the manubrium (Klima, 1978), although an additional, separately ossified element of uncertain identity seems to occur in at least some fetal specimens (Turner, 1870). This shortening of the sternum is accompanied by a reorganization of the rib cage: as in most mammals, the presence of several sternebrae in archaeocetes and odontocetes allows several ribs to articulate with the sternum, whereas in mysticetes this number is reduced to just one (the first rib). In both mysticetes and odontocetes, the anterior rib arches are separated into larger **vertebral ribs** that directly articulate with the vertebral column, and smaller **sternal ribs** that attach to the sternum. However, only some odontocetes (delphinids and phocoenids) develop fully ossified sternal ribs. In other toothed whales (e.g., physeteroids and ziphiids), they remain mostly cartilaginous, and in mysticetes they are rudimentary or absent, with the first vertebral rib effectively articulating directly with the manubrium (Eschricht and Reinhardt, 1866; Turner, 1870). No ossified sternal ribs have been reported for archaeocetes (Uhen, 2004).

Posterior to the thorax, the **lumbar** vertebrae retain the robust neural spines and transverse processes of the posterior thoracics, but do not attach to any ribs (Figure 3.17). In anterior view, the centrum of a lumbar vertebra is usually round or oval, and there is a tendency for the transverse processes to point slightly ventrolaterally. In archaic forms (e.g., basilosaurids and eomysticetids), this ventrolateral slope is often more pronounced (Kellogg, 1936; Sanders and Barnes, 2002). A characteristic, anteroposterior **carina** (keel) usually runs along the midline of the ventral surface of the centrum, but it may be variably developed and hence should not be seen as a reliable indicator of vertebral identity. In archaeocetes, the end of the lumbar series is marked by up to four **sacral vertebrae**, which in turn anchor the pelvic girdle to the vertebral column. In the most archaic forms, all of the four vertebrae are fused to each other and firmly connected to the pelvis (Moran *et al.*, 2015). By contrast, the number of fused vertebrae is increasingly reduced in more crownward lineages. Ultimately, the pelvis is left completely detached from the vertebral column in basilosaurid archaeocetes and neocetes, and the sacral vertebrae become morphologically and functionally similar to the posterior lumbars and anterior caudals (sections 5.1.2 and 8.2). In odontocetes and mysticetes, the sacrum can no longer be distinguished osteologically.

The vertebral column finishes with the **caudal** region, which extends all the way to the tip of the tail and, in all but the most archaic cetaceans, has become heavily involved in locomotion. The anteriormost caudal vertebra is traditionally determined based on the bifurcation of its ventral carina into two parallel ridges for the articulation of the first **chevron** bone (Figure 3.17). Chevron bones are located posteroventral to the vertebra with which they are associated. Together, they form a series of bony arches (the **hemal canal**) that protects the blood vessels supplying the tail and provides attachment surfaces for the hypaxial musculature (Rommel and Reynolds, 2009). In anterior view, the anteriormost caudal vertebrae have rounded centra and well-developed neural spines and transverse processes. Further posteriorly, the vertebrae decrease in size, and their various processes become increasingly smaller before eventually disappearing altogether. The beginning of the flukes is marked by a distinct change in vertebral proportions (section 3.4.6). Except for the anteriormost members of the series, the transverse processes or bodies of the caudal vertebrae are perforated by bilateral, vertical foramina for the transmission of vertebral arteries (Flower, 1885).

3.3.2 Forelimb

The forelimb is anchored to the body via the **scapula** (shoulder blade), which in turn rests on the vertebral column. The **clavicle**, which in many mammals forms a connection between the scapula and the sternum (thus forming the **pectoral girdle**), is absent in all cetaceans, although a rudimentary version of it still appears during odontocete (but not mysticete) embryogenesis (Klima, 1978). The scapula is broadly fan-shaped, with the base of the fan (**scapular neck**) carrying the articular cavity (**glenoid cavity**) for the top of the forelimb, as well as the anteriorly directed **coracoid process** (Figure 3.18). The dorsal margin of the blade is often continuous with a variably developed strip of **suprascapular cartilage**. The medial side of the scapula is somewhat concave and almost entirely occupied by the **subscapular fossa**. Laterally, the scapula is divided by the **scapular spine** into an anterior **supraspinous** and a posterior **infraspinous fossa**. In most cetaceans, the supraspinous fossa is reduced to a narrow groove, with the scapular

Figure 3.18 Right forelimb of (a) the protocetid archaeocete *Dorudon atrox*, (b) the common dolphin *Delphinus delphis*, and (c) the Bryde's whale *Balaenoptera edeni*, all in lateral view; (d) Carpus of *Dorudon atrox* in lateral view. Drawing of *Dorudon* adapted from Uhen (2004).

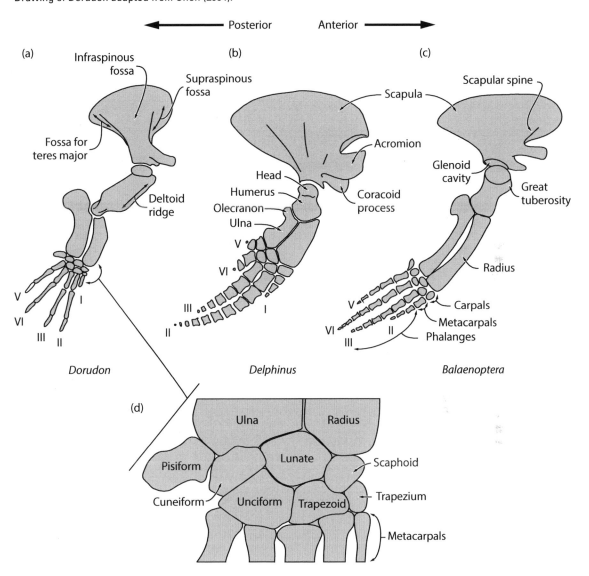

spine being almost coincident with the anterior margin of the bone. Posteriorly, the infraspinous fossa may be bounded by a dorsoventral ridge delimiting the area of insertion of the **teres major** muscle. Near its ventral base, the scapular spine gives rise to the anteriorly projected **acromion process**, which runs largely parallel to the more ventromedially located coracoid process.

As in terrestrial mammals, the forelimb of cetaceans consists of a single upper arm bone (humerus); two bones in the lower arm, namely, the anteriorly positioned radius and the more posteriorly located ulna, together also known as the **antebrachium**; and the hand, or **manus** (Figures 3.18 and 3.19). The latter can in turn be divided into the wrist (carpus); the metacarpus, a series of small

Figure 3.19 Manus and pes of the protocetid archaeocetes *Artiocetus clavis* and *Rodhocetus balochistanensis*. (a) Right astragalus and cuboid of *Artiocetus* in anterior view, (b) left pes, and (c) left manus of *Rodhocetus*. Nav. tr., navicular trochlea; tib. tr., tibial trochlea. Reproduced from Gingerich *et al.* (2001), with permission of the American Association for the Advancement of Science.

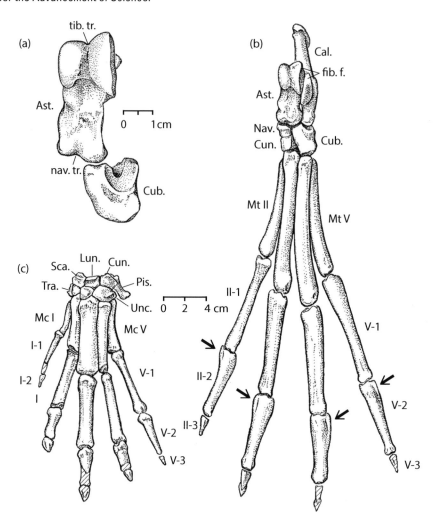

yet robust long bones forming the base of the fingers; and the fingers (digits) themselves, each of which comprises a series of individual phalanges. Distinct proximal and distal epiphyses that fuse to the shaft during ontogeny occur in the upper and lower arm bones, and frequently also the metacarpals and some of the phalanges (Flower, 1885; Rommel, 1990).

The **humerus** is generally robustly built but, at least in neocetes, often also somewhat shorter proximodistally than the antebrachium. In crown mysticetes and odontocetes, it is transversely flattened and distally flared in lateral view. Compared to most other mammals, its overall morphology is somewhat simplified. At its proximal end, the **head of the humerus** articulates with the glenoid cavity of the scapula. Anterior to the head, the variably developed **great** and **small tuberosities** serve as muscle attachment sites and, in some species, may fuse into a single, common tuberosity (Klima, 1980; Sanchez and Berta, 2010). Together, the head and tuberosities form the proximal epiphysis of the humerus. In archaeocetes and archaic neocetes, a well-developed **deltoid ridge** runs along most of the anterior face of the shaft (Kellogg, 1936; Okazaki, 2012; Sanders and Geisler, 2015). In later mysticetes and odontocetes, this ridge is usually either developed as a more or less prominent rugosity, or altogether absent. Distally, the elbow joint is rounded and troch-leated in archaeocetes, but divided into distinct, angled radial and ulnar facets (and thus effectively immobilized) in virtually all neocetes (Cooper et al., 2007b; Flower, 1885). The **radius** and **ulna** run parallel to each other and are comparatively featureless, except for the presence of a **radial tuberosity** and/or raised angle on the anterior surface of the radius in archaeocetes and archaic neocetes, and the **olecranon process** of the ulna, which is located just posterior to the elbow joint (Kellogg, 1936; Okazaki, 2012; Uhen, 2004). The relative size of these two bones is variable, with the ulna being equal to or bulkier than the radius in archaic forms, but somewhat more slender in modern mysticetes and odontocetes.

The morphology and composition of the **carpus** are variable (Figures 3.18 and 3.19). In mysticetes and odontocetes, the carpal bones are reduced to 5–6 polygonal, largely featureless elements that are arranged into two (proximal and distal) rows and often remain partially or entirely unossified. Owing to various patterns of fusion of some of the elements, the identity of individual carpals is sometimes disputed, and there exists a bewilder-ing range of terms for each element. However, most modern cetaceans seem to retain the **scaph-oid** (= radiale), **lunate** (= intermedium), **cuneiform** (=ulnare/triquetrum), and **pisiform** in the proximal row, and the **trapezoid** and **unciform** in the distal one (Cooper et al., 2007a; Flower, 1885). Archaeocetes may additionally retain the **trape-zium** anterior to the trapezoid, and have the trap-ezoid and unciform separated by the **magnum** and the **centrale** (Cooper et al., 2007a; Gingerich et al., 2009; Uhen, 2004). Distal to the carpus, the elongate **metacarpal** bones form the basis for the **digits**. Archaeocetes, odontocetes, and archaic mysticetes retain the primitive condition of five metacarpals and digits (i.e., they are **pentadactyl**). By contrast, most extant mysticetes (except right whales) only have four digits and metacarpals because of the loss of digit I, and hence have become **tetradactyl** (Figure 3.18). Besides this reduction in the number of digits, many modern cetaceans also have either a reduced (digits I and V) or increased (digits II–IV) number of phalanges in each finger, with the latter condition known as **hyperphalangy** (Cooper et al., 2007a).

3.3.3 Hind limb

As in all mammals, the hind limb of early archae-ocetes is anchored to the body via the **pelvic girdle**, also known as the **pelvis** or **innominate**, which attaches to the sacral region of the vertebral col-umn (Figures 3.1 and 3.2). The pelvic girdle forms a ring-shaped structure comprising three paired bones: the anterodorsally located **ilium**, which articulates with the sacral vertebrae; the more posteriorly pointing **ischium**; and the posteroven-trally oriented **pubis**, which closes the pelvic ring ventrally via the **pubic symphysis**. All three contribute to the **acetabulum**, a bowl-shaped fossa for the articulation of the hind limb proper. Posteroventral to the acetabulum, the pubis and ischium enclose a large **obturator foramen** (Madar, 2007; Madar et al., 2002). In basilosaurids, the pelvis has become detached from the vertebral column and is comparatively poorly developed, with the ilium, ischium, and obturator foramen being rather small. By contrast, the pubis is enlarged and still articulates with its bilateral counterpart, seemingly having moved to a posi-tion anteroventral (rather than posteroventral) to the acetabulum and obturator foramen. This interpretation is, however, disputed (Gingerich et al., 1990; Gol'din, 2014; Uhen and Gingerich, 2001).

Like the forelimb, the hind limb of early ceta-ceans consists of a single upper leg bone (**femur**), two lower leg bones (**tibia** and **fibula**), a complex assemblage of small bones forming the ankle (**tar-sus**), and a series of digital rays, each including a **metatarsal** and a series of **phalanges** (Figures 3.2 and 3.19). In addition, a small, rounded bone located anterior to the knee joint represents the kneecap (**patella**). Together, the ankle, metatarsus, and phalanges form the foot (**pes**). The posterior portion of the tarsus contains the **calcaneus**, or heel bone, and the **astragalus**, which together with the distal end of the tibia form the ankle joint. Both the proximal and the distal articular surfaces of the astragalus are distinctly trochleated, giving it the appearance of a "double pulley" (Gingerich *et al.*, 2001). This striking morphology is characteristic of both cetaceans and artiodactyls, and has played a major role in the recognition of their common origin (section 4.1). Besides the astragalus and calcaneus, the tarsus consists of the **cuboid**, the **navicular**, and three **cuneiform** bones (the **ecto-**, **meso-**, and **entocuneiform**), although the ecto- and mesocuneiform may be partially or entirely fused (Figure 3.19) (Madar, 2007; Madar *et al.*, 2002). Distal to the tarsus, there are four metatarsals supporting an equal number of digits (II–V). In addition, rudiments of the first digital ray persist in protocetids, and hence likely also earlier forms (Gingerich *et al.*, 2001).

In basilosaurids, all parts of the hind limb are comparatively poorly developed, with the tarsals largely fused into a single bony mass and digit II being almost entirely absent (section 8.1.2) (Gingerich *et al.*, 1990). In living cetaceans, no trace of an external hind limb remains. However, all species retain at least a small innominate, which functions as a point of attachment for abdominal muscles and the genitals (Dines *et al.*, 2014; Simões-Lopes and Gutstein, 2004). In addition, many mysticetes and sperm whales have a vestigial femur, although the latter often consists of little more than an ill-defined and sometimes entirely cartilaginous lump (Abel, 1908; Deimer, 1977; Omura, 1978). Right whales stand out for preserving the best-developed hind limb of any living cetacean, comprising not only the—often comparatively well-developed—pelvis and femur, but also a cartilaginous tibia, all of which articulate via recognizable, albeit presumably nonfunctional, synovial joints (Abel, 1908; Eschricht and Reinhardt, 1866).

3.4 Osteological correlates of soft tissue anatomy

3.4.1 Musculature

In order to be effective, muscles need to act on rigid structures, such as bone or cartilage. Depending on the size of the muscle and the strength of the action it performs, the site where it connects to the bone may become modified to increase its surface of attachment and provide firm anchorage on the skeleton. Typical examples of such modifications include bony crests (e.g., the external occipital crest), tuberosities (e.g., the radial tuberosity), and fossae—depressions in the bone surface which may house part of a muscle (e.g., the stapedial muscle fossa on the periotic).

On the dorsal surface of the rostrum, anterior to the external nares, is a well-developed **nasal fossa** in the premaxilla of mysticetes and some archaic odontocetes and archaeocetes (Figures 3.3 and 3.8). This fossa is homologous with the prenarial triangle of odontocetes (section 3.2.1) and mainly serves as attachment for the **nasal plug muscle,** or depressor alae nasi (Buono *et al.*, 2015; Mead and Fordyce, 2009). In its relaxed state, the nasal plug seals the narial passage, but it is pulled anterolaterally during respiration to allow air to enter the body (Heyning and Mead, 1990). Surrounding the nares in odontocetes is the expanded maxilla covering most of the supraorbital process of the frontal. The greatly enlarged surface area of the maxilla correlates with the presence of highly modified **maxillonasolabialis** muscles, which are involved in sound production and echolocation (section 5.2.2) (Geisler *et al.*, 2014; Huggenberger *et al.*, 2009). The muscles of the lips, tongue, and throat generally leave little trace on the associated bones, except maybe for a sulcus on the ventromedial portion of the mandible that may mark the attachment of the **mylohyoid** muscle in right whales. In addition, robustly built hyoid bones (e.g., in sperm whales) correlate with strongly developed gular musculature, such as the **geniohyoid**, **hyoglossus**, and **styloglossus** muscles, as well as a well-developed **sternohyoid muscle**, which connects the hyoid apparatus with the sternum (Bloodworth and Marshall, 2007; Reidenberg and Laitman, 1994; Werth, 2007).

Posterolateral to the rostrum, the temporal crest defines the often large temporal fossa. This structure has ancient origins, and its presence is one of the key features uniting cetaceans with all other mammals, and indeed all of their reptile-like synapsid forebears (Kemp, 2004). The size of the temporal fossa provides an indication of that of the **temporal** muscle, which extends from the temporal surface of the cranium to the coronoid process of the mandible and is chiefly responsible for raising the lower jaw (Figure 3.20). On the

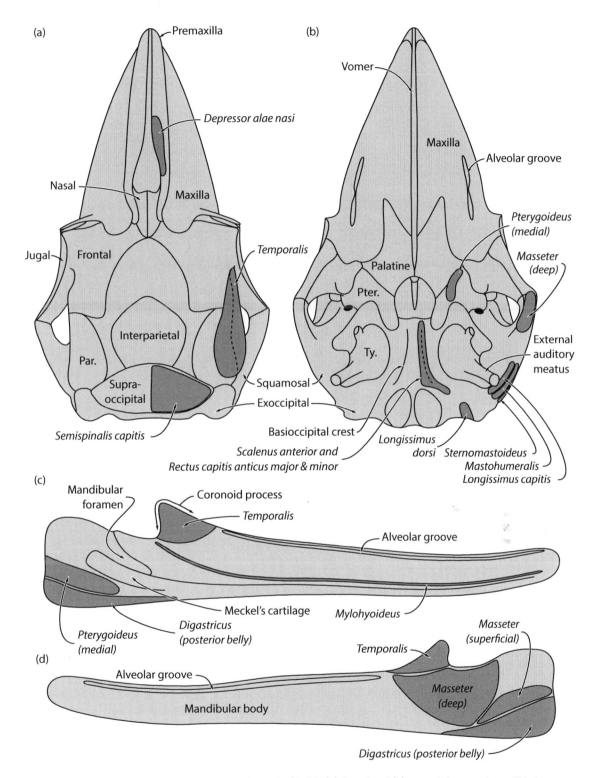

Figure 3.20 Fetus of the sei whale *Balaenoptera borealis*. Skull in (a) dorsal and (b) ventral views and mandible in (c) medial and (d) lateral views. Attachment sites of selected muscles are show in dark gray. par., parietal; pter., pterygoid; ty. tympanic bulla. Adapted from Schulte (1916).

mandible, additional fossae located between the coronoid process, the mandibular condyle, and the angular process sometimes provide details on the size of other jaw muscles, including: the **superficial** and **deep masseter** muscles, which connect the outer portion of the mandibular ramus to the zygomatic arch; the **medial pterygoid** muscle, which connects the medial surface of the mandibular ramus and the angular process to the anterior basicranium and posterior palate; and—in mysticetes—the posterior belly of the **digastric** muscle, which arises from the ventral surface of the angular process and inserts on to the skull just posterior to the external acoustic meatus (Figure 3.20) (El Adli and Deméré, 2015; El Adli et al., 2014; Schulte, 1916). In odontocetes, the posterior belly of the digastric either is absent or does not contact the mandible directly (Ito and Aida, 2005; Reidenberg and Laitman, 1994; Schulte and Smith, 1918).

Posterodorsally on the skull, the supraoccipital serves as the attachment site of the **semispinalis capitis** muscle, which stabilizes and extends the skull relative to the vertebral column (Pabst, 1990). The semispinalis capitis and, as a result, the supraoccipital shield are particularly well developed in mysticetes (Schulte, 1916), likely as a result of the elevated orientation of the skull (e.g., in right whales) and/or the need to raise the immense upper jaw during feeding (Arnold et al., 2005; Howell, 1930). A similar situation exists in early archaeocetes, such as protocetids, which have anterior thoracic neural spines that are distinctly taller than elsewhere along the vertebral column (Figure 3.2). As in terrestrial artiodactyls, the resulting hump likely anchored well-developed ligaments and muscles supporting the weight of the neck and skull (Gingerich et al., 1994; Kardong, 1998). Dorsal to the postglenoid process, the variably developed **sternomastoid fossa** houses the **splenius** and the **longissimus capitis**, both of which help to extend the neck. The actions of these muscles are opposed by the combined **scalenus** anterior and **rectus capitis anticus** (major and minor) muscles, which insert on to the basioccipital crest.

Posterior to the skull, the space defined by the well-developed neural spines and transverse processes of the vertebrae houses the extremely powerful **epaxial** muscles. These include mainly the **multifidus** and the **longissimus dorsi**, as well as their respective posterior extensions, the **extensor caudae medialis** (ECM) and **lateralis** (ECL). Together with the posteriorly located **intertransversarius caudae dorsalis** (ICD), these muscles act to stiffen the back and extend the caudal peduncle during upstroke (Pabst, 1990; Thewissen, 2009). In addition, when contracted unilaterally, the ICD also flexes the caudal peduncle sideways. In neocetes, the multifidus and the ECM originate from the lateral borders of the neural spines of the pre-caudal (multifidus) and caudal (ECM) vertebrae and insert on to the **deep tendon**, which connects the metapophyses of all vertebrae anterior to the fluke (Pabst, 1990). In archaeocetes, the multifidus in the thoracic region instead originated from well-developed fossae on the posterior margins of the neural spines, and may have inserted directly on to the metapophyses, as indicted by their robust nature and position farther away from the sagittal plane than in modern cetaceans (Uhen, 2004).

The longissimus and the ECL originate from the transverse processes and neural arches of the cervical, thoracic, and anterior lumbar vertebrae (longissimus) and the posterior lumbar and caudal vertebrae (ECL), respectively. In addition, the longissimus attaches to the exoccipital and, in some cases, the posterior portion of the squamosal. Besides its locomotory function, the longissimus thus assists the semispinalis in controlling the position of the head. Unlike the multifidus–ECM, both the longissimus–ECL and the ICD insert on to tendons without any clear osteological correlates—or lack correlates where they insert directly on to bone (Pabst, 1990; Uhen, 2004). Ventral to the transverse processes of the posterior thoracic, lumbar and caudal vertebrae, the **hypaxial** muscles comprise the **hypaxialis lumborum** and **intertransversarius caudae ventralis** (ICV). The hypaxial muscles jointly flex the caudal peduncle during downstroke. In addition, like its dorsal counterpart, the ICV can draw the caudal peduncle sideways when contracted just on one side (Pabst, 1990). There are generally no clear osteological correlates for the hypaxial muscles, but their strong development is indicated by the presence of numerous, robustly built chevron bones (Flower, 1885; Uhen, 2004).

Unlike in most mammals, the forelimb of modern cetaceans is relatively immobile (section 8.1.1). In general, all of the joints distal to the shoulder have become largely nonfunctional, although a limited degree of palmar flexion may be present in some mysticetes (Cooper *et al.*, 2007b). As a result, many of the muscles that normally effect specific forelimb movements are reduced or altogether absent, while others have shifted to assume a different role. The morphology of the scapula and humerus partially record this muscular rearrangement (Figure 3.21). Thus, the reduced importance of the **supraspinatus** muscle as an extensor of the forelimb is reflected in the small size of the supraspinous fossa on the scapula. By contrast, the enlarged infraspinous fossa, together with the scapular spine and the outside of the acromion process, mark the attachment site of the enlarged **deltoid** muscle. The latter attaches to the great tuberosity and deltoid ridge of the humerus, and serves as the major abductor and extensor of the flipper (Benke, 1993; Howell, 1930; Thewissen, 2009).

In archaeocetes, the medial side of the deltoid ridge may also anchor the **pectoralis**, which flexes and adducts the flipper; however, in neocetes the insertion of the pectoralis has shifted posteriorly on to the inside of the humeral shaft (Schulte, 1916; Uhen, 2004). The remainder of the infraspinous fossa is occupied by the **infraspinatus**, which inserts on the fringes of the great tuberosity of the humerus or, in odontocetes, a fossa located somewhat distal to the latter. Together with the **teres major** muscle, which originates from a variably developed fossa near the posterior border of the scapula, the infraspinatus acts to flex and rotate the flipper, and, at least in odontocetes, may also help to abduct it (Benke, 1993; Howell, 1930). A **teres minor** (also known as the **subdeltoid** muscle) may also occasionally insert on the scapula ventral to the deltoid; however, the identification of this muscle in cetaceans is contentious, and it often seems to be absent altogether (Klima, 1980; Murie, 1873; Strickler, 1978).

On the medial side of the humerus, the small tuberosity marks the site of attachment of the **subscapularis** (Figure 3.21) (Schulte, 1916). The subscapularis adducts the humerus and is generally well developed, which is reflected in both the generally broad subscapular fossa (where this muscle originates) and the conspicuously enlarged "small" tuberosity of many odontocetes. Note, however, that in mysticetes the small tuberosity is often smaller than might be expected, given the size of the associated muscle (Howell, 1930). Besides the subscapularis, the small **coracobrachialis** also inserts in the area of the small tuberosity. The coracobrachialis extends from the tip of the coracoid process and helps to flex and adduct the humerus (Benke, 1993). Other muscles involved in the movement of the flipper—including the **mastohumeralis** and the main adductor, the **latissimus dorsi**—also attach to the humerus, but usually do not leave distinct traces on the bone (Schulte, 1916; Uhen, 2004). Similarly, the positions of the muscles responsible for moving and fixing the scapula, including the **levator scapulae**, **serratus**, **rhomboideus**, and **omohyoideus**, can rarely be gauged based on the shape of the bone itself—either because their insertion is absent or small, or because they chiefly insert on the suprascapular cartilage (Howell, 1930; Klima, 1980). The **trapezius**, a large muscle present in most mammals, is absent or reduced in most cetaceans (Howell, 1930; Strickler, 1978), but seems to occur in some river dolphins and baleen whales (Klima, 1980; Schulte, 1916).

Distal to the humerus, the long head of the **triceps** connects the tip of the olecranon process with a variably developed tubercle on the posterior portion of the scapular neck, and helps to flex the shoulder. The medial head of the triceps, which normally flexes the elbow, also attaches to the olecranon; however, the locked elbow joint has made this muscle nonfunctional in all neocetes, which is likely reflected in the reduced size of the olecranon in many of the living species. Other muscles normally associated with the movement of the elbow, such as the **biceps** and the **brachialis**, are absent or vestigial in modern cetaceans, and consequently lack any distinct osteological correlates. By contrast, archaeocetes retain a distinct radial tuberosity for the insertion of the biceps, as well as tubercles on the humerus and radius likely associated with the brachialis (Uhen, 2004). The same holds true for the muscles of the wrist and digits, which still exist in modern cetaceans but

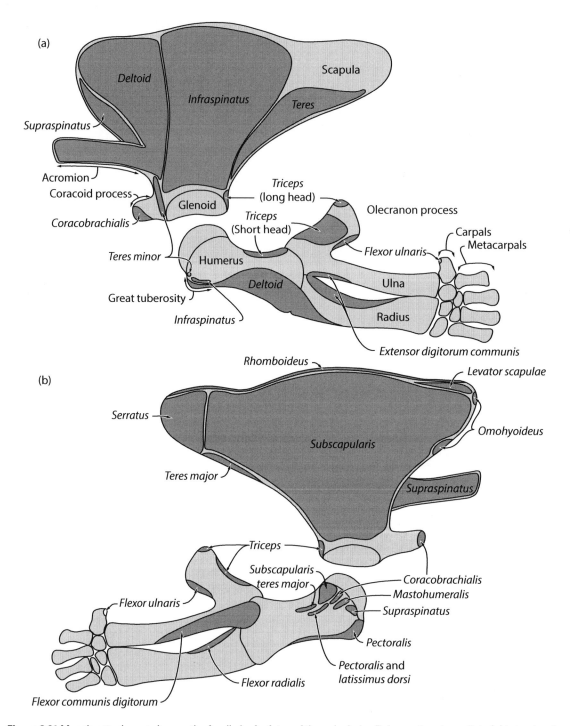

Figure 3.21 Muscle attachment sites on the forelimb of a fetus of the sei whale, *Balaenoptera borealis* in (a) lateral and (b) medial views. Adapted from Schulte (1916).

have largely lost their function. These muscles attach mostly on the inside of the radius and ulna, and thus within the space separating these two bones (Cooper *et al.*, 2007b; Schulte, 1916). The latter is reduced in many delphinoids, reflecting the loss of the distal flipper musculature in these taxa (Sanchez and Berta, 2010). Archaeocetes, on the other hand, sometimes possess distinct lateral crests on the phalanges which anchored well-developed digital abductors (Figure 3.19), and are likely indicative of a webbed foot (Gingerich *et al.*, 2001; Madar, 2007).

3.4.2 Baleen

All modern mysticetes use baleen to filter prey directly from the water and, as a result, have completely lost the use of teeth (sections 5.2.1 and 6.1.2). Baleen is a keratinous structure and, as such, is subject to much faster decay than bone. Nevertheless, in cases of exceptional preservation, which generally involve rapid burial of the remains, the proximal portion of entire racks of baleen may become fossilized (Plate 15b) (Brand *et al.*, 2004; Esperante *et al.*, 2008). Where such conditions are absent, the former presence of baleen can still be inferred from (1) the absence of functional teeth and (2) the presence of palatal sulci and grooves.

Incipient **tooth buds** still develop in mysticete fetuses but are quickly resorbed as the baleen begins to grow (Ishikawa *et al.*, 1999; Kükenthal, 1889–1893). All living mysticetes lack teeth as adults, and no teeth or tooth alveoli have ever been reported in any fossil mysticete except for members of the most archaic families (Aetiocetidae, Eomysticetidae, Llanocetidae, and Mammalodontidae; section 4.3.1). Nevertheless, interpreting a lack of teeth as an indicator of baleen may be problematic, because it assumes that the appearance of baleen, and hence filter feeding, makes the retention of a functioning dentition unnecessary. This ignores the potential for transitional feeding strategies employing both baleen and teeth at the same time, as may have been the case with some of the oldest families (Deméré *et al.*, 2008). In addition, several extant odontocete lineages adapted to suction feeding have also convergently lost much or all of their dentition, despite never having evolved baleen.

Nearly all mysticetes bear a series of **foramina** and associated **sulci** on their palate. Generally, these foramina fall into two distinct sets: a group of predominantly anteroposteriorly oriented sulci located near the midline of the skull, and a series of radially oriented sulci located near the lateral border of the rostrum (Figure 3.3). These two sets transmit the greater palatine artery and nerve and the superior alveolar artery and nerve, respectively. The superior alveolar artery supplies the tissues that give rise to baleen (Ekdale *et al.*, 2015; Walmsley, 1938), and the foramina and sulci that contain it appear to be unique to mysticetes (Deméré *et al.*, 2008). Because of this close association, laterally placed palatal foramina and sulci are often interpreted as an osteological correlate of baleen in fossil whales, and seem to indicate the concurrent presence of baleen and teeth in at least one family of archaic toothed mysticetes (Aetiocetidae; section 5.2.1) (Deméré and Berta, 2008; Deméré *et al.*, 2008). However, the foramina are weakly developed in both aetiocetids and at least some representatives of the earliest toothless mysticetes (Eomysticetidae) (Okazaki, 2012).

3.4.3 Air sinus system, air sacs and fat pads

Cetaceans differ markedly from terrestrial mammals in having ear bones that are physically and acoustically isolated from the remaining basicranium by an **accessory sinus system**, which is filled with an emulsion of mucus, oil, and air (section 5.1.2) (Nummela *et al.*, 2004; Oelschläger, 1986a). The sinus system is especially well developed in the skull of odontocetes, where it spreads: (1) around the tympanic bulla (**peribullary sinus**); (2) posteriorly along the paroccipital process of the exoccipital (**posterior sinus**); (3) laterally along the mandibular fossa and the postglenoid process of the squamosal, especially within the tympanosquamosal recess (**middle sinus**); (4) anteriorly on to the palatal portion of the pterygoid and the palatine (**pterygoid sinus**; present in all modern cetaceans); (5) dorsolaterally into the orbit (**pre- and postorbital lobes of the pterygoid sinus**); and (6) anterolaterally along the ventral surface of the rostrum (**anterior sinus**) (Figure 3.22) (Beauregard, 1894). These branches are variously developed in different odontocete lineages, and hence often used to diagnose taxa or establish evolutionary relationships. For example, ziphiids are characterized by an enlarged pterygoid sinus occupying the pterygoid hamulus; delphinids by an enlarged anterior sinus; monodontids by the lack of the orbital lobes; and phocoenids by the posterodorsal extension of the preorbital lobe. Bony fossae often record the minimum extent of a particular sinus, but they do not always form. Even when present, some fossae may be developed weakly and/or may be difficult to identify accurately. Thus, the lack of a well-defined bony cavity does not automatically indicate the absence of its associated branch of the sinus system.

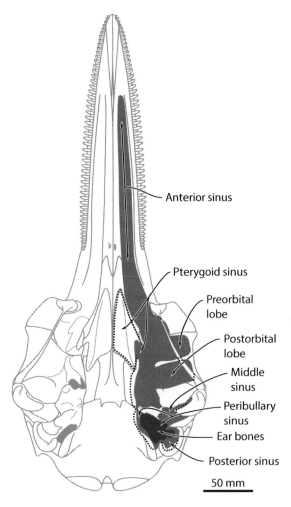

Figure 3.22 Accessory air sinus system of the common dolphin *Delphinus delphis*. Stippled lines indicate portions of the sinus system floored by bone.

A complex system of **nasal air sacs** and fat bodies occupies the forehead of extant odontocetes. The air sacs, which may have evolved from invaginations of originally external skin (Reidenberg and Laitman, 2008), are connected to the soft tissue nasal tracts, and may either serve to acoustically isolate the high-frequency sound production area (phonic lips) and/or act as air reservoirs during sound production (Cranford *et al.*, 1996; Heyning, 1989; Mead, 1975). Only one of the nasal sacs, the paired **premaxillary sac**, abuts the surface of the rostrum, where it forms the premaxillary sac fossa (Figure 3.7). Anterodorsal to the nasal air sacs lies the **melon**, a voluminous body composed of fat and connective tissue that likely serves to focus vocalizations (Cranford *et al.*, 1996; McKenna *et al.*, 2012). The presence of a melon strongly correlates with the tendency toward a concave facial area that is typical of most odontocetes. In the highly modified head of the extant sperm whale *Physeter*, the vast **supracranial basin** is occupied by two large fat bodies, the **junk** (which is homologous to the melon) and the **spermaceti organ**, both of which are likely involved in sound propagation (Cranford, 1999). A structure homologous to the spermaceti organ may also be present in the **prenarial basin** of some beaked whales (Cranford *et al.*, 2008). Notwithstanding its overall value as an osteological correlate, the supracranial basin is not necessarily a good indicator of the size of the soft tissue elements it contains. This is shown by pygmy and dwarf sperm whales (*Kogia* spp.), whose large basin only houses a proportionately small spermaceti organ. Finally, a further fat body is located in the posterior portion of the odontocete mandible, medial to the thin pan bone. This fat body is connected to the ear bones via the large mandibular foramen, and is likely involved in the reception of sounds (Cranford *et al.*, 2010). The size of the fat body may in some cases correlate with the size of the mandibular foramen; however, a considerable portion may also be located outside of the mandible, as in the minke whale *Balaenoptera acutorostrata* (Yamato *et al.*, 2012).

3.4.4 Brain anatomy and cranial nerves

Because it is entirely enclosed by bone, the size and external morphology of the mammalian brain can be gauged from the shape of the endocranial cavity. The latter is best observed either in the form of natural endocasts (Mitchell, 1989; Uhen, 2004) or as digital endocasts reconstructed via computed tomography (CT) scanning (Figure 3.23) (Colbert *et al.*, 2005; Marino *et al.*, 2000). Brain size and morphology have profoundly changed over the course of cetacean evolution, with most neocetes possessing an enlarged brain relative to their archaeocete forebears (Montgomery *et al.*, 2013). In addition, living cetaceans show a high degree of **gyrification**, that is, the development of deep folds and sulci in the cerebral cortex, as well as foreshortening of the brain as a result of facial telescoping (Morgane *et al.*, 1980). Like the facial bones, the brain of odontocetes may also be affected by a certain degree of asymmetry (Colbert *et al.*, 2005).

In cetaceans, the protective membranes (**meninges**) surrounding the brain are well developed and often partially ossify during ontogeny. This ossification helps to create a durable record of overall brain shape, but it may obscure the extent of gyrification and give a misleadingly smooth surface appearance to the resulting endocast (Colbert *et al.*, 2005; Racicot and Colbert, 2013). The dorsal surface of the endocranial cavity follows the outline of the **cerebral cortex** and traces the approximate outlines of its **frontal**, **parietal**, **temporal**, and **occipital lobes**. Anteroventrally, a separate chamber may trace the outline of the **olfactory bulb** (section 3.4.5) (Bajpai *et al.*, 2011; Godfrey *et al.*, 2013). The major meningeal ossifications are the bony **falx cerebri** (confluent with the **internal sagittal crest**), which dorsally separates the cerebral cortex into left and right hemispheres, and the **tentorium cerebelli**, which forms a horizontal shelf separating the main portion of the brain (**cerebrum**) from the **cerebellum** (Colbert *et al.*, 2005). Both the falx and the tentorium attach to the inside of the occipital bones and carry part of the vascular system that helps to drain the endocranial cavity. In particular, this includes the **dorsal sagittal sinus**, which runs along the falx cerebri, and the paired **transverse sinuses**, which follow the course of the tentorium. Both features can be identified based on the presence of grooves inside the cranial cavity. The sinuses converge posteriorly at the **confluence of the sinuses**, which in turn correlates with the

location of the **internal supraoccipital protuberance** (Colbert *et al.*, 2005; Mead and Fordyce, 2009). Other parts of the vascular system that can often be identified include the **spinal** and **middle meningeal arteries** (Figure 3.23).

Ventrally, a shallow **hypophyseal fossa** on the dorsal surface of the basisphenoid marks the position of the **pituitary gland**. Posterior to this fossa lies the cerebellum, the outline of which is traced by a corresponding concavity on the inside of the exoccipital. In mysticetes and archaeocetes, the cerebellum is largely enveloped in a complex vascular network (**caudal rete mirabile**) which largely obscures its shape (Geisler and Luo, 1998). Posteriorly, this network is connected to another, **epidural rete** running inside the neural canal (Vogl and Fisher, 1982). Ventral to the cerebellum, a variably developed fossa on the floor of the endocranial cavity may partially record the outline of the **brain stem**, including the **pons**. Anterior to and surrounding the hypophyseal fossa, the anatomy of the brain is largely obscured by the presence of the **rostral rete mirabile**, although the optic tract and olfactory lobe (not to be confused with the olfactory bulb) can be traced in at least some species (Racicot and Colbert, 2013).

Besides the shape of the brain itself, the morphology of the skull preserves information about the way it interfaces with the nearby sense organs, muscles, and skin. In particular, there are 13 **cranial nerves** (numbered 0 and I–XII) that originate directly from the brain and are highly conserved across all mammal lineages. Most of these nerves leave a series of osteological markers (highlighted in italics below) that can be used to trace them. In ascending order, the cranial nerves include:

0 and I: Terminal and olfactory nerves. The olfactory nerve conveys the sense of smell. It originates at the olfactory epithelium covering the ethmoturbinates, and from there runs to the olfactory bulb in the brain. On its way, it passes through the cribriform plate, where its course and development can be inferred from the presence of *cribriform foramina* (section 3.4.5) (Godfrey *et al.*, 2013; Thewissen *et al.*, 2011). The terminal nerve, also known as cranial nerve 0, is closely associated with the olfactory nerve and also passes through the cribriform plate toward the

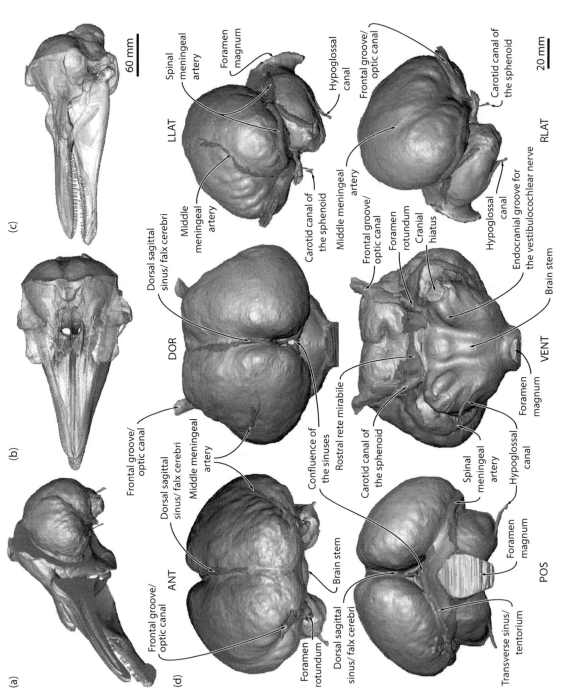

(a)

(b)

(c)

60 mm

(d)

ANT

Frontal groove/
optic canal

Dorsal sagittal
sinus/ falx cerebri

Middle meningeal
artery

Foramen
rotundum

Brain stem

Dorsal sagittal
sinus/ falx cerebri

Transverse sinus/
tentorium

POS

Foramen
magnum

Spinal
meningeal
artery

Hypoglossal
canal

DOR

Frontal groove/
optic canal

Dorsal sagittal
sinus/ falx cerebri

Middle
meningeal
artery

Confluence of
the sinuses

Rostral rete mirabile

Carotid canal of
the sphenoid

VENT

Foramen
magnum

Brain stem

Frontal groove/
optic canal

Foramen
rotundum

Cranial
hiatus

Hypoglossal
canal

Endocranial groove for
the vestibulocochlear nerve

LLAT

Spinal
meningeal
artery

Foramen
magnum

Hypoglossal
canal

Carotid canal of
the sphenoid

Middle meningeal
artery

RLAT

Carotid canal of
the sphenoid

20 mm

Figure 3.23 Brain anatomy of the harbor porpoise *Phocoena phocoena*, based on a digital endocast of the endocranial cavity reconstructed through CT scanning. (a) Size and position of the endocast within the skull, shown at a 45° angle; (b, c) partially transparent skull in (b) dorsal and (c) lateral views and; (d) endocast with main anatomical features labeled in anterior (ant), dorsal (dor), left lateral (llat), posterior (pos), ventral (vent), and right lateral (rlat) views. Adapted from Racicot and Colbert (2013).

nasal cavity. The terminal nerve is present in cetaceans, but its function remains unclear (Buhl and Oelschläger, 1986; Racicot and Rowe, 2014; Ridgway *et al.*, 1987). The vomeronasal nerve, which supplies the vomeronasal organ and is also commonly associated with the olfactory and terminal nerves, appears to be absent in cetaceans (Oelschläger and Buhl, 1985).

II: Optic nerve. The optic nerve connects the eyeball with the brain, and thus is responsible for vision. It originates in the orbit, where it is enveloped by fat and the retractor bulbi muscle (Zhu *et al.*, 2000). From here, the nerve enters the braincase via the *optic canal*. Inside the cranium, the two nerves cross and enter the brain within the ***chiasmatic groove*** on the presphenoid (Mead and Fordyce, 2009; Racicot and Colbert, 2013).

III, IV, and VI: Oculomotor, trochlear, and abducens nerves. Together, these nerves innervate the muscles that effect movements of the eye (Hosokawa, 1951). They enter the braincase via the (superior) ***orbital fissure***, which is located just posterior to the optic canal (Figure 3.6) (Mead and Fordyce, 2009).

V: Trigeminal nerve. The trigeminal is a complex nerve and divided into three major branches: the ophthalmic nerve (V_1), the maxillary nerve (V_2), and the mandibular nerve (V_3). The **ophthalmic nerve** is responsible for the sensory innervation of the forehead and, like cranial nerves III, IV, and VI, leaves the braincase via the (superior) *orbital fissure* (Figure 3.6). The **maxillary nerve** exits the braincase via the ***foramen rotundum***, which is located adjacent to, and often confluent with, the *orbital fissure* (e.g., Yamagiwa *et al.*, 1999); it provides sensory innervation for much of the rostrum, including the maxilla, palate, upper lip, and, where present, teeth. Because of the effects of telescoping, at least some cetaceans seem to show an inverse relationship between the size of the ophthalmic and maxillary nerves: whereas the former is reduced in line with the relatively small forehead, the latter is enlarged as a result of the posterior expansion of the rostral area (Rauschmann *et al.*, 2006). Within the maxilla, the maxillary nerve occupies the *infraorbital canal* and exits the rostrum via the *dorsal

infraorbital foramina* (Figure 3.3); consequently, this portion of the trigeminal is also known as the **infraorbital nerve**. On the palate, the maxillary nerve emerges as the **greater palatine nerve** via a series of foramina located along the maxilla–palatine suture, and from there runs anteriorly within a set of distinct *palatine sulci* (Figure 3.4). Finally, the **mandibular nerve** exits the skull via the *foramen ovale* and, where present, the *foramen pseudovale*. From here, it enters the mandible via the *mandibular foramen* and runs along the length of the *mandibular canal*, from which it exits via the *mental foramina*. The mandibular nerve provides sensory innervation to the lower teeth and lip, and motor innervation to the lower jaw musculature (Ito and Aida, 2005).

VII: Facial nerve. The facial nerve exits the braincase via the *cranial hiatus*, after which it enters the periotic via the *proximal opening of the facial canal* within the internal acoustic meatus (Figure 3.10). From here it travels through the pars cochlearis and, via the *distal opening of the facial canal*, on to the *facial sulcus*. As it passes through the periotic, the facial nerve gives rise to the **major petrosal nerve**, which anteriorly exits the pars cochlearis via the *hiatus Fallopii* (Figure 3.11). Within the middle ear, the facial nerve branches again to form the **chorda tympani**, which enters the malleus via a small foramen located just besides the articular facets for the incus, follows its eponymous *sulcus* and then goes on to innervate the tongue. Having emerged from the basicranium, the main branch of the facial nerve runs anteriorly toward the facial area, where it innervates the maxillonasolabialis muscles surrounding the blowhole and, in odontocetes, the main sound-producing apparatus (sections 3.4.3 and 5.2.2). On its way toward the face, it crosses in front of the orbit at the *antorbital notch* (Figure 3.3).

VIII: Vestibulocochlear nerve. This nerve follows a relatively short path from the brain to the inner ear. It exits the braincase via the *cranial hiatus* and then enters the pars cochlearis of the periotic via the *internal acoustic meatus*. Here, it splits into two subdivisions responsible for hearing (**cochlear nerve**) and balance (**vestibular nerve**), which connect to their respective organs via the *spiral cribriform tract* and the *foramen singulare* (section 3.4.5) (Mead and Fordyce, 2009).

IX–XI: Glossopharyngeal, vagus, and accessory nerves. Like cranial nerves VII and VIII, these nerves exit the braincase via the *cranial hiatus*, but then travel on toward the posterior border of the basicranium, where they pass through the *jugular notch* (Howell, 1930; Schulte and Smith, 1918). In older individuals of certain odontocete species, the cranial hiatus may become partially ossified, so that the path of nerves IX–XI is rendered distinct from that of nerves VII and VIII on the one hand, and from that of the jugular vein on the other; in this case, the exit for nerves IX–XI is known as the ***posterior lacerate foramen*** (Mead and Fordyce, 2009). While the glossopharyngeal nerve innervates the posterior portion of the tongue and the pharynx, the vagus, and accessory nerves interface with structures that are farther removed from the skull, including the viscera and the trapezius muscle (Oelschläger and Oelschläger, 2009; Schulte and Smith, 1918; Yamada *et al.*, 1998).

XII: Hypoglossal nerve. This hypoglossal nerve emerges from the *hypoglossal foramen*, located just inside the *jugular notch*. From here, it descends ventrally and finally turns anteriorly to innervate the muscles of the tongue and throat (Schulte and Smith, 1918).

3.4.5 Sensory organs

Information on cetacean sensory organs can be gleaned from osteological correlates relating to the eye, nose, and ear, and thus the senses of sight, smell, hearing, and balance. Anatomical structures related to sight are mainly the bony orbit and the optic canal, although the orbit may not always be a reliable predictor of eyeball size (Racicot and Colbert, 2013; Thewissen and Nummela, 2008). In contrast to this relatively limited information, the bony traces left by the olfactory system are more diverse. Neural signals related to smell are conveyed by the **olfactory nerve** (cranial nerve I), which extends from the olfactory epithelium in the nasal cavity to the olfactory bulb in the brain (section 3.4.4). To reach the brain, the olfactory

nerve passes through the cribriform plate of the ethmoid, which is perforated by a series of cribriform foramina. In modern odontocetes, the sense of smell is generally thought to be reduced or absent, which is reflected in (1) the absence of ethmoturbinates, which support the olfactory epithelium; (2) a largely or entirely imperforate cribriform plate, indicating the reduction or loss of the olfactory nerve; and (3) the absence of a distinct olfactory bulb in adults (Mead and Fordyce, 2009; Oelschläger and Oelschläger, 2009). By contrast, all of these features are preserved in archaeocetes (Godfrey *et al.*, 2013; Uhen, 2004), mysticetes (Thewissen *et al.*, 2011) and, presumably, archaic odontocetes (Fordyce, 2002; Godfrey *et al.*, 2013), all of which thus seem to possess a functioning olfactory system.

As in other mammals, the auditory apparatus of cetaceans comprises the **outer ear**, which includes the ear canal (external auditory meatus) and, where present, the externally visible pinna; the **middle ear**, which contains a chain of three auditory ossicles (malleus, incus, and stapes) and is separated from the outer ear by the eardrum, or tympanic membrane; and the **inner ear**, which houses a complex set of spaces inside the periotic (the **osseous labyrinth**), filled with soft tissue ducts and sacs (Figure 3.24). All three of these subdivisions leave clear traces in the bones of the skull. Starting from the outer ear, the external acoustic meatus is represented by the eponymous groove located on the ventral surface of the squamosal. Proximally, the meatus meets the tympanic bulla at the **tympanic ring**, consisting of the sigmoid process, the conical process, and the (outer) posterior pedicle (Ekdale *et al.*, 2011; Ridewood, 1923). As in other mammals, this ring supports the tympanic membrane, whose attachment is marked by the variably developed tympanic sulcus (often only noticeable as a scar or crest) originating on the posterior face of the sigmoid process. Hearing underwater has rendered the tympanic membrane largely nonfunctional in modern cetaceans (section 5.1.2). In mysticetes, it has been transformed into a sac-like structure protruding into the external auditory meatus, and is commonly referred to as the **glove finger** (Figure 3.24) (Beauregard, 1894; Lillie, 1910). In odontocetes, the tympanic membrane is

Figure 3.24 Location and morphology of the tympanic membrane (glove finger) in the minke whale, *Balaenoptera acutorostrata*. (a) Overview of the right half of the basicranium in ventral view, with tympanic bulla *in situ*; note how the sac-like glove finger extends into the proximal portion of the external acoustic meatus; the distal portion of the meatus of balaenopterids is filled with a waxy earplug that can be used to determine the age of the individual (section 6.7). (b) Detailed view of the glove finger and its attachment to the malleus via the tympanic ligament, with the bulla removed. c., conical process; s., sigmoid process. Adapted from Ekdale *et al.* (2011) under a Creative Commons Attribution license.

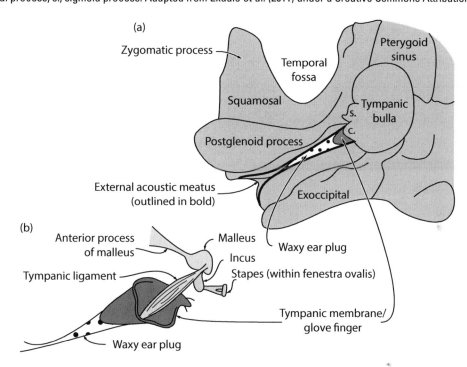

reduced or even absent (Fleischer, 1975; Mead and Fordyce, 2009).

Medial to the tympanic ring, the eardrum is attached to the often ill-defined manubrium of the malleus via the **tympanic ligament**. With the malleus begins the chain of auditory ossicles, and thus the middle ear. The movement of the ossicular chain is controlled by two muscles (Purves, 1966): (1) the tensor tympani, which originates on the anterior process of the periotic and/or the accessory ossicle, and inserts on the malleus via an often clearly defined scar; via the malleus, this muscle can temporarily affect the tension of the tympanic membrane, although this function may be largely lost in cetaceans (Fraser and Purves, 1960); and (2) the stapedius muscle, which originates from a distinct fossa just posterior to the fenestra ovalis and inserts on to the neck of the stapes (sections 3.2.4 and 3.2.6). The chain terminates with

the stapes, whose footplate is lodged in the fenestra ovalis and directly contacts the inner ear. Two sets of sensory organs—the **cochlea**, responsible for sound reception, and the **vestibular apparatus**, or organ of balance—are located here, and can be digitally reconstructed in both extant and fossil species via micro CT scans (Ekdale and Racicot, 2015; Spoor *et al.*, 2002).

The cochlea is coiled like a spiral staircase and consists of a **spiral canal** winding itself around a bony core (the **modiolus**). Inside the spiral canal, the cochlea is divided into three ducts, namely: (1) the **scala tympani**, which communicates with the fenestra rotunda; (2) the **scala vestibuli**, which is in indirect contact with the fenestra ovalis (via the vestibule; discussed in the next paragraph), and thus the footplate of the stapes; and (3) the **cochlear duct**, or scala media (Figure 3.25). The former two are filled with a fluid known as **perilymph**,

Figure 3.25 Schematic reconstruction of the bony labyrinth of the bottlenose dolphin *Tursiops truncatus*. The vestibular apparatus is shown in dark gray. Adapted from Ekdale (2013) under a Creative Commons Attribution license.

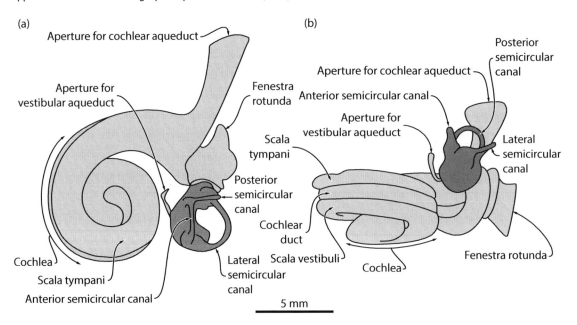

whereas the cochlear duct is filled with **endo-lymph**. The scala tympani is separated from the cochlear duct by the primary and secondary **spiral laminae**, which are connected to each other by the **basilar membrane**. The latter bears the hair cells for the **organ of Corti**, which transduces mechanical signals into neural responses (Ekdale, 2013; Luo and Eastman, 1995). The base of the scala tympani gives rise to the **perilymphatic duct**, which leaves the periotic via the cochlear aqueduct (section 3.2.4).

The vestibular apparatus is connected to the cochlea via the chamber-like **vestibule** and provides information on linear motions, as well as tilts and rotational movements of the head. Linear motions and head tilts are detected by two **otolith organs**, the **saccule** and the **utricle**, both of which are located inside the vestibule. By contrast, rotational movements are picked up by three **semicircular canals** (anterior, lateral, and posterior) and their corresponding ampullae (Figure 3.25). The vestibule furthermore gives rise to the **endolymphatic duct**, which ascends toward the brain via the vestibular aqueduct. All parts of the vestibular apparatus are filled with endolymph (Ekdale, 2013; Spoor and Thewissen, 2008).

In addition to the major sense organs, numerous pits and/or extensive vascularization in some taxa may correlate with the presence of vibrissae (Thewissen and Bajpai, 2009; Thewissen and Nummela, 2008) or tactile upper (dorsal infraorbital foramina) and lower (mental foramina) lips (de Muizon and Domning, 2002; Fitzgerald, 2010; Racicot *et al.*, 2014). These structures may thus be indicative of well-developed **mechanoreception**, but they have rarely been studied in detail.

3.4.6 Flukes

Cetacean **flukes** are composed of five component parts: a cutaneous layer; a thin, subcutaneous blubber layer; a ligamentous layer; a core of tough, dense fibrous tissue; and a pattern of superficial veins between the subcutaneous and ligamentous layers, along with a deep arterial system surrounded by veins (Felts, 1966). These soft tissue structures insert on to the caudalmost vertebrae, which are anteroposteriorly short and dorsoventrally compressed. These fluke vertebrae are posterior to the **peduncular region** of modern cetaceans, where the vertebral bodies are instead laterally compressed. The transition between these two regions is indicated by the presence of a

ball vertebra (Fish *et al.*, 2006; Watson and Fordyce, 1993), which is convex both anteriorly and posteriorly, and forms the main locus of the bending motion in the tail (Fish, 1998). Pakicetids, ambulocetids, remingtonocetids, and protocetids lack any osteological indicators of flukes (Gingerich *et al.*, 2009; Uhen, 2008a). By contrast, dorsoventrally compressed fluke vertebrae are present in basilosaurids, but peduncular vertebrae are not, suggesting that these early whales possessed flukes but no peduncle (Uhen, 2004).

3.5 Suggested readings

Flower, W. H. 1885. An Introduction to the Osteology of the Mammalia. Macmillan and Co., London.

Fraser, F. C., and P. E. Purves. 1960. Hearing in cetaceans: Evolution of the accessory air sacs and the structure of the outer and middle ear in recent cetaceans. Bulletin of the British Museum (Natural History), Zoology 7:1–140.

Mead, J. G. 1975. Anatomy of the external nasal passages and facial complex in the Delphinidae (Mammalia: Cetacea). Smithsonian Contributions to Zoology 207:1–67.

Mead, J. G., and R. E. Fordyce. 2009. The therian skull: a lexicon with emphasis on the odontocetes. Smithsonian Contributions to Zoology 627.

Uhen, M. D. 2004. Form, function, and anatomy of *Dorudon atrox* (Mammalia, Cetacea): an archaeocete from the middle to late Eocene of Egypt. University of Michigan Papers on Paleontology 34:1–222.

References

Abel, O. 1908. Die Morphologie der Hüftrudimente der Cetaceen. Denkschriften der Kaiserlichen Akademie der Wissenschaften 81:139–195.

Armfield, B. A., Z. Zheng, S. Bajpai, Christopher J. Vinyard, and J. G. M. Thewissen. 2013. Development and evolution of the unique cetacean dentition. PeerJ 1:e24.

Arnold, P. W., R. A. Birtles, S. Sobtzick, M. Matthews, and A. Dunstan. 2005. Gulping behaviour in rorqual whales: underwater observations and functional interpretation. Memoirs of the Queensland Museum 51:309–332.

Bajpai, S., J. G. M. Thewissen, and R. W. Conley. 2011. Cranial anatomy of Middle Eocene *Remingtonocetus* (Cetacea, Mammalia) from Kutch, India. Journal of Paleontology 85:703–718.

Barroso, C., T. W. Cranford, and A. Berta. 2012. Shape analysis of odontocete mandibles: functional and evolutionary implications. Journal of Morphology 273:1021–1030.

Beauregard, H. 1894. Recherches sur l'appareil auditif chez les mammifères III. Journal de l'Anatomie et de la Physiologie 30:366–413.

Benke, H. 1993. Investigations on the osteology and functional morphology of the flipper of whales and dolphins. Investigations on Cetacea 24:9–252.

Berta, A., E. G. Ekdale, and T. W. Cranford. 2014. Review of the cetacean nose: form, function, and evolution. The Anatomical Record 297:2205–2215.

Bianucci, G., and W. Landini. 2006. Killer sperm whale: a new basal physeteroid (Mammalia, Cetacea) from the Late Miocene of Italy. Zoological Journal of the Linnean Society 148:103–131.

Bloodworth, B. E., and C. D. Marshall. 2007. A functional comparison of the hyolingual complex in pygmy and dwarf sperm whales (*Kogia breviceps* and *K. sima*), and bottlenose dolphins (*Tursiops truncatus*). Journal of Anatomy 211:78–91.

Boessenecker, R. W., and R. E. Fordyce. 2015. A new eomysticetid (Mammalia: Cetacea) from the Late Oligocene of New Zealand and a re-evaluation of '*Mauicetus*' *waitakiensis*. Papers in Palaeontology 1:107–140.

Bouetel, V., and C. de Muizon. 2006. The anatomy and relationships of *Piscobalaena nana* (Cetacea, Mysticeti), a Cetotheriidae s.s. from the early Pliocene of Peru. Geodiversitas 28:319–396.

Brand, L. R., R. Esperante, A. V. Chadwick, O. P. Porras, and M. Alomía. 2004. Fossil whale preservation implies high diatom accumulation rate in the Miocene–Pliocene Pisco Formation of Peru. Geology 32:165–168.

Buchholtz, E. A. 2007. Modular evolution of the cetacean vertebral column. Evolution & Development 9:278–289.

Buchholtz, E. A. 2011. Vertebral and rib anatomy in *Caperea marginata*: Implications for evolutionary patterning of the mammalian vertebral column. Marine Mammal Science 27:382–397.

Buchholtz, E. A., and S. A. Schur. 2004. Vertebral osteology in Delphinidae (Cetacea). Zoological Journal of the Linnean Society 140:383–401.

Buhl, E. H., and H. A. Oelschläger. 1986. Ontogenetic development of the nervus terminalis in toothed whales. Anatomy and Embryology 173:285–294.

Buono, M. R., M. S. Fernández, R. E. Fordyce, and J. S. Reidenberg. 2015. Anatomy of nasal complex in the southern right whale, *Eubalaena australis* (Cetacea, Mysticeti). Journal of Anatomy 226:81–92.

Colbert, M., R. Racicot, and T. Rowe. 2005. Anatomy of the cranial endocast of the bottlenose dolphin, *Tursiops truncatus*, based on HRXCT. Journal of Mammalian Evolution 12:195–207.

Cooper, L. N., A. Berta, S. D. Dawson, and J. S. Reidenberg. 2007a. Evolution of hyperphalangy and digit reduction in the cetacean manus. The Anatomical Record 290:654–672.

Cooper, L. N., S. D. Dawson, J. S. Reidenberg, and A. Berta. 2007b. Neuromuscular anatomy and evolution of the cetacean forelimb. The Anatomical Record 290:1121–1137.

Cranford, T. W. 1999. The sperm whale's nose: sexual selection on a grand scale? Marine Mammal Science 15:1133–1157.

Cranford, T. W., M. Amundin, and K. S. Norris. 1996. Functional morphology and homology in the odontocete nasal complex: implications for sound generation. Journal of Morphology 228:223–285.

Cranford, T. W., P. Krysl, and M. Amundin. 2010. A new acoustic portal into the odontocete ear and vibrational analysis of the tympanoperiotic complex. PLoS One 5:e11927.

Cranford, T. W., M. F. McKenna, M. S. Soldevilla, S. M. Wiggins, J. A. Goldbogen, R. E. Shadwick, P. Krysl, J. A. St. Leger, and J. A. Hildebrand. 2008. Anatomic geometry of sound transmission and reception in Cuvier's beaked whale (*Ziphius cavirostris*). The Anatomical Record 291:353–378.

de Muizon, C., and D. P. Domning. 2002. The anatomy of *Odobenocetops* (Delphinoidea, Mammalia), the walrus-like dolphin from the Pliocene of Peru and its palaeobiological implications. Zoological Journal of the Linnean Society 134:423–452.

Deimer, P. 1977. Der rudimentäre hintere Extremitätengürtel des Pottwals (*Physeter macrocephalus* Linnaeus, 1758), seine Variabilität und Wachstumsallometrie. Zeitschrift für Säugetierkunde 42:88–101.

Deméré, T. A., and A. Berta. 2008. Skull anatomy of the Oligocene toothed mysticete *Aetiocetus weltoni* (Mammalia; Cetacea): implications for mysticete evolution and functional anatomy. Zoological Journal of the Linnean Society 154:308–352.

Deméré, T. A., M. R. McGowen, A. Berta, and J. Gatesy. 2008. Morphological and molecular evidence for a stepwise evolutionary transition from teeth to baleen in mysticete whales. Systematic Biology 57:15–37.

Dines, J. P., E. Otárola-Castillo, P. Ralph, J. Alas, T. Daley, A. D. Smith, and M. D. Dean. 2014. Sexual selection targets cetacean pelvic bones. Evolution 68:3296–3306.

Ekdale, E. G. 2013. Comparative anatomy of the bony labyrinth (inner ear) of placental mammals. PLoS One 8:e66624.

Ekdale, E. G., A. Berta, and T. A. Deméré. 2011. The comparative osteology of the petrotympanic complex (ear region) of extant baleen whales (Cetacea: Mysticeti). PLoS One 6:e21311.

Ekdale, E. G., T. A. Deméré, and A. Berta. 2015. Vascularization of the gray whale palate (Cetacea, Mysticeti, *Eschrichtius robustus*): soft tissue evidence for an alveolar source of blood to baleen. The Anatomical Record:691–702.

Ekdale, E. G., and R. A. Racicot. 2015. Anatomical evidence for low frequency sensitivity in an archaeocete whale: comparison of the inner ear of *Zygorhiza kochii* with that of crown Mysticeti. Journal of Anatomy 226:22–39.

El Adli, J. J., and T. A. Deméré. 2015. On the anatomy of the temporomandibular joint and the muscles that act upon it: observations on the gray whale, *Eschrichtius robustus*. The Anatomical Record 298:680–690.

El Adli, J. J., T. A. Deméré, and R. W. Boessenecker. 2014. *Herpetocetus morrowi* (Cetacea: Mysticeti), a new species of diminutive baleen whale from the Upper Pliocene (Piacenzian) of California, USA, with observations on the evolution and relationships of the Cetotheriidae. Zoological Journal of the Linnean Society 170:400–466.

Ellenberger, W., H. Baum, H. Dittrich, and G. Münch. 1911. Handbuch der Anatomie der Tiere für Künstler Bd. V: Anatomie des Hundes. 20 pp. Dieterich, Leipzig.

Eschricht, D. F., and J. Reinhardt. 1866. On the Greenland right whale (*Balaena mysticetus* Linn.); with especial reference to its geographical distribution and migrations in times past and present, and to its external and internal characteristics.; pp. 1–150 in W. H. Flower (ed.), Recent Memoirs on the Cetacea by Professors Eschricht, Reinhardt, and Lilljeborg. Robert Hardwicke, London.

Esperante, R., L. Brand, K. E. Nick, O. Poma, and M. Urbina. 2008. Exceptional occurrence of fossil baleen in shallow marine sediments of the Neogene Pisco Formation, Southern Peru. Palaeogeography Palaeoclimatology Palaeoecology 257:344–360.

Fahlke, J. M., P. D. Gingerich, R. C. Welsh, and A. R. Wood. 2011. Cranial asymmetry in Eocene archaeocete whales and the evolution of directional hearing in water. Proceedings of the National Academy of Sciences of the United States of America 108:14545–14548.

Felts, W. J. L. 1966. Some functional and structural characteristics of cetacean flippers and flukes; pp. 255–276 in K. S. Norris (ed.), Whales, Dolphins and Porpoises. University of California Press, Berkeley.

Fish, F. E. 1998. Biomechanical perspective on the origin of cetacean flukes; pp. 303–324 in J. G. M. Thewissen (ed.), The Emergence of Whales. Plenum Press, New York.

Fish, F. E., M. K. Nusbaum, J. T. Beneski, and D. R. Ketten. 2006. Passive cambering and flexible propulsors: cetacean flukes. Bioinspiration & Biomimetics 1:S42.

Fitzgerald, E. M. G. 2006. A bizarre new toothed mysticete (Cetacea) from Australia and the early evolution of baleen whales. Proceedings of the Royal Society B 273:2955–2963.

Fitzgerald, E. M. G. 2010. The morphology and systematics of *Mammalodon colliveri* (Cetacea: Mysticeti), a toothed mysticete from the Oligocene of Australia. Zoological Journal of the Linnean Society 158:367–476.

Fitzgerald, E. M. G. 2012. Archaeocete-like jaws in a baleen whale. Biology Letters 8:94–96.

Fleischer, G. 1975. Über das spezialisierte Gehörorgan von *Kogia breviceps* (Odontoceti). Zeitschrift für Säugetierkunde 40:89–102.

Flower, W. H. 1885. An Introduction to the Osteology of the Mammalia. Macmillan and Co., London.

Fordyce, R. E. 1994. *Waipatia maerewhenua*, new genus and new species (Waipatiidae, new family), an archaic Late Oligocene dolphin (Cetacea: Odontoceti: Platanistoidea) from New Zealand. Proceedings of the San Diego Society of Natural History 29:147–176.

Fordyce, R. E. 2002. *Simocetus rayi* (Odontoceti: Simocetidae, new family): a bizarre new archaic Oligocene dolphin from the eastern North Pacific. Smithsonian Contributions to Paleobiology 93:185–222.

Fordyce, R. E., P. G. Quilty, and J. Daniels. 2002. *Australodelphis mirus*, a bizarre new toothless ziphiid-like fossil dolphin (Cetacea: Delphinidae) from the Pliocene of Vestfold Hills, East Antarctica. Antarctic Science 14:37–54.

Fraser, F. C., and P. E. Purves. 1960. Hearing in cetaceans. Bulletin of the British Museum (Natural History) 7:1–140.

Geisler, J. H., M. W. Colbert, and J. L. Carew. 2014. A new fossil species supports an early origin for toothed whale echolocation. Nature 508:383–386.

Geisler, J. H., and Z.-X. Luo. 1996. The petrosal and inner ear of *Herpetocetus* sp. (Mammalia; Cetacea) and their implications for the phylogeny and hearing of archaic mysticetes. Journal of Paleontology 70:1045–1066.

Geisler, J. H., and Z.-X. Luo. 1998. Relationships of Cetacea to terrestrial ungulates and the evolution of cranial vasculature in Cete; pp. 163–212 in J. G. M. Thewissen (ed.), The Emergence of Whales. Plenum Press, New York.

Geisler, J. H., and A. E. Sanders. 2003. Morphological evidence for the phylogeny of Cetacea. Journal of Mammalian Evolution 10:23–129.

Gingerich, P. D., M. u. Haq, I. S. Zalmout, I. H. Khan, and M. S. Malakani. 2001. Origin of whales from early artiodactyls: hands and feet of Eocene Protocetidae from Pakistan. Science 293:2239–2242.

Gingerich, P. D., S. M. Raza, M. Arif, M. Anwar, and X. Zhou. 1994. New whale from the Eocene of Pakistan and the origin of cetacean swimming. Nature 368:844–847.

Gingerich, P. D., B. H. Smith, and E. L. Simons. 1990. Hind limbs of Eocene *Basilosaurus*: evidence of feet in whales. Science 249:154–157.

Gingerich, P. D., M. Ul-Haq, W. von Koenigswald, W. J. Sanders, B. H. Smith, and I. S. Zalmout. 2009. New protocetid whale from the Middle Eocene of Pakistan: birth on land, precocial development, and sexual dimorphism. PLoS One 4:1–20.

Godfrey, S. J., J. H. Geisler, and E. M. G. Fitzgerald. 2013. On the olfactory anatomy in an archaic whale (Protocetidae, Cetacea) and the minke whale *Balaenoptera acutorostrata* (Balaenopteridae, Cetacea). The Anatomical Record 296:257–272.

Gol'din, P. 2014. Naming an innominate: pelvis and hindlimbs of Miocene whales give an insight into evolution and homology of cetacean pelvic girdle. Evolutionary Biology 41:473–479.

Hampe, O., H. Franke, C. A. Hipsley, N. Kardjilov, and J. Müller. 2015. Prenatal cranial ossification of the humpback whale (*Megaptera novaeangliae*). Journal of Morphology 276:564–582.

Heyning, J. E. 1989. Comparative facial anatomy of beaked whales (Ziphiidae) and a systematic revision among the families of extant Odontoceti. Contributions in Science, Natural History Museum of Los Angeles County 405:1–64.

Heyning, J. E., and J. G. Mead. 1990. Evolution of the nasal anatomy of cetaceans; pp. 67–79 in J. Thomas, and R. Kastelein (eds.), Sensory Abilities of Cetaceans. Springer, New York.

Hosokawa, H. 1951. On the extrinsic eye muscles of the whale with special remarks on the innervation and function of the musculus retractor bulbi. Scientific Reports of the Whales Research Institute Tokyo 6:1–33.

Howell, A. B. 1930. Aquatic Mammals: Their Adaptations to Life in the Water. Charles C. Thomas, Springfield, IL.

Huggenberger, S., M. A. Rauschmann, T. J. Vogl, and H. H. A. Oelschläger. 2009. Functional morphology of the nasal complex in the harbor porpoise (*Phocoena phocoena* L.). The Anatomical Record 292:902–920.

Ichishima, H. 2011. Do cetaceans have the mesethmoid? Memoir of the Fukui Prefectural Dinosaur Museum 10:63–75.

Ishikawa, H., H. Amasaki, H. Dohguchi, A. Furuya, and K. Suzuki. 1999. Immunohistological distributions of fibronectin, tenascin, type I, III and IV collagens, and laminin during tooth development and degeneration in fetuses of minke whale, *Balaenoptera acutorostrata*. Journal of Veterinary Medical Science 61:227–232.

Ito, H., and K. Aida. 2005. The first description of the double-bellied condition of the digastric muscle in the finless porpoise *Neophocaena phocaenoides* and Dall's porpoise *Phocoenoides dalli*. Mammal Study 30:83–87.

Jefferson, T. A. 1988. *Phocoenoides dalli*. Mammalian Species 319:1–7.

Kardong, K. V. 1998. Vertebrates: Comparative Anatomy, Function, Evolution. WCB/McGraw-Hill, Boston.

Kasuya, T. 1973. Systematic consideration of recent toothed whales based on the morphology of tympano-periotic bone. Scientific Reports of the Whales Research Institute Tokyo 25:1–103.

Kellogg, R. 1936. A review of the Archaeoceti. Carnegie Institution of Washington Publication 482. Carnegie Institution, Washington, DC.

Kemp, T. S. 2004. The Origin and Evolution of Mammals. Oxford University Press, Oxford.

Klima, M. 1978. Comparison of early development of sternum and clavicle in striped dolphin and in humpback whale. Scientific Reports of the Whales Research Institute Tokyo 30:253–269.

Klima, M. 1980. Morphology of the pectoral girdle in the Amazon dolphin *Inia geoffrensis* with special reference to the shoulder joint and the movements of the flippers. Zeitschrift für Säugetierkunde 45:288–309.

Kükenthal, W. 1889–1893. Vergleichend-anatomische und entwicklungsgeschichtliche Untersuchung an Walthieren. Denkschriften der Medicinisch-Naturwissenschaftlichen Gesellschaft zu Jena 3:1–448.

Lambertsen, R., N. Ulrich, and J. Straley. 1995. Frontomandibular stay of Balaenopteridae: a mechanism for momentum recapture during feeding. Journal of Mammalogy 76:877–899.

Lillie, D. G. 1910. Observations on the anatomy and general biology of some members of the larger Cetacea. Proceedings of the Zoological Society of London 1910:769–792.

Luo, Z.-X., and E. R. Eastman. 1995. Petrosal and inner ear of a squalodontoid whale: implications for evolution of hearing in odontocetes. Journal of Vertebrate Paleontology 15:431–442.

Luo, Z.-X., and P. D. Gingerich. 1999. Terrestrial Mesonychia to aquatic Cetacea: transformation of the basicranium and evolution of hearing in whales. University of Michigan Papers on Paleontology 31:1–98.

Luo, Z.-X., and K. Marsh. 1996. Petrosal (periotic) and inner ear of a Pliocene kogiine whale (Kogiinae, Odontoceti): implications on relationships and hearing evolution of toothed whales. Journal of Vertebrate Paleontology 16:328–348.

Madar, S. I. 2007. The postcranial skeleton of Early Eocene pakicetid cetaceans. Journal of Paleontology 81:176–200.

Madar, S. I., J. G. M. Thewissen, and S. T. Hussain. 2002. Additional holotype remains of *Ambulocetus natans* (Cetacea, Ambulocetidae), and their implications for locomotion in early whales. Journal of Vertebrate Paleontology 22:405–422.

Marino, L., M. D. Uhen, B. Frohlich, J. M. Aldag, C. Blane, D. Bohaska, and F. C. Whitmore, Jr. 2000. Endocranial volume of Mid-Late Eocene archaeocetes (Order: Cetacea) revealed by computed tomography: implications for cetacean brain evolution. Journal of Mammalian Evolution 7:81–94.

McDonald, M., N. Vapniarsky-Arzi, F. J. M. Verstraete, C. Staszyk, D. M. Leale, K. D. Woolard, and B. Arzi. 2015. Characterization of the temporomandibular joint of the harbour porpoise (*Phocoena phocoena*) and Risso's dolphin (*Grampus griseus*). Archives of Oral Biology 60:582–592.

McKenna, M. F., T. W. Cranford, A. Berta, and N. D. Pyenson. 2012. Morphology of the odontocete melon and its implications for acoustic function. Marine Mammal Science 28:690–713.

Mead, J. G. 1975. Anatomy of the external nasal passages and facial complex in the Delphinidae (Mammalia: Cetacea). Smithsonian Contributions to Zoology 207:1–72.

Mead, J. G., and R. E. Fordyce. 2009. The therian skull: a lexicon with emphasis on the odontocetes. Smithsonian Contributions to Zoology 627. Smithsonian Institution Scholarly Press, Washington, DC.

Miller, G. S. 1923. The telescoping of the cetacean skull. Smithsonian Miscellaneous Collections 76:1–70.

Mitchell, E. D. 1989. A new cetacean from the late Eocene La Meseta Formation, Seymour Island, Antarctic Peninsula. Canadian Journal of Fisheries and Aquatic Science 46:2219–2235.

Montgomery, S. H., J. H. Geisler, M. R. McGowen, C. Fox, L. Marino, and J. Gatesy. 2013. The evolutionary history of cetacean brain and body size. Evolution 67:3339–3353.

Moore, J. C. 1968. Relationships among the living genera of beaked whales with classifications, diagnoses and keys. Fieldiana Zoology 53:209–298.

Moran, M., S. Bajpai, J. C. George, R. Suydam, S. Usip, and J. G. M. Thewissen. 2015. Intervertebral and epiphyseal fusion in the postnatal ontogeny of cetaceans and terrestrial mammals. Journal of Mammalian Evolution 22:93–109.

Morgane, P. J., M. S. Jacobs, and W. L. McFarland. 1980. The anatomy of the brain of the bottlenose dolphin (*Tursiops truncatus*). Surface configurations of the telencephalon of the bottlenose dolphin with comparative anatomical observations in four other cetacean species. Brain Research Bulletin 5:1–107.

Muller, J. 1954. Observations on the orbital region of the skull of the Mystacoceti. Zoologische Mededelingen (Leiden) 32:279–290.

Murie, J. 1873. On the organization of the caaing whale, *Globicephalus melas*. Transactions of the Zoological Society of London 8:235–301.

Narita, Y. 2005. Evolution of the vertebral formulae in mammals: a perspective on developmental constraints. Journal of Experimental Zoology 304B:91–106.

Nummela, S., S. T. Hussain, and J. G. M. Thewissen. 2006. Cranial anatomy of Pakicetidae (Cetacea, Mammalia). Journal of Vertebrate Paleontology 26:746–759.

Nummela, S., J. G. M. Thewissen, S. Bajpai, S. T. Hussain, and K. Kumar. 2004. Eocene evolution of whale hearing. Nature 430:776–778.

O'Leary, M. A. 2010. An anatomical and phylogenetic study of the osteology of the petrosal of extant and extinct artiodactylans (Mammalia) and relatives. Bulletin of the American Museum of Natural History:1–206.

Oelschläger, H. A. 1986a. Comparative morphology and evolution of the otic region in toothed whales (Cetacea, Mammalia). American Journal of Anatomy 177:353–368.

Oelschläger, H. A. 1986b. Tympanohyal bone in toothed whales and the formation of the tympano-periotic complex (Mammalia: Cetacea). Journal of Morphology 188:157–165.

Oelschläger, H. A., and E. H. Buhl. 1985. Development and rudimentation of the peripheral olfactory system in the harbor porpoise *Phocoena phocoena* (Mammalia: Cetacea). Journal of Morphology 184:351–360.

Oelschläger, H. A., and J. S. Oelschläger. 2009. Brain; pp. 134–149 in W. F. Perrin, B. Würsig, and J. G. M. Thewissen (eds.), Encyclopedia of Marine Mammals. Academic Press, Burlington, MA.

Okazaki, Y. 2012. A new mysticete from the upper Oligocene Ashiya Group, Kyushu, Japan and its significance to mysticete evolution. Bulletin of the Kitakyushu Museum of Natural History and Human History, Series A (Natural History) 10:129–152.

Omura, H. 1964. A systematic study of the hyoid bones in the baleen whales. Scientific Reports of the Whales Research Institute Tokyo 18:149–170.

Omura, H. 1978. Preliminary report on morphological study of pelvic bones of the minke whale from the Antarctic. Scientific Reports of the Whales Research Institute Tokyo 30:271–279.

Pabst, D. A. 1990. Axial muscles and connective tissues of the bottlenose dolphin; pp. 51–67 in S. Leatherwood and R. Reeves (eds.), The Bottlenose Dolphin. Academic Press, San Diego.

Purves, P. E. 1966. Anatomy and physiology of the outer and middle ear in cetaceans; pp. 320–380 in K. S. Norris (ed.), Whales, Dolphins, and Porpoises. California University Press, Berkeley.

Racicot, R. A., and M. W. Colbert. 2013. Morphology and variation in porpoise (Cetacea: Phocoenidae) cranial endocasts. The Anatomical Record 296:979–992.

Racicot, R. A., T. A. Deméré, B. L. Beatty, and R. W. Boessenecker. 2014. Unique feeding morphology in a new prognathous extinct porpoise from the Pliocene of California. Current Biology 24:774–779.

Racicot, R. A., and T. Rowe. 2014. Endocranial anatomy of a new fossil porpoise (Odontoceti, Phocoenidae) from the Pliocene San Diego Formation of California. Journal of Paleontology 88:652–663.

Rauschmann, M. A., S. Huggenberger, L. S. Kossatz, and H. H. A. Oelschläger. 2006. Head morphology in perinatal dolphins: a window into phylogeny and ontogeny. Journal of Morphology 267:1295–1315.

Reichert, C. 1837. Über die Viceralbögen der Wirbeltiere im Allgemeinen und deren Metamorphosen bei den Vögeln und Säugetieren. Archiv für Anatomie, Physiologie und Wissenschaftliche Medicin 1837:120–220.

Reidenberg, J. S., and J. T. Laitman. 1994. Anatomy of the hyoid apparatus in Odontoceti (toothed whales): specializations of their skeleton and musculature compared with those of terrestrial mammals. The Anatomical Record 240:598–624.

Reidenberg, J. S., and J. T. Laitman. 2008. Sisters of the sinuses: cetacean air sacs. The Anatomical Record 291:1389–1396.

Reidenberg, J. S., and J. T. Laitman. 2011. Moving forward with the mysticete hyoid: biomechanics of the whale's hyoid in body wave locomotion. Journal of the Federation of American Societies for Experimental Biology 25:867.4.

Ridewood, W. G. 1923. Observations on the skull in foetal specimens of whales of the genera *Megaptera* and *Balaenoptera*. Philosophical Transactions of the Royal Society of London B 211:209–272.

Ridgway, S. H., L. S. Demski, T. H. Bullock, and M. Schwanzel-Fukuda. 1987. The terminal nerve in odontocete cetaceans. Annals of the New York Academy of Sciences 519:201–212.

Rommel, S. 1990. Osteology of the bottlenose dolphin; pp. 29–49 in S. Leatherwood, and R. Reeves, R. (eds.), The Bottlenose Dolphin. Academic Press, San Diego.

Rommel, S., and J. E. Reynolds III. 2009. Skeleton, postcranial; pp. 1021–1033 in W. F. Perrin, B. Würsig, and J. G. M. Thewissen (eds.), Encyclopedia of Marine Mammals. Academic Press, Burlington, MA.

Sanchez, J. A., and A. Berta. 2010. Comparative anatomy and evolution of the odontocete forelimb. Marine Mammal Science 26:140–160.

Sanders, A. E., and L. G. Barnes. 2002. Paleontology of the Late Oligocene Ashley and Chandler Bridge Formations of South Carolina, 3: Eomysticetidae, a new family of primitive mysticetes (Mammalia: Cetacea). Smithsonian Contributions to Paleobiology 93:313–356.

Sanders, A. E., and J. H. Geisler. 2015. A new basal odontocete from the upper Rupelian of South Carolina, U.S.A., with contributions to the systematics of *Xenorophus* and *Mirocetus* (Mammalia, Cetacea). Journal of Vertebrate Paleontology 35:e890107.

Schenkkan, E. J. 1973. On the comparative anatomy and function of the nasal tract in odontocetes (Mammalia, Cetacea). Bijdragen tot de Dierkunde 43:127–159.

Schulte, H. v. W. 1916. Anatomy of a foetus of *Balaenoptera borealis*. Memoirs of the American Museum of Natural History 1:389–502.

Schulte, H. v. W. 1917. The skull of *Kogia breviceps* Blainv. Bulletin of the American Museum of Natural History 37:361–404.

Schulte, H. v. W., and M. d. F. Smith. 1918. The external characters, skeletal muscles, and peripheral nerves of *Kogia breviceps* (Blainville). Bulletin of the American Museum of Natural History 38:7–72.

Simões-Lopes, P. C., and C. S. Gutstein. 2004. Notes on the anatomy, positioning and homology of the pelvic bones in small cetaceans (Cetacea, Delphinidae, Pontoporiidae). Latin American Journal of Aquatic Mammals 3:157–162.

Slijper, E. J. 1936. Die Cetaceen, vergleichend-anatomisch und systematisch. M. Nijhoff, Amsterdam.

Spoor, F., S. Bajpai, S. T. Hussain, K. Kumar, and J. G. M. Thewissen. 2002. Vestibular evidence for the evolution of aquatic behaviour in early cetaceans. Nature 417:163–166.

Spoor, F., and J. G. M. Thewissen. 2008. Comparative and functional anatomy of balance in aquatic mammals; pp. 257–284 in J. G. M. Thewissen, and S. Nummela (eds.), Sensory Evolution on the Threshold. University of California Press, Berkeley.

Strickler, T. L. 1978. Myology of the shoulder of *Pontoporia blainvillei*, including a review of the literature on shoulder morphology in the Cetacea. American Journal of Anatomy 152:419–431.

Thewissen, J. G. M. 2009. Musculature; pp. 744–747 in W. F. Perrin, B. Würsig, and J. G. M. Thewissen (eds.), Encyclopedia of Marine Mammals. Academic Press, Burlington, MA.

Thewissen, J. G. M., and S. Bajpai. 2001. Dental morphology of Remingtonocetidae (Cetacea, Mammalia). Journal of Paleontology 75:463–465.

Thewissen, J. G. M., and S. Bajpai. 2009. New skeletal material of *Andrewsiphius* and

Kutchicetus, two Eocene cetaceans from India. Journal of Paleontology 83:635–663.

Thewissen, J. G. M., J. George, C. Rosa, and T. Kishida. 2011. Olfaction and brain size in the bowhead whale (*Balaena mysticetus*). Marine Mammal Science 27:282–294.

Thewissen, J. G. M., and S. Nummela. 2008. Towards an integrative approach; pp. 333–340 in J. G. M. Thewissen, and S. Nummela (eds.), Sensory Evolution on the Threshold. University of California Press, Berkeley.

Todd, T. W. 1922. Numerical significance in the thoracicolumbar vertebrae of the mammalia. The Anatomical Record 24:260–286.

Tsai, C.-H., and R. E. Fordyce. 2015. The earliest gulp-feeding mysticete (Cetacea: Mysticeti) from the Oligocene of New Zealand. Journal of Mammalian Evolution 22:535–560.

Turner, S. W. 1870. On the sternum and ossa innominata of the Longniddry whale (*Balaenoptera sibbaldii*). Journal of Anatomy and Physiology 4:271–281.

Uhen, M. D. 2004. Form, function, and anatomy of *Dorudon atrox* (Mammalia: Cetacea): an archaeocete from the Middle to Late Eocene of Egypt. University of Michigan Papers on Paleontology 34:1–222.

Uhen, M. D. 2008a. New protocetid whales from Alabama and Mississippi, and a new cetacean clade, Pelagiceti. Journal of Vertebrate Paleontology 28:589–593.

Uhen, M. D. 2008b. A new *Xenorophus*-like odontocete cetacean from the Oligocene of North Carolina and a discussion of the basal odontocete radiation. Journal of Systematic Palaeontology 6:433–452.

Uhen, M. D., and P. D. Gingerich. 2001. New genus of dorudontine archaeocete (Cetacea) from the middle-to-late Eocene of South Carolina. Marine Mammal Science 17:1–34.

Van Beneden, P.-J., and P. Gervais. 1880. Ostéographie des cétacés vivants et fossiles. Bertrand, Paris.

Vogl, A. W., and H. D. Fisher. 1982. Arterial retia related to supply of the central nervous system in two small toothed whales—narwhal (*Monodon monoceros*) and beluga (*Delphinapterus leucas*). Journal of Morphology 174:41–56.

Walmsley, R. 1938. Some observations on the vascular system of a female fetal finback. Contributions to Embryology 27:109–178.

Walsh, B. M., and A. Berta. 2011. Occipital ossification of balaenopteroid mysticetes. The Anatomical Record 294:391–398.

Watson, A. G., and R. E. Fordyce. 1993. Skeleton of two minke whales, *Balaenoptera acutorostrata*, stranded on the south-east coast of New Zealand. New Zealand Natural Sciences 20:1–14.

Werth, A. J. 2007. Adaptations of the cetacean hyolingual apparatus for aquatic feeding and thermoregulation. The Anatomical Record 290:546–568.

Whitmore, F. C., Jr., and L. G. Barnes. 2008. The Herpetocetinae, a new subfamily of extinct baleen whales (Mammalia, Cetacea, Cetotheriidae). Virginia Museum of Natural History Special Publication 14:141–180.

Yamada, T. K., H. Ito, and H. Takakura. 1998. On the dolphin shoulder muscles, with special reference to their nerves of supply. World Marine Mammal Science Conference, Monaco, 20–24 January 1998. Abstracts:152.

Yamagiwa, D., H. Endo, I. Nakanishi, A. Kusanagi, M. Kurohmaru, and Y. Hayashi. 1999. Anatomy of the cranial nerve foramina in the Risso's dolphin (*Grampus griseus*). Annals of Anatomy—Anatomischer Anzeiger 181:293–297.

Yamato, M., D. R. Ketten, J. Arruda, S. Cramer, and K. Moore. 2012. The auditory anatomy of the minke whale (*Balaenoptera acutorostrata*): a potential fatty sound reception pathway in a baleen whale. Anatomical Record 295:991–998.

Yamato, M., and N. D. Pyenson. 2015. Early development and orientation of the acoustic funnel provides insight into the evolution of sound reception pathways in cetaceans. PLoS One 10:e0118582.

Zhu, Q., D. J. Hillmann, and W. G. Henk. 2000. Observations on the muscles of the eye of the bowhead whale, *Balaena mysticetus*. The Anatomical Record 259:189–204.

(a)

(b)

Plate 1 Reconstructions of (a) one of the earliest known cetaceans, *Pakicetus*, and (b) the remingtonocetid archaeocete *Remingtonocetus*, hunting for prey in freshwater and the sea, respectively. Both drawings © C. Buell.

Cetacean Paleobiology, First Edition. Felix G. Marx, Olivier Lambert, and Mark D. Uhen.
© 2016 John Wiley & Sons, Ltd. Published 2016 by John Wiley & Sons, Ltd.

Plate 2 Reconstruction of the "walking, swimming" whale, *Ambulocetus natans*. © C. Buell.

(a)

(b)

Plate 3 Reconstructions of (a) the protocetid *Georgiacetus* and (b) the basilosaurid *Dorudon*, both on the hunt for fish. Both drawings © C. Buell.

(a)

(b)

Plate 4 Reconstructions of the mammalodontids (a) *Janjucetus* and (b) *Mammalodon*. Note how *Mammalodon* may have been able to manipulate its well-developed lips for use during benthic suction feeding. © C. Buell.

Plate 5 Reconstruction/drawing of (a) *Aetiocetus*, using its combination of small teeth and proto-baleen to feed on a school of small fish; and (b) the extant bowhead whale, *Balaena mysticetus*, swimming toward a cloud of prey. Both drawings © C. Buell.

(a)

(b)

Plate 6 Drawings of (a) the extant blue whale, *Balaenoptera musculus*, about to engulf a swarm of krill; and (b) the extant giant sperm whale, *Physeter macrocephalus*, attacking a giant squid. Note the large, expandable throat pouch of *B. musculus*. Both drawings © C. Buell.

Plate 7 Reconstruction of the macropredatorial stem physeteroid *Livyatan*, feeding on a small mysticete. © C. Letenneur.

Plate 8 Reconstruction of the extinct ziphiid *Ninoziphius* preying upon fish on the seafloor. © C. Letenneur.

Plate 9 The four extant "river dolphins." From top to bottom: *Inia geoffrensis* (Amazon River dolphin), *Platanista gangetica* (South Asian river dolphin), *Lipotes vexillifer* (baiji, Yangtze River dolphin), and *Pontoporia blainvillei* (Franciscana). © C. Buell.

Plate 10 Extant oceanic dolphins. Clockwise from upper left: *Delphinus delphis* (common dolphin), *Pseudorca crassidens* (false killer whale), *Stenella coeruleoalba* (striped dolphin), *Steno bredanensis* (rough-toothed dolphin), and *Orcinus orca* (killer whale) © C. Buell.

Plate 11 Reconstruction of the bizarre fossil delphinoid *Odobenocetops* bottom-feeding like a walrus. © Mary Parrish.

(a)

(b)

Plate 12 Field photographs. (a) A large, articulated mysticete skeleton exposed at Cerro Colorado, Pisco Formation, Pisco Basin, Peru; courtesy of G. Bianucci. (b) Chilean and Smithsonian paleontologists studying several fossil whale skeletons at Cerro Ballena, next to the Pan-American Highway in the Atacama Region of Chile, 2011; courtesy of N. D. Pyenson, Smithsonian Institution; photo by A. Metallo.

(a)

(b)

Plate 13 Field photographs. (a) A view south of the Sharktooth Hill National Natural Landmark and other localities exposing the Middle Miocene Round Mountain Silt, which includes the Sharktooth Hill bone bed; courtesy of N. D. Pyenson. (b) The excavation of a Miocene dolphin skull south of Plum Point, along Calvert Cliffs, Calvert County, Maryland, USA. The lower two-thirds of the cliff include a portion of the Plum Point Member of the Calvert Formation, whereas the upper third of the cliff face exposes the Middle Miocene Choptank Formation; courtesy of the Calvert Marine Museum; photo by S. J. Godfrey.

(a)

(b)

Plate 14 Field photographs of (a) Eocene outcrops of the Samlat Formation, near Ad-Dakhla, Western Sahara; and (b) the excavation of a fossil cetacean from the Otekaike Limestone, exposed in Hakataramea Valley, South Canterbury, New Zealand; (b) courtesy of R. E. Fordyce.

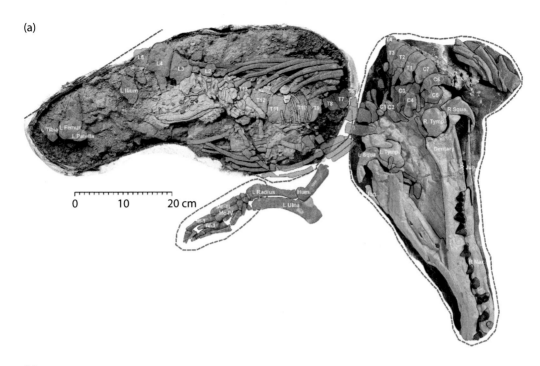

(a)

(b)

Plate 15 (a) Adult female (GSP-UM 3475a, type) and fetal skeletons (GSP-UM 3475b) of the protocetid *Maiacetus inuus*. The skeleton of the adult is shown in beige (skull) and red (postcrania), and that of the fetus in blue. Reproduced from Gingerich *et al.* (2009; PLoS 4:e4366) under a Creative Commons Attributions license. (b) Permineralized baleen of a large balaenopteroid from the Pisco Formation, Pisco Basin, Peru.

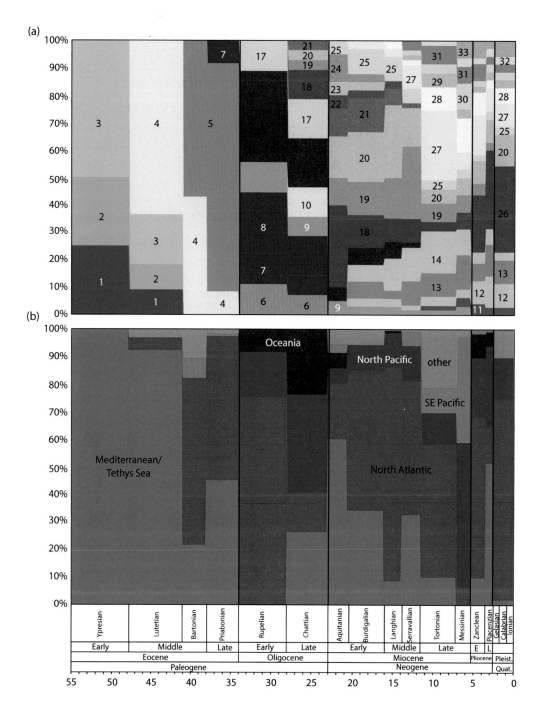

Plate 16 (a) Plot of proportional, family-level cetacean diversity over time. Note the dramatic change in the number of families from the Eocene, to the Oligocene, to the Neogene. Archaeocetes are shown in shades of green, Mysticeti in shades of blue, and Odontoceti in shades of red and orange. Groups are identified as follows: 1, Ambulocetidae; 2, Pakicetidae; 3, Remingtonocetidae; 4, Protocetidae; 5, Basilosauridae; 6, Kekenodontidae; 7, Llanocetidae + Mammalodontidae; 8, Aetiocetidae; 9, Cetotheriopsidae; 10, Eomysticetidae; 11, Eschrichtiidae; 12, Balaenidae; 13, Balaenopteridae; 14, Cetotheriidae; 15, Agorophiidae + Simocetidae + Waipatiidae; 16, Squalodontidae; 17, Xenorophidae; 18, Eurhinodelphinidae; 19, Kentriodontidae + Albireonidae; 20, Physeteroidea minus Kogiidae; 21, Squalodelphinidae; 22, Allodelphinidae; 23, Dalpiazinidae; 24, Eoplatanistidae; 25, Platanistidae; 26, Delphinidae; 27, Ziphiidae; 28, Phocoenidae; 29, Iniidae; 30, Kogiidae; 31, Pontoporiidae; 32, Lipotidae; 33, Monodontidae + Odobenocetopsidae. (b) Proportional representation of fossil cetaceans by ocean basin. Ocean basins are represented by different color blocks. The "other" category includes the South Atlantic, Arctic, and Indian oceans. Note that the proportion of fossil cetaceans from different regions changes over time. In particular, the contribution from the Mediterranean–Tethys Sea region disappears during times of low sea level in the Rupelian, as well as during the Messinian Salinity Crisis.

4 | Phylogeny and Taxonomy

4.1 Cetacean origins

For many decades, the relationship of whales to other mammals remained something of a mystery. Nineteenth- and early 20th-century workers variably suspected the origin of cetaceans among early mammals, marsupials, insectivores, creodonts, pinnipeds, sirenians, and perissodactyls, or even thought that they were derived from aquatic reptiles (Kellogg, 1936). Nevertheless, as early as 1787, the famous surgeon and scientist John Hunter recognized that cetaceans resemble "ruminants" in certain aspects of their anatomy but, working outside an evolutionary framework, did not go so far as to propose any actual relationship between the two (Hunter, 1787). Almost 100 years later, William H. Flower drew on Hunter's findings in an insightful lecture at the Royal Institution of Great Britain, titled "On Whales, Past and Present, and Their Probable Origin." In it, Flower suggested that whales might have arisen from some early form of archaic hoofed mammal, based on such similarities as "the complex stomach, simple liver, respiratory organs, and especially the reproductive organs and structures relating to the development of the young" (Flower, 1883: p. 229).

At least initially, Flower's ideas did not find much purchase in the scientific community, possibly because they relied mainly on soft tissue characters, rather than osteological data or fossils. Throughout the first half of the 20th century, scientists either clung to the idea of an insectivore and/or creodont origin (Kellogg, 1936), or simply gave up on trying to decipher the relationships of cetaceans altogether, as summarized by Simpson (1945: p. 213) in his *Classification of Mammals*:

> Because of their perfected adaptation to a completely aquatic life, with all its attendant conditions of respiration, circulation, dentition, locomotion, etc., the cetaceans are on the whole the most peculiar and aberrant of mammals. Their place in the sequence of cohorts and orders is open to question and is indeed quite impossible to determine in any purely objective way.

Simpson's statement of abject uncertainty was the culmination of decades of seemingly fruitless speculation. Ironically, however, it also triggered what would ultimately evolve into our modern understanding of cetacean evolution. Prompted by his comments, Boyden and Gemeroy (1950) conducted a series of immune-reactivity tests that compared cetacean proteins with those of several other living orders of mammals. Their results confirmed artiodactyls as the closest match, thus reviving Flower's (1883) earlier hypothesis. Over the next 50 years, their findings were largely backed by a series of

Cetacean Paleobiology, First Edition. Felix G. Marx, Olivier Lambert, and Mark D. Uhen.
© 2016 John Wiley & Sons, Ltd. Published 2016 by John Wiley & Sons, Ltd.

additional studies focusing on proteins and, increasingly from about 1990 onward, DNA (Gatesy, 1998). Some of these analyses even went so far as to place cetaceans within Artiodactyla, with hippopotamids as their closest living relatives (e.g., Irwin and Árnason, 1994). Both concepts were eventually formalized, and the proposed clades became known as **Cetartiodactyla** and **Whippomorpha**, respectively (Figure 4.1) (Montgelard *et al.*, 1997; Waddell *et al.*, 1999).

Meanwhile, morphologists had developed some novel ideas of their own. Although most accepted the association of cetaceans and artiodactyls implied by the molecular data, they also followed an idea by Van Valen (1966), who suggested that the closest relatives of whales were the **Mesonychia**—a group of carnivorous condylarths (ancient hoofed mammals) that resemble early cetaceans in the morphology of their teeth. Paleontologists in particular remained skeptical about the nesting of cetaceans (or cetaceans plus mesonychians) within Artiodactyla, whose monophyly they regarded as well-supported by morphological data (Geisler and Luo, 1998; O'Leary and Uhen, 1999). This perception was largely founded on the morphology of the ankle bone, or astragalus, which in artiodactyls assumes a highly characteristic "double-pulley" shape not found in any other

mammal, including mesonychians (Figure 4.2). There was only one problem: along with their hind limbs, all living cetaceans have lost the astragalus, and until the end of the 20th century this bone either remained entirely unknown in their fossil relatives or else was much too reduced and modified to be of much phylogenetic value (Gingerich *et al.*, 1990).

All of this changed in 2001, when two publications provided dramatic evidence that some of the most archaic cetaceans—the Eocene archaeocetes *Ichthyolestes*, *Pakicetus*, *Artiocetus*, and *Rodhocetus*—had astragali exactly matching those of artiodactyls, but not mesonychians (Figure 4.2). The morphology of the artiodactyl ankle bone is probably an adaptation for running (section 5.1.1), making it highly unlikely that its presence in an aquatic mammal arose by convergence. Thus, the new fossil evidence overwhelmingly supported a direct relationship between cetaceans and artiodactyls, which finally brought paleontological and molecular evidence back into alignment. With few exceptions, this view has prevailed to the present day (O'Leary and Gatesy, 2008; Spaulding *et al.*, 2009).

The final twist in the story of the origin of cetaceans came in 2007, with the publication of new material of the raoellid artiodactyl *Indohyus*. Even

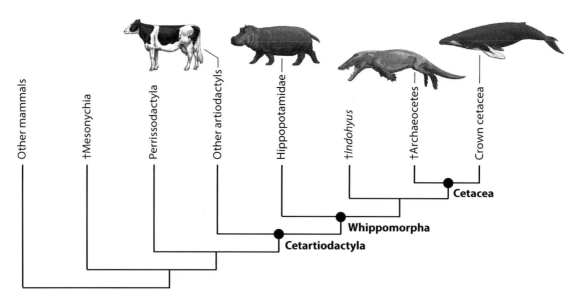

Figure 4.1 Simplified phylogeny placing cetaceans within the context of Mammalia. Life reconstructions © C. Buell.

Figure 4.2 Comparison of the astragalus of (a) the mesonychian *Pachyaena*; (b) the archaic artiodactyl *Diacodexis*; (c) the pakicetid archaeocete *Pakicetus*; and (d) the protocetid archaeocete *Rodhocetus*. Drawing of *Rodhocetus* adapted from Gingerich *et al.* (2009), under a Creative Commons Attribution license.

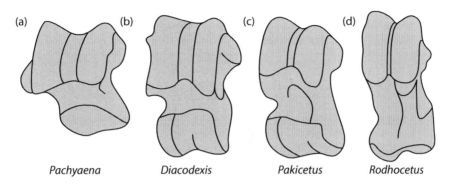

(a)   (b)   (c)   (d)

Pachyaena   *Diacodexis*   *Pakicetus*   *Rodhocetus*

though the discovery of artiodactyl-type ankle bones in archaic cetaceans demonstrated the close affinity of these two groups, the question of whether cetaceans were nested *within* (crown) Artiodactyla remained contentious (Geisler and Uhen, 2003, 2005; Thewissen *et al.*, 2001). The new specimens of *Indohyus* showed some characteristics otherwise typical of archaic cetaceans, such as a tympanic bulla with an incipient involucrum. These observations firmly established not just that cetaceans and artiodactyls were related, but also that whales had indeed arisen from artiodactyl stock. Initially, it appeared that cetaceans and raoellids might together form the sister group to all other artiodactyls, which would have allowed crown Artiodactyla to remain monophyletic. However, subsequent analysis including more morphological and molecular data found that cetaceans and *Indohyus* together were most closely related to hippopotamids, in line with the results of most molecular studies (Figure 4.1) (Geisler and Theodor, 2009; Spaulding *et al.*, 2009).

4.2 The earliest whales: archaeocetes

Archaeocetes, or ancient whales, are those early cetaceans that made the transition from land to water. Thus, the earliest archaeocetes look essentially terrestrial, whereas the youngest ones are fully aquatic and in many ways resemble living cetaceans. Though originally considered a formal

suborder (Archaeoceti), it is now known that archaeocetes are a paraphyletic assemblage of stem cetaceans defined entirely by their retention of many plesiomorphic characteristics, such as a differentiated (heterodont) dentition, tooth replacement, well-developed external hind limbs, and little or no cranial telescoping (sections 5.1 and 8.3). Archaeocetes first appeared in the Early Eocene, around 53 Ma, and within as little as 12 million years evolved first into amphibious and, ultimately, fully aquatic taxa capable of spreading across all of the world's oceans. In this form, archaeocetes persisted until the Late Oligocene (ca 25 Ma), well past the origin of their living descendants, the **Neoceti** (baleen and toothed whales), around 36 Ma (sections 4.3 and 4.4).

4.2.1 Pakicetids, ambulocetids, and remingtonocetids

At the very base of the cetacean tree of life, there is a phylogenetically unequivocal succession of three clades of ancient whales almost exclusively known from the Eocene of Indo-Pakistan (Figure 4.3). The oldest and most archaic of these, the **Pakicetidae**, occurred during the latest Early to early Middle Eocene (ca 47–53 Ma) (Plate 1a). With the possible exception of *Himalayacetus* (discussed further here), pakicetids only occur in freshwater deposits, and likely subsisted on freshwater prey (Clementz *et al.*, 2006; Gingerich *et al.*, 1983). They probably were capable of normal terrestrial locomotion but had nevertheless already adapted to a semiaquatic existence, such

Figure 4.3 Simplified phylogeny of archaeocetes and their relationship to Neoceti. Note that both Protocetidae and Basilosauridae are likely paraphyletic. After Uhen (2010).

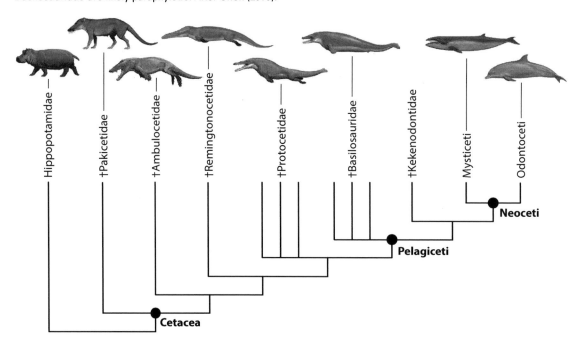

as bottom walking on river beds (section 5.1.1) (Madar, 2007). Compared to most other early archaeocetes, pakicetids stand out for their extremely narrow skull and closely spaced, dorsally oriented eyes (Nummela *et al.*, 2006) (Figure 4.4). In accordance with their age and archaic nature, they retain many features typical of terrestrial artiodactyls, including a small mandibular foramen, long and gracile limb bones, no signs of cranial telescoping, and a solidly fused sacrum (Bajpai and Gingerich, 1998; Thewissen *et al.*, 2001). On the other hand, their pachyosteosclerotic tympanic bulla, their shearing-adapted dentition, and the anteroposterior alignment of the anterior teeth clearly mark them out as cetaceans (Uhen, 2007). To date, the family includes at least three genera: *Ichthyolestes*, *Nalacetus*, and *Pakicetus*. *Himalayacetus* is also commonly included, but its affinities are still under debate.

Morphologically intermediate between the pakicetids and later, more aquatic cetaceans are the **Ambulocetidae**, which lived during the early Middle Eocene (ca 46–48 Ma) (Plate 2). Ambulocetids comprise two genera, *Ambulocetus* and *Gandakasia*, but

the latter is extremely poorly known. The material preserved for *Ambulocetus* is considerably more extensive, but fragmentary (Madar *et al.*, 2002; Thewissen *et al.*, 1996). *Ambulocetus* resembles pakicetids in having a transversely narrow skull, but it is larger and more obviously adapted to an amphibious lifestyle in terms of its relatively smaller femur, robust tail, and enlarged, paddle-like hands and feet. At the same time, *Ambulocetus* retains a well-developed hind limb and a completely fused sacrum, and hence was clearly capable of supporting its weight on land (Thewissen *et al.*, 1994). Its eyes face sideways, yet are located unusually far dorsally. Together with isotopic evidence indicating a freshwater habitat (section 5.1.1), this may suggest that *Ambulocetus* was a crocodile-like ambush predator that lay in wait in shallow water, perhaps at the mouths of rivers, ready to lunge out at (potentially terrestrial) prey (Thewissen *et al.*, 1996).

The first appearance of ambulocetids coincides with that of the **Remingtonocetidae** (named after Remington Kellogg), which survived until the beginning of the Late Eocene around 41 Ma (Plate 1b). Like pakicetids and ambulocetids, this family only occurs

Figure 4.4 Skull of a representative member of Pakicetidae, *Pakicetus inachus* in (a) dorsal and (b) lateral views. ali, alisphenoid; C, canine; eoc, exoccipital; fr, frontal; I, incisor; jug, jugal; lac, lacrimal; M, molar; man, mandible; mx, maxilla; nas, nasal; os, orbitosphenoid; P, premolar; par, parietal; per, periotic; pmx, premaxilla; soc, supraoccipital; sq, squamosal; ty, tympanic bulla. Adapted from Nummela *et al.* (2006).

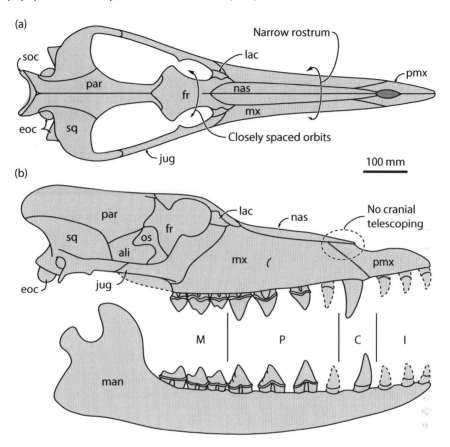

in Indo-Pakistan, with the exception of a single specimen from Egypt (Bebej *et al.*, 2012b). Together with protocetids (section 4.2.2), remingtonocetids were the first truly marine cetaceans, and likely foraged near the coast (Clementz *et al.*, 2006). A well-developed sacrum and hind limb suggest that they were capable of some degree of terrestrial locomotion (Bebej *et al.*, 2012a). Remingtonocetid hallmarks include a long body with relatively short limbs; an elongate skull, rostrum, and mandibular symphysis; lower molars consisting of a series of linearly arranged cusps; an enlarged mandibular foramen (shared with all later cetaceans); and, in some taxa, small eye sockets (Figure 4.5) (Bebej *et al.*, 2012a; Thewissen and Bajpai, 2001, 2009). Superficially,

these traits give remingtonocetids an even more crocodile-like appearance than *Ambulocetus*. As currently defined, the family comprises five genera in two subfamilies, including (1) *Andrewsiphius* and *Kutchicetus* (Andrewsiphiinae); and (2) *Attockicetus*, *Dalanistes*, and *Remingtonocetus* (Remingtonocetinae).

4.2.2 Protocetidae and basal Pelagiceti

Like its predecessors, the family **Protocetidae** originated in Indo-Pakistan during the early Middle Eocene (ca 48 Ma), but then rapidly diversified and spread across North Africa and North America, surviving until as late as 37 Ma (section 7.6.1) (Gingerich *et al.*, 2014). All protocetids

Figure 4.5 Skull of a representative member of Remingtonocetidae, *Remingtonocetus harudiensis*, in (a) dorsal and (b) lateral views. Lower jaw reconstructed. Abbreviations as in Figure 4.4; adapted from Bajpai *et al.* (2011).

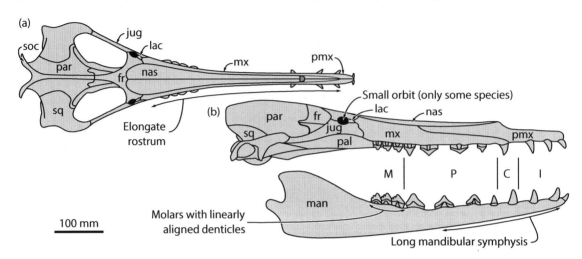

have been found in deposits indicative of a coastal marine environment, where they spent most of their time, foraged for invertebrates and fish and, possibly, even hunted seabirds and other mammals (Plate 3a) (Clementz *et al.*, 2006; Fahlke *et al.*, 2013). Like all archaeocetes up to this point, they retain a well-developed hind limb and pelvis; however, the latter is attached to an increasingly weak sacrum comprising ever fewer vertebrae, which suggests only limited terrestrial abilities. At the same time, increased uniformity in the morphology of the vertebrae points to the emergence of a modern, undulatory swimming mode, which makes protocetids the major transitional stage connecting modern, ocean-going cetaceans with their terrestrial ancestors (section 5.1.2) (Bianucci and Gingerich, 2011). Protocetids are still characterized by the primitive mammalian dental formula of 3.1.4.3/3.1.4.3. Their cheek teeth retain a distinct trigonid and talonid and, with few exceptions, lack accessory denticles. However, they resemble more crownward cetaceans in having eyes that are roofed by broad supraorbital processes (Figure 4.6).

The family is large by cetacean standards, and currently comprises 19 genera: *Aegyptocetus, Artiocetus, Babiacetus, Carolinacetus, Crenatocetus, Dhedacetus, Eocetus, Gaviacetus, Georgiacetus, Indocetus, Kharodacetus, Maiacetus,* *Makaracetus, Natchitochia, Pappocetus, Protocetus, Qaisracetus, Rodhocetus,* and *Takracetus* (Uhen, 2015). Of these, *Artiocetus* and *Rodhocetus* have featured prominently in establishing cetacean origins (section 4.1); *Georgiacetus* in tracing the evolution of swimming styles (section 5.1.3); and *Maiacetus* in determining that protocetids likely still gave birth on land (section 6.3). Though generally treated as a family, protocetids are likely paraphyletic and fall into a series of successive stem branches leading up to fully aquatic whales (Pelagiceti) (Figure 4.3). *Artiocetus* and *Maiacetus* may be the most archaic representatives of the group, whereas *Georgiacetus, Babiacetus,* and *Eocetus* may be the most crownward. The phylogenetic position of the remaining genera remains poorly resolved (Uhen, 2014).

Pelagiceti is a clade of fully aquatic cetaceans comprising modern whales and dolphins, collectively known as **Neoceti**, as well as two families of archaeocetes: basilosaurids and kekenodontids. Pelagicetes are characterized by an extremely small and rotated pelvis; the marked reduction or even absence of the hind limb; the presence of more than 10 lumbar vertebrae; and rectangular, short, and dorsoventrally compressed posterior caudal vertebrae. In addition, archaic pelagicetes differ from archaeocetes in the presence of multiple accessory

Figure 4.6 Skull of a representative member of Protocetidae, *Artiocetus clavis*, in (a) dorsal and (b) lateral views. Lower jaw reconstructed from *Rodhocetus kasranii*. Abbreviations as in Figure 4.4; adapted from Gingerich *et al.* (2001).

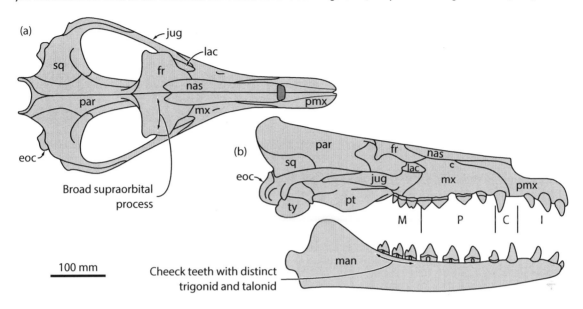

denticles on the cheek teeth, although this trait is later lost in toothless baleen whales and homodont odontocetes (Uhen, 2008a).

Basilosauridae existed from the late Middle Eocene to the Late Eocene and are found throughout the world, including Africa, Asia, Europe, North and South America, New Zealand, and even Antarctica (Plate 3b). An unusually old, unnamed basilosaurid has been reported from the Ypresian (late Early Eocene) portion of the La Meseta Formation of Antarctica (Reguero *et al.*, 2012), but both the identification and age of this specimen have yet to be confirmed. Basilosaurids show many anatomical features indicating that they were fully aquatic, such as a well-developed pterygoid sinus acoustically isolating the ear, a relatively short neck (compared to other archaeocetes), a flattened forelimb that likely took the shape of a flipper, and a relatively immobile wrist (Uhen, 1998, 2004). Most importantly, the hind limbs of basilosaurids are minuscule, and the pelvis, itself greatly reduced in size, is completely detached from the vertebral column. This means that basilosaurids were no longer able to support their weight on land, and thus were the first cetaceans to forage, mate, and, crucially, give birth in water

(Gingerich *et al.*, 1990). Their vertebral column has become largely homogenized, which enabled them to swim via undulatory and oscillatory movements of the torso and tail (section 5.1), and they are the oldest cetaceans to show marked posterior migration of the external nares, as well as evidence of tail flukes (Figure 4.7) (Buchholtz, 1998). Like protocetids, they likely subsisted on both invertebrate and vertebrate prey, including other marine mammals (Fahlke *et al.*, 2013).

Beyond those features directly related to becoming fully aquatic, basilosaurids are characterized by the lack of an upper third molar, the presence of accessory denticles on the cheek teeth, and, in some species, extremely elongated trunk and tail vertebrae that give rise to a somewhat snake-like appearance (Figure 4.7) (Uhen, 1998). This elongation is particularly noticeable in *Basilosaurus*, which could reach a total body length of as much as 17 m. There are currently 13 recognized genera in this family: *Ancalecetus, Basilosaurus, Basiloterus, Basilotritus, Chrysocetus, Cynthiacetus, Dorudon, Masracetus, Ocucajea, Saghacetus, Stromerius, Supayacetus,* and *Zygorhiza* (Uhen, 2015). Like protocetids, basilosaurids are likely paraphyletic and include among them the ancestor of neocetes

Figure 4.7 Skull of a representative member of Basilosauridae, *Dorudon atrox*, in (a) dorsal and (b) lateral views. Abbreviations as in Figure 4.4; adapted from Uhen (2004).

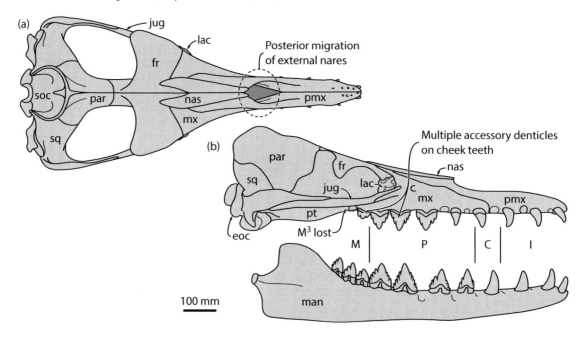

(Figure 4.3). However, the degree of this paraphyly is debated, with some analyses recovering a nearly monophyletic Basilosauridae excluding only *Ocucajea* (Gol'din and Zvonok, 2013; Martínez Cáceres and de Muizon, 2011), whereas others envisage a gradual, step-by-step succession of individual stem taxa leading up to modern cetaceans (Uhen, 2004). In either case, *Basilotritus* and maybe *Supayacetus* are among not only the oldest but likely also the basal-most members of the group, since both retain features that are to some degree reminiscent of protocetids (Gol'din and Zvonok, 2013; Uhen *et al.*, 2011).

Kekenodontidae are a small, enigmatic family known only from the Oligocene, which starkly sets them apart from all other, exclusively Eocene archaeocetes. At present there is just a single kekenodontid genus, the enigmatic *Kekenodon* from the Late Oligocene (Duntroonian) of New Zealand, although some fragmentary material from the Early Oligocene of Europe (*Phococetus vasconum*) has at times also been referred to this family (Mitchell, 1989). Not much is known of *Kekenodon* besides some cranial fragments and

teeth, which has resulted in the family variably being placed within or outside Neoceti. In either case, however, they are likely to be more closely related to neocetes than to any other archaeocetes, including basilosaurids. New, better preserved material from New Zealand—including a virtually complete skull—will help to settle this question, but has yet to be formally described (Clementz *et al.*, 2014).

4.3 Filter-feeding whales: Mysticeti

Mysticetes, or baleen whales, are one of the two groups of living cetaceans (**Neoceti**). Baleen whales are highly distinctive inhabitants of the modern oceans and include the largest animals that have ever lived, ranging in length from about 6.5 to as much as 30 m and weighing up to 190 tonnes. With few exceptions, all mysticetes are characterized by a dorsoventrally flattened snout, a well-developed infraorbital plate and thickened basioccipital crests. All mysticetes furthermore exhibit some degree of

Figure 4.8 Simplified phylogeny of toothed mysticetes and their relationship to chaeomysticetes. A and L denote alternative placements of Aetiocetidae and Llanocetidae (see text for details). Life reconstructions © C. Buell.

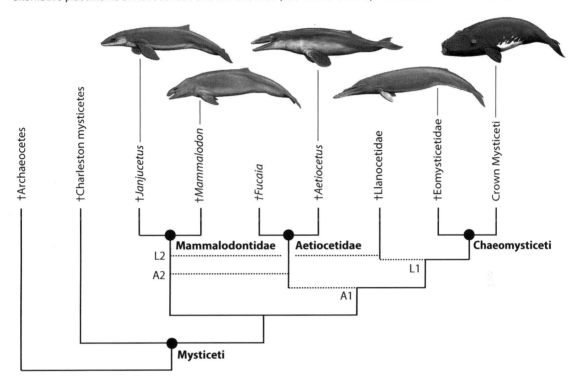

telescoping of the rostral bones (premaxillae, maxillae, and nasals) and, except for the most archaic taxa, the supraoccipital. As their name suggests, modern baleen whales have no teeth. Instead, they use their hallmark characteristic, **baleen**, to filter vast amounts of prey from the water, thereby sustaining their impressive bulk (section 5.2.1). By contrast, many of the earliest mysticetes defy all notions of what a baleen whale ought to be: unlike their gigantic descendants, they were small, bore well-developed teeth, and followed a variety of feeding strategies. Mysticetes first appeared during the latest Eocene, about 34 Ma, and during the Early Oligocene evolved into several disparate lineages, including both toothed and early toothless forms (Figure 4.8).

4.3.1 Toothed mysticetes

The existence of toothed baleen whales may seem counterintuitive, but it makes perfect sense considering that they are derived from tooth-bearing ancestors. Toothed mysticetes range from the latest Eocene (ca 34 Ma) to the latest Oligocene (23 Ma), and possibly even the late Early Miocene (17–19 Ma), thus implying a long period of coexistence with their toothless cousins. Taxonomically, toothed mysticetes fall into three main groups, the relationships of which are still a matter of debate (Figure 4.8). The oldest and least known of these are the **Llanocetidae**, which currently only include *Llanocetus* from the Late Eocene of Antarctica (ca 34 Ma). Only fragments of the relatively complete holotype have so far been formally described (Mitchell, 1989). Additional, as-yet-unpublished material possibly referable to *Llanocetus* or Llanocetidae, including a diminutive form from the Early Oligocene, has been reported from Antarctica, New Zealand, California, and Peru (Fordyce, 2003; Martínez Cáceres *et al.*, 2011; Rivin, 2010). Preliminary descriptions of the main portion of the holotype specimen of *Llanocetus* indicate that it was a relatively large

Figure 4.9 Skull of a representative member of Mammalodontidae, *Janjucetus hunderi*, in (a) dorsal and (b) lateral views. Abbreviations as in Figure 4.4; adapted in part from Fitzgerald (2010).

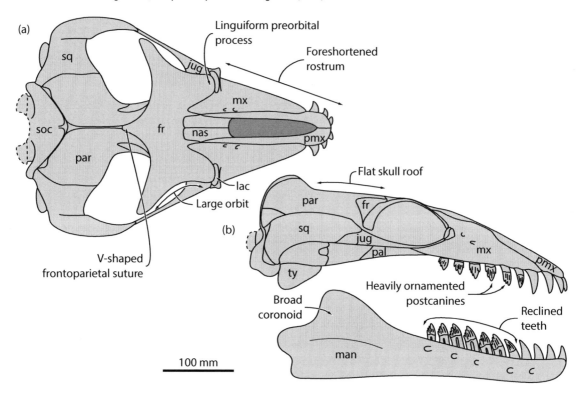

animal (ca 8–9 m long) bearing relatively robust, heterodont teeth inside a dorsoventrally flattened snout (Fordyce, 2003). Fine palatal grooves surrounding the upper teeth may indicate the presence of an enhanced blood supply, but it is unclear whether these structures are homologous with the palatal nutrient foramina of toothless mysticetes. The phylogenetic position of llanocetids remains unresolved, with current proposals including a sister group relationship with either mammalodontids (discussed further here) or toothless mysticetes (Fitzgerald, 2010; Marx and Fordyce, 2015; Steeman, 2007).

Mammalodontidae are an extinct family of toothed mysticetes currently only known from the Late Oligocene (24–27 Ma) of Australia, New Zealand, and, possibly, Malta (Bianucci *et al.*, 2011; Fitzgerald, 2010) (Plate 4a,b). Members of this clade stand out for their small size (<5 m long), short snout, large and anteriorly directed eyes,

tongue-shaped anterior border of the supraorbital process, flattened skull roof, heavily ornamented tooth enamel, and posteriorly reclined lower cheek teeth (Figure 4.9). In common with archaeocetes and early odontocetes, mammalodontids also retain many archaic characteristics, such as distinctly heterodont teeth, nasal openings placed halfway along the snout, a non-telescoped supraoccipital, a large temporal fossa, an unfused compound posterior process of the tympanoperiotic, a distinct median furrow on the non-rotated tympanic bulla, a fused (albeit small) mandibular symphysis, a broad and plate-like coronoid process, a large mandibular foramen, and no evidence of baleen in the form of palatal nutrient foramina. Only two genera, *Mammalodon* and *Janjucetus*, have so far been recognized.

The inclusion of mammalodontids within Mysticeti is widely agreed upon (Figure 4.8), but questions persist regarding the monophyly of the family (Deméré et al., 2008; Fitzgerald, 2006), as well as the existence of a potential relationship with either llanocetids or aetiocetids (Fitzgerald, 2010; Marx, 2011). The dental and cranial morphology of mammalodontids suggest that they occupied ecological niches—suction feeding (*Mammalodon*) and macrophagy (*Janjucetus*)—fundamentally different from those of most other living or fossil baleen whales. Pedomorphism may be among the reasons for the rather disparate nature of these animals, and could plausibly account for their small size, large eyes, and large occipital condyles (Fitzgerald, 2010; Sanders and Barnes, 2002). In addition, it might explain the absence of any sign of baleen, with the juvenile condition (i.e., teeth) being retained into adulthood instead.

Aetiocetidae are a group of extinct mysticetes that may have possessed both teeth and baleen at the same time (Plate 5a). To date, aetiocetid fossils have only been found around the North Pacific, in particular the Early–Late Oligocene (23–34 Ma) of Japan and the western coast of North America (Barnes et al., 1995; Emlong, 1966). Additional specimens from the Early Oligocene of South Australia

and the Early Miocene of California are either fragmentary or still awaiting description (Pledge, 2005; Rivin, 2010). Aetiocetids are generally small—less than 5 m long, although some may have reached as much as 8 m (Tsai and Ando, 2015). They are characterized by a large orbit, an enlarged lacrimal, a centrally constricted zygomatic process and mandible, a finger-like coronoid process, an unfused mandibular symphysis, and, at least in some cases, small palatal nutrient foramina (Figure 4.10). The teeth are clearly heterodont in some species, but more widely spaced and incipiently homodont in others. On the other hand, aetiocetids retain somewhat anteriorly placed nasal openings, a non-telescoped supraoccipital, a large temporal fossa, an unfused compound posterior process of the tympanoperiotic, and a distinct median furrow on the large, non-rotated tympanic bulla (Barnes et al., 1995; Deméré and Berta, 2008; Marx, 2011). Six genera have been described so far: *Aetiocetus*, *Ashorocetus*, *Chonecetus*, *Fucaia*, *Morawanocetus*, and *Willungacetus*. Of these, *Ashorocetus* and *Willungacetus* are only known from poorly preserved material and in need of reassessment (Marx et al., 2015).

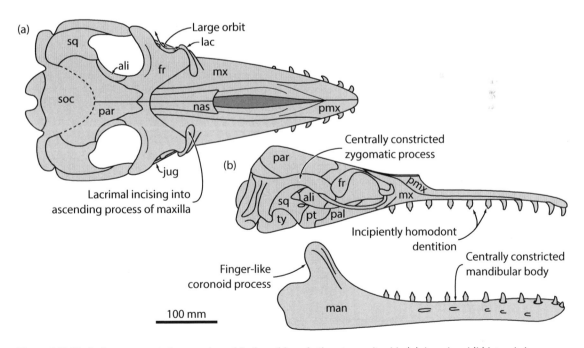

Figure 4.10 Skull of a representative member of Aetiocetidae, *Aetiocetus weltoni*, in (a) dorsal and (b) lateral views. Abbreviations as in Figure 4.4; adapted in part from Deméré and Berta (2008).

Aetiocetids have been known since the 1960s (Emlong, 1966), but their true significance was only discovered more than 40 years later with the discovery of palatal nutrient foramina reminiscent of those of living baleen whales (Deméré et al., 2008). Remarkably, these palatal foramina occur alongside teeth, which might suggest the concurrent presence of an early form of baleen. Aetiocetids thus seem to be morphologically, and possibly ecologically, intermediate between the toothed archaeocetes and their toothless, baleen-bearing mysticete descendants. The presence of an unfused mandibular symphysis, often associated with filter feeding, further adds to this impression, as does the phylogenetic position of aetiocetids at or near the base of toothless mysticetes (Deméré et al., 2008). However, the appearance of features not necessarily associated with filter feeding (e.g., the progressive development of near-homodonty) and the long persistence of aetiocetids alongside toothless mysticetes add a further level of complexity that is not yet fully understood (Marx et al., 2015). As in the case of mammalodontids, pedomorphism has been invoked as a possible explanation for the small size of aetiocetids, as well as their com-

paratively large eye sockets, occipital condyles, and tympanic bullae (Fitzgerald, 2010; Sanders and Barnes, 2002).

4.3.2 Toothless mysticetes

Toothless, "true" mysticetes (**Chaeomysticeti**) first appeared during the Early Oligocene and subsequently radiated into at least five, but probably considerably more, distinct lineages (Figure 4.11). Of these, four (right whales, the pygmy right whale, the gray whale, and rorquals) are still extant. As their name suggests, toothless mysticetes have lost any trace of a dentition except for fetal tooth buds, which in living baleen whales are resorbed before birth (Deméré et al., 2008). All chaeomysticetes furthermore share the presence of a palatal keel running anteriorly along the center of the rostrum, the presence of palatal nutrient foramina, and an unfused mandibular symphysis, with the latter two features also occurring in aetiocetids (as mentioned in this chapter).

The extinct **Eomysticetidae**, or "dawn mysticetes," are among the oldest and most archaic chaeomysticetes; they are known from the late Early–Late Oligocene (ca 25–29 Ma) of Japan, New Zealand, and the East Coast of the United

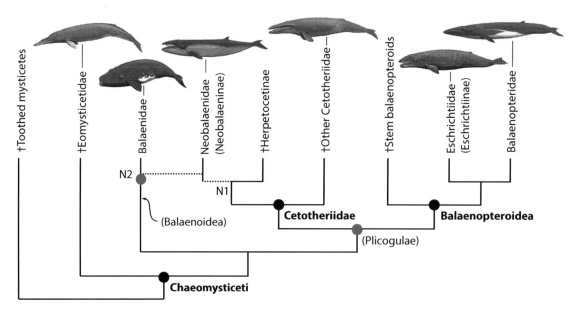

Figure 4.11 Simplified phylogeny of Chaeomysticeti. N denotes alternative placements of Neobalaenidae (see text for details). Life reconstructions © C. Buell.

States (Boessenecker and Fordyce, 2014; Okazaki, 2012; Sanders and Barnes, 2002). The taxonomic identity of additional material from the Early Miocene of California (17–19 Ma) needs to be confirmed (Rivin, 2010). Eomysticetids share the loss of functional teeth with all other chaeomysticetes, and furthermore are distinguished by the presence of a relatively long rostrum and markedly elongated nasal, a large and oval temporal fossa, and a robustly built zygomatic process of the squamosal (Figure 4.12). However, they also retain a plethora of primitive features, such as poorly developed palatal nutrient foramina, a little-telescoped supraoccipital, an unfused compound posterior process of the tympanoperiotic, a distinct median furrow on the (non-rotated) tympanic bulla, a broad and plate-like coronoid process, a large mandibular foramen, and shallow, much-reduced alveoli along the anterior portion of the rostrum housing small (presumably nonfunctional) teeth (Boessenecker and Fordyce, 2015b; Okazaki, 2012; Sanders and Barnes, 2002).

At present, the family comprises five genera (*Eomysticetus*, *Tohoraata*, *Tokarahia*, *Waharoa*, and *Yamatocetus*), but more material from New Zealand and the United States still awaits description. *Micromysticetus* from the East Coast of the United States and *Cetotheriopsis* from Europe may constitute a family of their own, **Cetotheriopsidae**, but more likely also represent eomysticetids or—in the case of *Cetotheriopsis*—are too poorly known to be sure of familial affinities (Boessenecker and Fordyce, 2015a, 2015b; Marx and Fordyce, 2015). In general, eomysticetids seem to have been larger than most coeval toothed mysticetes. They likely represent the basal-most chaeomysticetes and, as such, capture a "morphological snapshot" of the early evolution of baleen-assisted filter feeding (Deméré *et al.*, 2008; Marx and Fordyce, 2015; Steeman, 2007). Nevertheless, their feeding strategy remains something of a mystery, with no living mysticete showing the eomysticetid combination of a long and narrow rostrum, a relatively gracile mandible

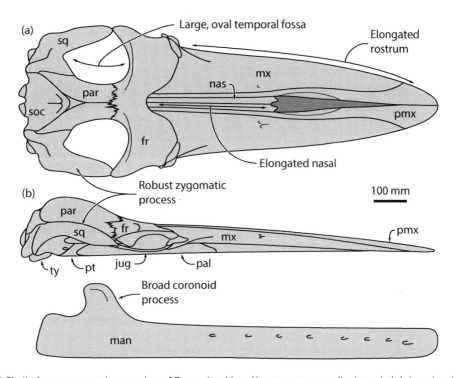

Figure 4.12 Skull of a representative member of Eomysticetidae, *Yamatocetus canaliculatus*, in (a) dorsal and (b) lateral views. Abbreviations as in Figure 4.4

(e.g., in *Waharoa*), and powerful jaw adductor muscles, as implied by the large temporal fossa.

Balaenomorpha comprises all toothless mysticetes except eomysticetids, and therefore all of the living species (Figure 4.11) (Geisler and Sanders, 2003). Balaenomorphs are characterized by a distinctly telescoped supraoccipital, a supraorbital process that laterally descends from the skull vertex (as opposed to being nearly horizontal), well-developed palatal nutrient foramina and sulci, the fusion of the posterior processes of the tympanoperiotic, a poorly defined mallear fossa, the reduction or loss of the ventral border of the sigmoid process, and a general trend toward the reduction of the mandibular foramen. A further, particularly striking feature of all balaenomorphs is the gradual rotation of the tympanic bulla along its anteroposterior axis, so that the lateral lobe becomes positioned ventrally, and the medial one dorsally. A variably developed median furrow of the tympanic bulla may still be present, but tends to become reduced in most lineages.

Balaenidae (right whales) are among the earliest diverging balaenomorphs, and are still represented in the modern ocean by the globally distributed genus *Eubalaena* and the Arctic bowhead whale, *Balaena mysticetus* (Plate 5b). The earliest reliable occurrence of the family comes from the late Early Miocene (ca 20 Ma) of Argentina (Cabrera, 1926). The balaenid affinities of some considerably older (ca 28 Ma), undescribed material from the early Late Oligocene of New Zealand are doubtful (Marx and Fordyce, 2015). Fossil remains of this family have been recovered from virtually every continent, yet mostly cluster in the Late Miocene and Pliocene, thus leaving much of the Miocene history of balaenids in the dark (Bisconti, 2003; Fitzgerald, 2004).

The skull morphology of balaenids is highly distinctive, and dominated by the pronounced dorsoventral arching of the narrow rostrum. This arching may in turn have necessitated the development of a long and narrow supraorbital process, a dorsoventrally high parietal and squamosal, a short and laterally oriented zygomatic process, and a large and steeply ascending supraoccipital (Figure 4.13). In addition, balaenids stand out for their hypertrophied anterior process and lateral tuberosity of the periotic, box-shaped, and posteriorly

diverging tympanic bullae, long and U-shaped postglenoid process, greatly reduced coronoid process, markedly twisted anterior portion of the mandible, mylohyoidal sulcus running along the ventromedial surface of the mandible, and fused cervical vertebrae (Bisconti, 2005; Churchill *et al.*, 2011; Marx, 2011). Unlike all other living mysticetes, balaenids retain five digits inside the flipper. Externally, living balaenids are more bulky than other baleen whales, with highly arched lower lips, broad flippers, and no dorsal fin. On the snout and above the eyes, *Eubalaena* bears an individually unique pattern of thickened patches of skin, or **callosities**, often heavily infested with whale lice (cyamids). The highly arched rostrum is matched by long and slender, finely fringed baleen plates employed in continuous skim feeding (Pivorunas, 1979).

In addition to the extant *Eubalaena* and *Balaena*, four extinct genera—*Balaenella*, *Balaenula*, *Idiocetus*, and *Morenocetus*—have so far been described. The status of a fifth, *Balaenotus*, currently remains uncertain. Whereas both *Balaena* and *Eubalaena* tend to be extremely large (up to 18 m long), and have been so since at least the latest Miocene (Kimura, 2009), *Balaenella*, *Balaenula*, and *Morenocetus* were likely no larger than 5–6 m each. The evolutionary relationships of balaenids have been a matter of some debate, with morphological and total evidence analyses either supporting a closely knit mysticete crown group to the exclusion of most fossil chaeomysticetes (Bouetel and de Muizon, 2006; Churchill *et al.*, 2011; Deméré *et al.*, 2008; Geisler *et al.*, 2011), or interpreting balaenids as the earliest diverging balaenomorphs (Figure 4.11) (Bisconti *et al.*, 2013; Marx and Fordyce, 2015; Steeman, 2007). Most morphological analyses furthermore recover balaenids as the sister group of the extant pygmy right whale, *Caperea* (e.g., Bisconti, 2014; El Adli *et al.*, 2014), a result contradicted by the vast majority of molecular and total evidence studies, as well as some morphological analyses (e.g., Deméré *et al.*, 2008; Fordyce and Marx, 2013; Hassanin *et al.*, 2012; Marx and Fordyce, 2015; Steeman *et al.*, 2009). See the discussion of Neobalaenidae and Cetotheriidae that follows next.

For most of its existence, the family **Cetotheriidae** served as a wastebasket taxon for a variety

Figure 4.13 Skull of a representative member of Balaenidae, *Eubalaena glacialis*, in (a) dorsal and (b) lateral views. Abbreviations as in Figure 4.4.

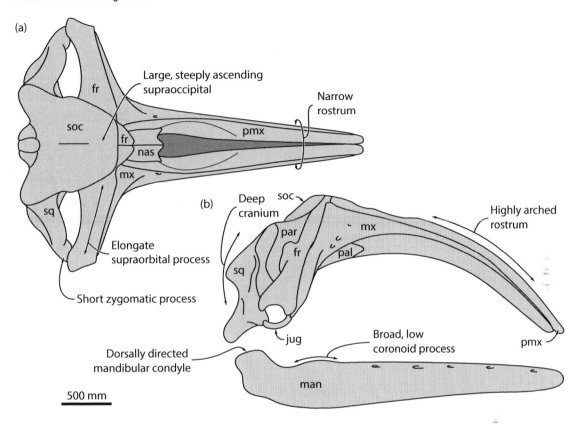

of extinct chaeomysticetes that could not be reliably referred to any of the living families. This resulted in a polyphyletic assemblage of both stem and crown mysticetes, ranging from the Oligocene to the Pliocene. Over the past decade, detailed comparative and phylogenetic studies have started to untangle this confusion by (1) recognizing the distinctiveness of Eomysticetidae (discussed in this chapter); and (2) sorting the remaining species into those related to *Cetotherium rathkii* (**Cetotheriidae *sensu stricto***; here simply referred to as Cetotheriidae) and those whose affinities remain uncertain ("cetotheres" *sensu lato*). The latter include *Aglaocetus*, *Cophocetus*, *Diorocetus*, *Hibacetus*, *Horopeta*, *Isanacetus*, *Mauicetus*, *Parietobalaena*, *Pelocetus*, *Thinocetus*, and *Titanocetus*, all of which have variously been placed as sister to cetotheriids

proper or as stem balaenopteroids (discussed further here), and have been both included (Bisconti, 2014; Marx and Fordyce, 2015; Steeman, 2007) and excluded (Deméré *et al.*, 2008; Geisler *et al.*, 2011) from crown Mysticeti as a whole. Until a more stable picture of their affinities emerges, all of these former "cetotheres" should thus be considered Chaeomysticeti *incertae sedis*.

Most phylogenetic analyses envisage cetotheriids as a group of relatively small-bodied taxa (4–6 m) with a highly distinctive cranial architecture, known from the late Middle Miocene–Early Pleistocene of North and South America, Europe, and Japan (e.g., Bouetel and de Muizon, 2006; Brandt, 1873; Whitmore and Barnes, 2008). This view remains the most widely accepted interpretation (but see the discussion of Neobalaenidae in this section). So defined, cetotheriids are characterized

Figure 4.14 Skull of a representative member of Cetotheriidae, *Piscobalaena nana*, in (a) dorsal and (b) lateral views. Abbreviations as in Figure 4.4.

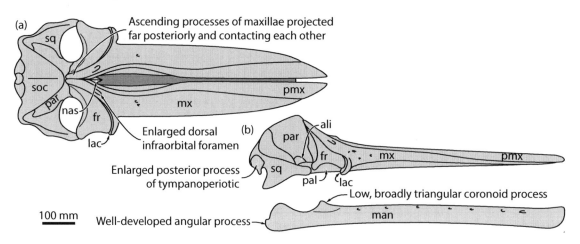

by an X-shaped skull vertex formed by the nuchal crests and the far posteriorly telescoped maxillae. Likely as a result of this telescoping, the anterior-most point of the supraoccipital in cetotheres is located much further posteriorly than in all other balaenomorphs, except gray whales (Figure 4.14). In some species, the posterior ends of the ascending processes of the maxillae approximate each other along the midline, which causes the intervening nasals and premaxillae to become transversely compressed, or even to disappear entirely (Bouetel and de Muizon, 2006; Gol'din *et al.*, 2014). Besides this unique pattern of cranial telescoping, cetotheriid hallmarks include an extremely enlarged compound posterior process of the tympanoperiotic, an often prominent external occipital crest, the presence of an enlarged dorsal infraorbital foramen aligned with the ascending process of the maxilla, a reduced coronoid process and a variably enlarged angular process. Some species furthermore show a squamosal cleft. Although their tympanic bullae are fully rotated, cetotheriids primitively retain a well-developed median lobe (dorsal posterior prominence) as well as a well-developed median furrow (Bouetel and de Muizon, 2006).

Besides *Cetotherium*, Cetotheriidae appear to comprise at least a further 10 genera: *Cephalotropis, Joumocetus, Herentalia, Herpetocetus, Kurdalagonus, Metopocetus, Nannocetus, Piscobalaena, Vampalus,* and *Zygiocetus* (Bouetel and de Muizon, 2006; Gol'din *et al.*, 2014; Marx and Fordyce, 2015). The affinities of "*Cetotherium*" *megalophysum*, "*Metopocetus*" *vandelli*, "*Aulocetus*" *latus, Mesocetus, Mixocetus,* and *Tranatocetus* remain uncertain, but they have recently been proposed to form a separate family, **Tranatocetidae** (Gol'din and Steeman, 2015). *Herpetocetus* and *Nannocetus* are generally grouped into their own subfamily, **Herpetocetinae**, which is characterized by an externally exposed, plug-like compound posterior process of the tympanoperiotic, a medially oriented postglenoid process and a continuous posterolateral skull wall in dorsal view (Whitmore and Barnes, 2008). Broader definitions of this subfamily including *Vampalus, Piscobalaena,* "*C.*" *megalophysum*, and "*M.*" *vandelli* have also been proposed (El Adli *et al.*, 2014; Tarasenko and Lopatin, 2012). Phylogenetic analyses have variably placed cetotheriids basal to all crown mysticetes (Bouetel and de Muizon, 2006; Deméré *et al.*, 2008) or grouped them with gray whales (Bisconti, 2008; Steeman, 2007), gray whales and rorquals (Bisconti *et al.*, 2013; Marx, 2011), or pygmy right whales (Fordyce and Marx, 2013; Gol'din and Steeman, 2015; Marx and Fordyce, 2015). None of these hypotheses are as yet widely accepted.

The **Neobalaenidae** (pygmy right whales) are maybe the most enigmatic of all mysticetes; they are represented by *Caperea*, the smallest living mysticete (ca 6.5 m long), as well as a limited fossil record of similarly sized animals stretching back to the early Late Miocene (Bisconti, 2012; Buono *et al.*, 2014). However, molecular divergence estimates for this lineage range from 18 to 27 million years, and thus considerably predate the oldest available fossil evidence (Hassanin *et al.*, 2012; McGowen *et al.*, 2009; Steeman *et al.*, 2009). Both *Caperea* and all of the described fossil material occur exclusively in the Southern Hemisphere, with the living species being limited to a relatively narrow, circumpolar distribution (ca 30–55°S) (Kemper, 2009).

Neobalaenids have the most disparate skull and postcranial morphology of any living or extinct baleen whale. In particular, they stand out for their arched rostrum, far-anteriorly projected supraoccipital shield, pointed nasals, longitudinal maxillary fenestra on the palate, elongate squamosal fossa, medially shifted and oriented post-glenoid process, reduced pterygoid hamulus, conical and externally exposed compound posterior process of the tympanoperiotic, highly arched mandible, strongly reduced coronoid process, fused cervical vertebrae, fused first and second ribs, large number of thoracic vertebrae (generally

18) followed by just one or two lumbar vertebrae, and flattened and overlapping posterior ribs (Figure 4.15) (Beddard, 1901; Bisconti, 2012; Buchholtz, 2011; Fordyce and Marx, 2013). In addition, extant *Caperea* is unique in having an almost completely detached anterior process of the periotic and in having the foramen pseudovale located entirely within the pterygoid (Bisconti, 2012; Fordyce and Marx, 2013). Like rorquals, *Caperea* has a squamosal cleft, only four digits in each flipper and a distinct dorsal fin (Kemper, 2009), but it more closely matches right whales in having long, finely fringed baleen (Sekiguchi *et al.*, 1992).

The fossil record of neobalaenids is extremely patchy. Besides the living *Caperea*, the family only includes the somewhat more archaic, yet structurally similar, extinct *Miocaperea*. The affinities of *Balaena simpsoni*, previously considered to be a neobalaenid, are controversial (Canto *et al.*, 2010). Other fossil material referred to this family includes a fragmentary ear bone from Australia (Fitzgerald, 2012) and an isolated mandible from Argentina (Buono *et al.*, 2014), although neither of these specimens can be identified with complete confidence.

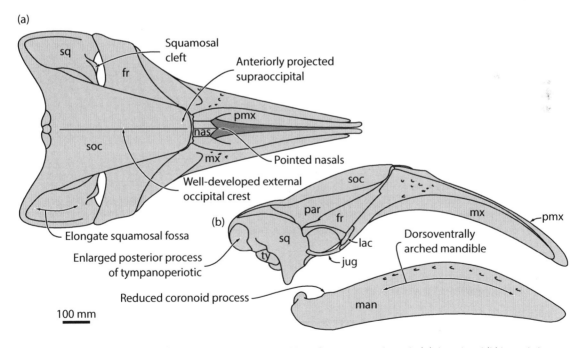

Figure 4.15 Skull of a representative member of Neobalaenidae, *Caperea marginata*, in (a) dorsal and (b) lateral views. Abbreviations as in Figure 4.4.

As their name suggests, pygmy right whales have, on morphological grounds, traditionally been regarded as distant relatives of right whales, and were united with them in the superfamily **Balaenoidea** (e.g., Bisconti, 2012, 2014; El Adli *et al.*, 2014). However, this interpretation is contradicted by molecular evidence, which generally allies *Caperea* with rorquals and gray whales in the clade **Plicogulae**, to the exclusion of right whales (e.g., Deméré *et al.*, 2008; Hassanin *et al.*, 2012; McGowen *et al.*, 2009). Some morphological analyses contradict the traditional interpretation of *Caperea* and potentially resolve this apparent molecular–morphological discrepancy by nesting neobalaenids, as the subfamily **Neobalaeninae**, within Cetotheriidae (Figure 4.11) (Fordyce and Marx, 2013; Marx and Fordyce, 2015). However, in the absence of transitional fossils revealing more about ancestral neobalaenid morphology, as well as the obvious lack of molecular data for cetotheriids, these hypotheses currently remain difficult to test.

Balaenopteroidea traditionally comprise rorquals (Balaenopteridae) and gray whales (Eschrichtiidae). A range of distantly related Oligocene and Miocene taxa—"cetotheres" *sensu lato*—may also fall into this group (Figure 4.11), but their relationships remain a matter of debate (Marx and Fordyce, 2015; Steeman, 2007). Within balaenopteroids, the clade formed by balaenopterids and eschrichtiids is defined by the presence of an abruptly depressed (relative to the skull vertex), broad and horizontally oriented supraorbital process, a squamosal cleft, long and narrow ascending processes of the maxilla, a cranially elongated pars cochlearis and sharply triangular anterior process of the periotic, a semi-synovial or non-synovial (fibrocartilaginous) mandibular joint, and the presence of just four digits in the flipper (Deméré *et al.*, 2005).

Gray whales (**Eschrichtiidae**) are small- to medium-sized (ca 7–15 m) mysticetes including the extant *Eschrichtius* and a small number of extinct taxa from the Late Miocene and Pliocene of Italy and the United States (Bisconti, 2008; Ekdale and Racicot, 2015; Whitmore and Kaltenbach, 2008). *Eschrichtius* is today restricted to the North Pacific, but until very recently it also occurred in the North Atlantic (Mead and Mitchell, 1984). Within the last few years, stray animals presumably originating from the Pacific population have been sighted as far afield as the

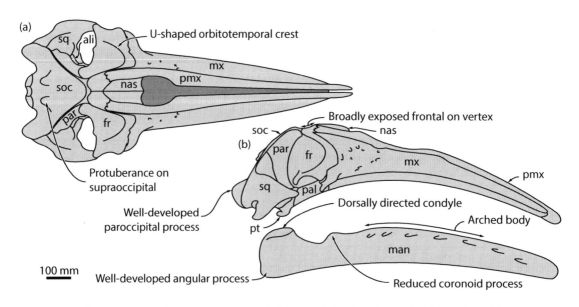

Figure 4.16 Skull of a representative member of Eschrichtiidae, *Eschrichtius robustus*, in (a) dorsal and (b) lateral views. Abbreviations as in Figure 4.4.

Mediterranean (Scheinin *et al.*, 2011). Unlike all other living mysticetes, which filter their prey from the water column, gray whales are mostly bottom feeders and sift their food from sediments deposited on the seabed (Werth, 2000). The skull of gray whales is characterized by a well-developed paroccipital process, a U-shaped orbitotemporal crest, and the virtual absence of telescoping of the supraoccipital, which allows the frontal to be broadly exposed on the vertex (Figure 4.16). The mandible is highly arched and bears a dorsally directed condyle, as well as a robust angular process. The coronoid is reduced and paralleled on its medial side by a secondary "satellite" process (Bisconti and Varola, 2006). In the periotic, the fenestra rotunda and the aperture for the cochlear aqueduct are generally confluent. To varying degrees, eschrichtiids also display a pair of protuberances on the supraoccipital (best developed in *Eschrichtius*), although a similar feature may also occur in other taxa clearly outside the family. Externally, gray whales are easily identified by their name-giving body coloration, their short and coarse baleen, and the presence of a series of knobs on their back in lieu of a dorsal fin (Barnes and McLeod, 1984).

Besides the living *Eschrichtius*, gray whales include the extinct *Archaeschrichtius*, *Eschrichtioides*, *Gricetoides*, and, possibly, "*Balaenoptera*" *cortesii* var. *portisi* (Marx and Fordyce, 2015). Cladistic analyses generally support the monophyly of this family and its affinities with rorquals, but differ on its exact placement within Balaenopteroidea. Most molecular studies nest gray whales inside crown Balaenopteridae, usually as the sister group of the fin (*Balaenoptera physalus*) and humpback (*Megaptera novaeangliae*) whales (Hassanin *et al.*, 2012; McGowen *et al.*, 2009). By contrast, morphological analyses, supported by some molecular studies, generally envisage eschrichtiids either as sister to a monophyletic Balaenopteridae (e.g., Ekdale *et al.*, 2011; Steeman *et al.*, 2009), as nested within balaenopterids, but outside crown Balaenopteridae (Marx and Fordyce, 2015), or as a more basal lineage allied with cetotheriids (Bisconti, 2008; Steeman, 2007). In light of the potential inclusion of eschrichtiids within rorquals, it has been suggested that the clade should be reduced to the rank of a subfamily, **Eschrichtiinae** (Marx and Fordyce, 2015).

The **Balaenopteridae** are the most prominent of the living baleen whales (Plate 6a). They represent more than half of the extant species, are found throughout all of the world's oceans, and include the largest animal ever known. Balaenopterids range in length from around 8–9 m to as much as 33 m and, in the case of the blue whale, *Balaenoptera musculus*, can in extreme cases reach weights of up to 190 tonnes (Sears and Perrin, 2009). Rorquals also stand out for their unique strategy of bulk filter feeding, which involves the engulfment of vast quantities of prey-laden water via an expandable throat cavity. The earliest balaenopterids are known from the late Middle Miocene (Serravallian) of Japan (Kohno *et al.*, 2007), followed by early Late Miocene (Tortonian) specimens from North America and Italy (Bisconti, 2010; Hanna and McLellan, 1924). However, it is possible that at least some of these early species may fall along the balaenopteroid stem lineage leading to both eschrichtiids and crown Balaenopteridae (Marx and Fordyce, 2015).

In addition to the traits typical of all balaenopteroids, rorquals are characterized by a highly telescoped skull with far posteriorly projected rostral bones (premaxilla, maxilla, and nasal) and a well-developed, anteriorly projected supraoccipital shield overhanging the lateral skull wall (Figure 4.17). Most rorquals furthermore have a distinctive bony pocket between the long, parallel-sided ascending process of the maxillae and the anteromedial corner of the supraorbital process, a trapezoidal postglenoid process (in posterior view), a bony shelf along the anterolateral corner of the tympanic bulla, and a subcondylar furrow extending to the lateral side of the mandible. Finally, all of the extant members of this group, except the humpback whale *Megaptera novaeangliae*, display a distinctive fold (the **squamosal crease**) in the posteroventral border of the temporal fossa (Deméré *et al.*, 2005). Externally, rorquals stand out for having an enormous ventral throat pouch lined by longitudinal folds (**ventral throat pleats**), which allow the extension of the pouch during feeding. All living species furthermore bear at least one external, longitudinal ridge running along the midline of the dorsal side of the rostrum,

Figure 4.17 Skull of a representative member of Balaenopteridae, *Balaenoptera acutorostrata*, in (a) dorsal and (b) lateral views. Abbreviations as in Figure 4.4.

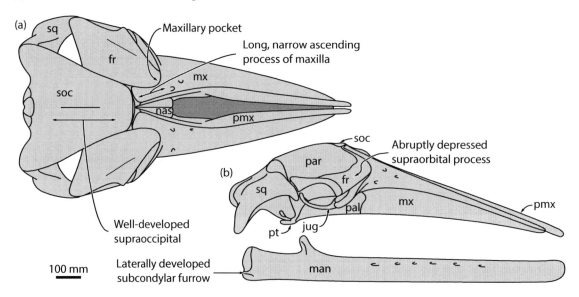

as well as a well-developed dorsal fin. The morphology of the baleen varies from species to species, but it is generally much shorter and coarser than in balaenids and *Caperea* (Pivorunas, 1979).

Besides the extant *Balaenoptera* and *Megaptera*, five extinct balaenopterid genera have been described so far: *Archaebalaenoptera*, *Diunatans*, *Parabalaenoptera*, *Plesiobalaenoptera*, and *Protororqualus*. In addition, several species currently referred to either *Balaenoptera* or *Megaptera*, including "*B.*" *ryani*, "*M.*" *miocaena*, and possibly "*M.*" *hubachi*, likely represent distinct genera that are yet to be redescribed and formally named (Deméré *et al.*, 2005). In terms of their size, most of these fossil taxa are similar to or smaller than the smallest extant rorqual, the 8–9 m long minke whale *Balaenoptera acutorostrata* (Bisconti, 2007; Dathe, 1983). Phylogenetically, the placement of balaenopterids within Balaenopteroidea, and thus as the closest relatives of eschrichtiids, is widely supported (e.g., Deméré *et al.*, 2008; Marx and Fordyce, 2015; McGowen *et al.*, 2009). However, morphological and molecular analyses frequently disagree on exactly how rorquals and gray whales are related, with the former often seen as potentially paraphyletic. See the discussion on Eschrichtiidae earlier in this section.

4.4 Echolocating whales: Odontoceti

Odontocetes (toothed whales) are the most diverse group of marine mammals, including at least 72 living species in 10 families. Their size ranges from less than 2 m for some porpoises and the Franciscana (*Pontoporia blainvillei*) to as much as 18 m in adult male sperm whales (*Physeter macrocephalus*). The key characteristic of all odontocetes is their ability to **echolocate** (i.e., to detect prey, congeners, and topographical features using high-frequency sounds) (section 5.2.2). Morphological correlates for this specialized biosonar system are found all across the forehead, and include: (1) a concave facial area accommodating the melon and the hypertrophied facial muscles; (2) the covering of most of the dorsal exposure of the frontal by the maxilla; (3) the presence of a premaxillary foramen and a premaxillary sac fossa; and (4) facial asymmetry (generally skewed toward the left). In addition, all modern odontocetes are both homodont and polydont, and, unlike baleen whales, only have a single external blowhole. The oldest, though as yet undescribed, fossil odontocete comes from the earliest

Oligocene of Washington State, United States (Barnes *et al.*, 2001). In addition, several slightly younger taxa are known from the Early Oligocene of the eastern and western coasts of North America (Uhen, 2008b).

4.4.1 Stem odontocetes

The early history of odontocetes during the Oligocene and Early Miocene is dominated by a variety of archaic forms that seemingly were capable of echolocation, yet retained a suite of primitive features inherited from their archaeocete ancestors. Most salient among the latter are the retention of a heterodont dentition, limited or no polydonty, a marked intertemporal constriction, a poorly developed sinus system, and relatively far anteriorly placed external nares. Most of these archaic taxa are represented by very few specimens, and their phylogenetic relationships are still unsettled—especially with regards to the question of which, if any, of them form part of the crown group (Figure 4.18).

The **Ashleycetidae**, which are currently limited to *Ashleycetus* from the Early Oligocene (ca 29 Ma)

of South Carolina, United States, may be the most archaic of all known toothed whales (Sanders and Geisler, 2015). *Ashleycetus* is clearly an odontocete, based on its enlarged maxilla covering much of the supraorbital process. Nevertheless, it also retains several archaeocete-like features, including a sizeable, dorsally open temporal fossa and anteriorly placed external nares, with the latter being roofed by well-developed nasals. Seemingly unique to this genus are its enlarged parietal, which extends along the posterior border of the supraorbital process, as well as a distinctly angled posteromedial corner of the ascending process of the maxilla (Figure 4.19). Because of the scarcity of available material, extremely little is known about the biology of this animal. Nevertheless, the only cladistic analysis to date which included *Ashleycetus* recovered it as the basal-most stem odontocete (Sanders and Geisler, 2015).

Much better known than ashleycetids are the **Xenorophidae**, although they, too, have so far only been recorded from the Oligocene of the United States East Coast (Geisler *et al.*, 2014; Uhen,

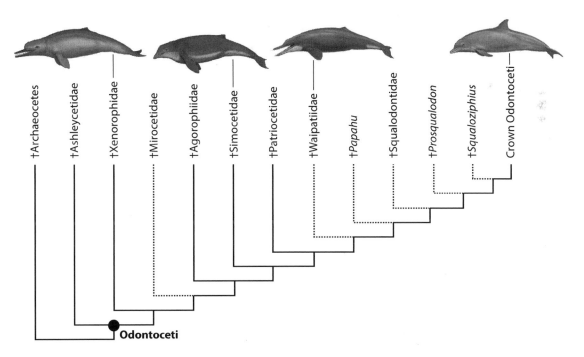

Figure 4.18 Simplified phylogeny of stem odontocetes. Stippled lines indicate taxa whose placement is highly preliminary (Mirocetidae), or that appear as part of the crown group in some analyses. Life reconstructions © C. Buell.

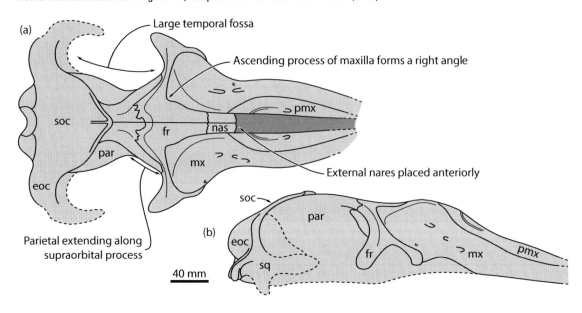

Figure 4.19 Skull of the only known member of Ashleycetidae, *Ashleycetus planicapitis*, in (a) dorsal and (b) lateral views. Abbreviations as in Figure 4.4; adapted from Sanders and Geisler (2015).

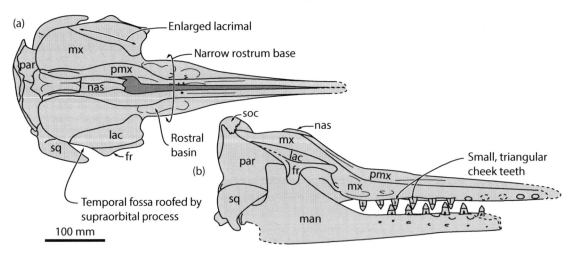

Figure 4.20 Skull of a representative member of Xenorophidae, *Cotylocara macei*, in (a) dorsal and (b) lateral views. Lateral view has been flipped horizontally. Abbreviations as in Figure 4.4; adapted from Geisler *et al.* (2014).

2008b). Xenorophid hallmarks include a greatly enlarged lacrimal partially covering the supraorbital process, an extremely narrow rostrum, rostral basins in the maxilla, and the partial or nearly complete roofing of the temporal fossa (Figure 4.20). The dentition is only weakly polydont and, like that of archaeocetes, remains heterodont,

with small, triangular, denticulate cheek teeth. Compared to that of other odontocetes, the xenorophid nasal is generally elongate, but tends to be retracted far posteriorly (along with external nares), thus pushing the parietal and supraoccipital backward (Geisler *et al.*, 2014). This peculiar style of facial telescoping clearly sets xenorophids apart from other archaic odontocetes, in which the retraction of the nares happens at the expense of the nasals, and the supraoccipital roughly maintains its relative position. At present, the family includes *Albertocetus*, *Archaeodelphis*, *Cotylocara*, and *Xenorophus* itself (Geisler *et al.*, 2014; Uhen, 2008b).

Xenorophidae are one of the most basal branches of the odontocete tree, second only to ashleycetids (Figure 4.18). Current interpretations envisage the family as having, up to a point, evolved in parallel with most other odontocetes, especially as regards the marked elaboration of their echolocation apparatus. The evolution of the xenorophid biosonar foreshadowed some similar, convergent developments that took place along the lineage to crown odontocetes (e.g., the roofing of the temporal fossa), but also led to the emergence of traits entirely unique to this family, such as the appearance of rostral basins in the maxilla (Geisler *et al.*, 2014).

Like ashleycetids, the **Mirocetidae** are a recent creation, and based on just a single genus—in this case, *Mirocetus* from the late Early or early Late Oligocene of Azerbaijan (Sanders and Geisler, 2015). The single available specimen is poorly preserved, but seems to unite a bizarre array of primitive and derived features. The former include a tooth-bearing infraorbital process of the maxilla, as well as an elongate humerus bearing a well-developed deltoid crest and a potentially mobile elbow joint. All of these features are typical of basilosaurid archaeocetes, but so far unique among odontocetes—which, at least in part, may simply reflect a lack of described (postcranial) material. On the other hand, *Mirocetus* is clearly identifiable as an odontocete, based on its partially roofed temporal fossa and broad ascending process of the maxilla (Figure 4.21). The phylogenetic position of Mirocetidae is somewhat questionable because of the lack of informative material, but it may be just crownward of Xenorophidae (Figure 4.18) (Sanders and Geisler, 2015).

Agorophiidae are a family of archaic odontocetes known from the Early Oligocene of the eastern United States and, possibly, Denmark, Japan, and the western coast of North America (Godfrey *et al.*, in press; Goedert *et al.*, 1995; Sawamura *et al.*, 1996). Judging from the dimensions of the skull, agorophiids were small animals, about the size of a harbor porpoise. They are characterized by a broad-based

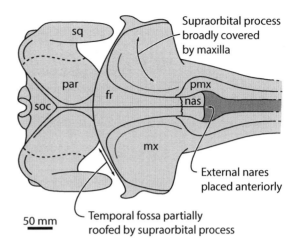

Figure 4.21 Skull of the only known member of Mirocetidae, *Mirocetus riabinini*, in dorsal view. Abbreviations as in Figure 4.4; adapted from Sanders and Geisler (2015).

rostrum, a deep premaxillary cleft, a flattened skull roof, a robust hamular process, and relatively large teeth bearing well-developed accessory denticles (Figure 4.22). The anteriormost portion of the temporal fossa is roofed by the supraorbital process, which in turn is broadly covered by the ascending process of the maxilla; however, neither feature seems as far developed as in the putatively more basal Mirocetidae.

Like so many other early odontocete families, agorophiids are currently represented by just a single genus, *Agorophius*. Prior to the discovery of ashleycetids and mirocetids, and the establishment of Xenorophidae, agorophiids were widely held to be among the most, if not *the* most, archaic lineage of odontocetes. Testing this claim was long hampered by the loss of the type specimen of *Agorophius pygmaeus*, on which the family is based (Fordyce, 1981). Drawing on additional material, more recent analyses have confirmed the basal position of agorophiids within Odontoceti, albeit somewhat removed from the origin of the group as a whole (Figure 4.18).

Simocetidae (pug-nosed whales) are currently a monogeneric family restricted to *Simocetus*, which in turn was founded on a single specimen from the Early Oligocene (ca 32 Ma) of Oregon, United States (Fordyce, 2002). At present, this specimen is the

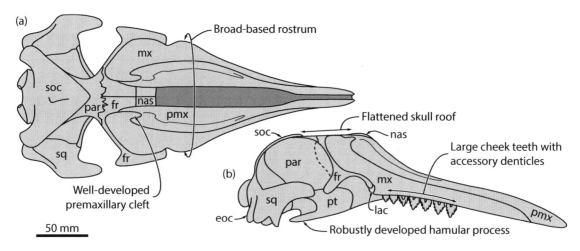

Figure 4.22 Skull of a representative member of Agorophiidae, *Agorophius pygmaeus*, in (a) dorsal and (b) lateral views. Abbreviations as in Figure 4.4; dorsal view adapted from Godfrey *et al.* (in press), lateral view from Fordyce (1981).

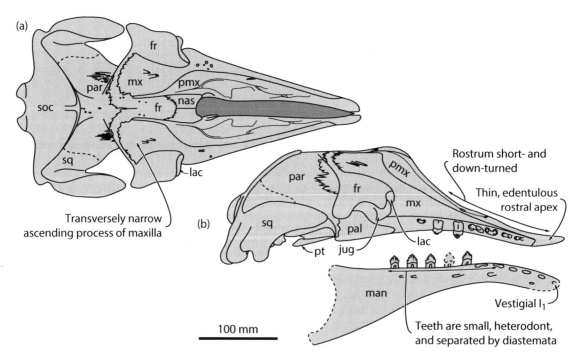

Figure 4.23 Skull of the only known member of Simocetidae, *Simocetus rayi*, in (a) dorsal and (b) lateral views. Abbreviations as in Figure 4.4; adapted from Fordyce (2002).

oldest odontocete material that has been formally described. *Simocetus* is a small-bodied cetacean characterized by a relatively short rostrum with an edentulous and ventrally deflected apex, delicate posterior cheek teeth occluding into opposing diastemata, and a vestigial I$_1$. At the same time, *Simocetus* retains such primitive traits as a broadly exposed parietal on the skull vertex, a comparatively narrow ascending process of the maxilla, anteriorly placed external nares, heterodonty, and the absence of polydonty (Figure 4.23). Phylogenetically, *Simocetus* seems to have branched off the main odontocete lineage either just before or just after agorophiids (Geisler *et al.*, 2011, 2014). Its rather unique rostral morphology and delicate teeth suggest that it may have been a bottom feeder that specialized on soft-bodied, benthic invertebrates (Fordyce, 2002).

Patriocetidae are a group of relatively poorly known species all belonging to the single genus *Patriocetus*, known from the Late Oligocene of Europe, Central Asia, and, possibly, Japan (Dubrovo and Sanders, 2000; Okazaki, 1987). *Patriocetus* is about the size of a bottlenose dolphin, and is defined by its narrow exposure of the parietal on the vertex, the pistol-like arrangement of the zygomatic and postglenoid processes, heterodonty, and a moderate degree of polydonty (Figure 4.24). Because of the poor state of preservation of the material available at the time, patriocetids were once regarded as potential mysticete ancestors (Abel, 1914). However, subsequent reinterpretations of the type species, *Patriocetus ehrlichii*, and the discovery of additional material made it clear that patriocetids are indeed stem odontocetes

(Dubrovo and Sanders, 2000; Geisler *et al.*, 2011; Rothausen, 1968).

4.4.2 Potential crown odontocetes

As in the case of mysticetes, the contents of the odontocete crown group are controversial. Much of this lingering disagreement hinges on the proposed grouping of the extant South Asian river dolphin, *Platanista*, with a variable range of extinct families in the superfamily **Platanistoidea** (Figure 4.25) (de Muizon, 1991, 1994). Platanistoids are primarily defined by the reduction of the coracoid process of the scapula and the disappearance of the supraspinous fossa. At its most comprehensive, the group potentially includes the extant platanistids, as well as the extinct allodelphinids, squalodelphinids, squalodontids, waipatiids, and, possibly, dalpiazinids (discussed in the remainder of this section and section 4.4.3) (Barnes, 2006; de Muizon, 1994; Fordyce, 1994). While the platanistoid affinities of squalodelphinids are largely agreed upon, those of the remaining families have either been questioned or only rarely been tested. In particular, both squalodontids and waipatiids have at times been excluded from Platanistoidea, and indeed from crown Odontoceti as a whole (Figure 4.18).

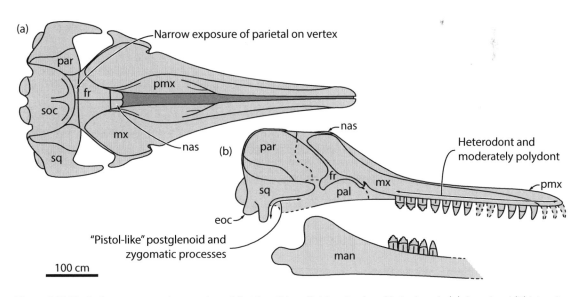

Figure 4.24 Skull of a representative member of Patriocetidae, *Patriocetus kazakhstanicus*, in (a) dorsal and (b) lateral views. Dorsal view adapted from Dubrovo and Sanders (2000).

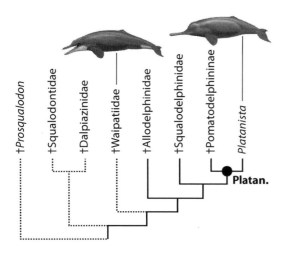

Figure 4.25 Simplified phylogeny of Platanistoidea. Stippled lines indicate taxa that instead appear as part of the odontocete stem group in some analyses. Platan., Platanistidae. Life reconstructions © C. Buell.

The **Squalodontidae** (shark-toothed dolphins) are widely distributed in the Late Oligocene–Middle Miocene of Argentina, Europe, New Zealand, and eastern North America (Tanaka and Fordyce, 2014). The youngest, unquestionably identified material dates from the Serravallian. This family has long served as a wastebasket for fragmentary, seemingly archaic cetacean remains, consisting mostly of double-rooted cheek teeth with accessory denticles. As a result, several proposed records based largely on plesiomorphic dental features are in need of reexamination (Dooley, 2005). Squalodontids are medium-sized odontocetes characterized by an elongate rostrum housing large, procumbent incisors, frequently within a robust and transversely expanded rostral apex. The development of the latter varies within species, and may thus be subject to sexual dimorphism or ontogenetic variation (Dooley, 2005). The squalodontid dentition is moderately polydont, but also still clearly heterodont, with double-rooted cheek teeth bearing triangular, often heavily ornamented crowns. The facial region is long, with a large ascending process of the maxilla covering almost the entire frontal, a dorsally open temporal fossa, and a weakly elevated vertex (Figure 4.26). There are currently three genera represented by diagnostic material: *Eosqualodon*,

Phoberodon, and *Squalodon* (Cabrera, 1926; Dooley, 2005; Rothausen, 1968).

Depending on the analysis, squalodontids have been interpreted either as basal platanistoids or as stem odontocetes, albeit relatively derived ones (Figures 4.18 and 4.25) (Fordyce, 1994; Geisler *et al.*, 2014; Tanaka and Fordyce, 2015). In either case, it is likely that squalodontids were among the major marine predators of their time, preying on large fish and, possibly, even other marine mammals, although it is questionable whether their elongate rostrum would have allowed for a powerful bite (Loch *et al.*, 2015). Possibly related to squalodontids are the enigmatic **Dalpiazinidae**, which are based on some fragmentary specimens belonging to a single genus, *Dalpiazina*, from the Early Miocene of Italy. Some of the material originally referred to *Dalpiazina* is questionable, and what remains is too poorly preserved to determine its evolutionary relationships with any degree of certainty (de Muizon, 1994). More complete material from New Zealand has been mentioned, but not yet described (Clementz *et al.*, 2014).

Members of the **Waipatiidae** are principally known from the Late Oligocene (ca 25 Ma) of New Zealand. Other potential records come from the Late Oligocene–Middle Miocene of Malta, the Caucasus, and the Northwest Pacific (Bianucci *et al.*, 2011; Fordyce, 1994; Tanaka and Fordyce, 2014). Although most of these referrals are based on fragmentary remains or specimens in need of revision, they suggest a moderately diversified family and a wide geographic distribution. Waipatiids have a relatively long, anteriorly attenuated rostrum bearing delicate, procumbent incisors and a moderately polydont series of small cheek teeth in each quadrant (Figure 4.27). The external nares are far posterior to the level of the antorbital notch, and the facial region shows a weak, but noticeable, degree of asymmetry. The mandibular symphysis is proportionally short and not ankylosed (Fordyce, 1994).

Pending the reassessment of the uncertainly referred material mentioned in this chapter, which includes the poorly known genera *Sachalinocetus* and *Sulakocetus*, the family is currently restricted to its type genus, *Waipatia*—as well as, possibly, the closely related *Otekaikea* (Tanaka and Fordyce, 2014, 2015). Starting with their original description, waipatiids have often been included

Figure 4.26 Skull of a representative member of Squalodontidae, *Squalodon bellunensis* in (a) dorsal and (b) lateral views. Abbreviations as in Figure 4.4.

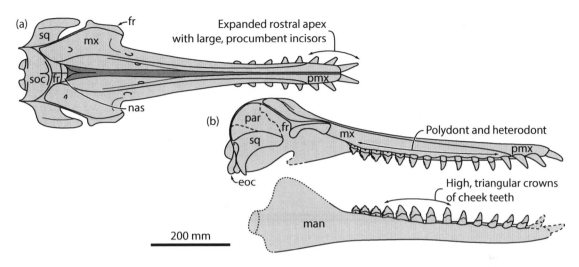

Figure 4.27 Skull of a representative member of Waipatiidae, *Waipatia maerewhenua*, in (a) dorsal and (b) lateral views. Abbreviations as in Figure 4.4; adapted from Fordyce (1994).

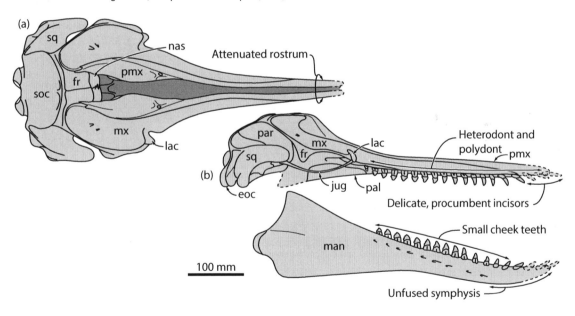

in the superfamily Platanistoidea, and thus crown Odontoceti (Fordyce, 1994; Murakami *et al.*, 2014; Tanaka and Fordyce, 2015). However, other analyses have interpreted them either as located elsewhere within the crown group or, more commonly, as stem odontocetes (Figure 4.18) (Aguirre-

Fernández and Fordyce, 2014; Geisler *et al.*, 2011; Lambert, 2005a). With their elongate snouts and gracile teeth, waipatiids likely were snap feeders (section 6.1.3).

Besides these established families, there are several well-preserved, yet enigmatic, potential stem odontocetes whose higher level affinities remain uncertain. They include *Prosqualodon*, a robustly built, short-snouted, heterodont taxon known primarily from the Early Miocene of Argentina and Australia, and sometimes considered to represent a family (**Prosqualodontidae**) in its own right (Cozzuol, 1996; Flynn, 1948; Lydekker, 1894); the small-bodied, homodont dolphin *Papahu* from the Early Miocene of New Zealand (Aguirre-Fernández and Fordyce, 2014); and the medium-sized *Squaloziphius*, which is known from the Early Miocene of the western United States and characterized by an elevated vertex with transverse premaxillary crests, as well as a large hamular process of the pterygoid (de Muizon, 1991). Although sometimes thought to represent a basal platanistoid (de Muizon, 1994; Fordyce, 1994; Murakami *et al.*, 2014), several phylogenetic analyses have instead interpreted *Prosqualodon* as a stem odontocete (Geisler *et al.*, 2014; Tanaka and Fordyce, 2015). *Papahu* and *Squaloziphius* have likewise been placed both within and outside the crown group (Figure 4.18): *Papahu* as either a (phylogenetically unstable) stem taxon or as a relatively basal member of the crown group, just apical to platanistoids; and *Squaloziphius* as either sister to the crown group, or as a basal crown odontocete or beaked whale (Ziphiidae) (section 4.4.3) (Aguirre-

Fernández and Fordyce, 2014; Geisler *et al.*, 2014; Tanaka and Fordyce, 2015).

4.4.3 Basal crown odontocetes

Most crown odontocetes can broadly be categorized into four major groups: (1) physeteroids (sperm whales), (2) platanistoids (the South Asian river dolphin, *Platanista*, and relatives), (3) ziphiids (beaked whales), and (4) delphinidans (Figure 4.28). With the exception of some fossil platanistoids and iniids (see this section and sections 4.4.2 and 4.4.4), all known crown odontocetes are effectively homodont, although some taxa show secondary variations, such as the development of tusks.

The **Physeteroidea** were the first of the major odontocete crown lineages to diverge, and comprise the living pygmy, dwarf, and giant sperm whales (families Kogiidae and Physeteridae). Additional physeteroid stem taxa are relatively diverse, and include what may be the largest known macropredator, *Livyatan melvillei* from the Middle-Late Miocene of Peru (Plate 7). Physeteroids are easily recognizable, thanks to their highly asymmetrical facial region and the presence of a vast **supracranial basin** occupying much of the dorsal surface of the skull. In life, this basin houses the **spermaceti organ** and other soft tissue structures related to sound production and transmission (Cranford, 1999). In addition, sperm

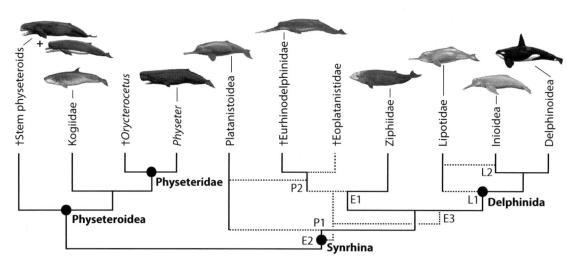

Figure 4.28 Simplified phylogeny of crown Odontoceti. E, L, and P denote alternative placements of Eurhinodelphinidae, Lipotidae, and Platanistoidea, respectively (see text for details). Life reconstructions © C. Buell.

whales are characterized by proportionally large dental roots (owing to continuous deposition of cementum on the outer surface of the root), a massive lacrimal–jugal complex, a large accessory ossicle on the periotic, a bilobate involucrum, an enlarged posterior process of the tympanic bulla, and, for the most part, the loss of at least one nasal bone. In stem physeteroids, functional teeth generally line both the upper and lower jaws and, in most taxa, bear short crowns covered in ornamented enamel. By contrast, functional teeth only occur in the lower jaw and largely or entirely lack enamel in all of the extant taxa. Although extant sperm whales, like other odontocetes, have a single blowhole, their left and right nasal tracts are separated along most of their length. This condition harks back to the ancestral condition of having two separate external narial openings, as seen in all other mammals, and reaffirms the basal position of physeteroids among odontocetes.

The oldest stem physeteroid may be *Ferecetotherium* from the Late Oligocene of the Caucasus, but the available material is fragmentary and its affinities within Physeteroidea are uncertain (Mchedlidze, 1976). The youngest occurrence of a stem physeteroid comes from the Pleistocene of the Southwest Pacific (Fitzgerald, 2011). During the Miocene, physeteroids attained a worldwide distribution, with fossils known from Europe, Japan, Australasia, and the Atlantic and Pacific coasts of both North and South America (Bianucci and Landini, 2006; Fitzgerald, 2004; Lambert *et al.*, 2010). Stem physeteroids range in size from medium (ca 6–7 m long) to very large—up to possibly 17.5 m in the case of *Livyatan* (Bianucci and Landini, 2006; Lambert *et al.*, 2010). Besides *Ferecetotherium* and *Livyatan*, there are five more commonly recognized stem genera: *Acrophyseter*, *Brygmophyseter*, *Diaphorocetus*, *Eudelphis*, and *Zygophyseter*. *Idiorophus*, *Orycterocetus*, *Placoziphius*, and "*Ontocetus*" *oxymycterus* may also fall into this category or, alternatively, might represent extinct physeterids (Bianucci and Landini, 2006; Lambert *et al.*, 2010; Vélez-Juarbe *et al.*, 2015). The genera *Hoplocetus*, *Preaulophyseter*, and *Scaldicetus* have all been used to refer to a variety of isolated, enamel-bearing physeteroid teeth and ear bones; however, they are based on

largely non-diagnostic-type material, and thus may have to be abandoned.

Kogiidae are today represented by the extant, globally distributed pygmy and dwarf sperm whales (*Kogia*). Extant kogiids are relatively small (ca 3 m), deep-diving odontocetes that rely on suction feeding to prey on squid, benthic fish, and crabs (section 6.1.3) (Bloodworth and Marshall, 2005). The earliest well-dated members of this family come from the Late Miocene of Malta (ca 9 Ma), but one record possibly dates from the late Early or Middle Miocene of Belgium (Bianucci *et al.*, 2011; Lambert, 2008b). Other, mostly Pliocene, records are known from other parts of Europe, Australia, North America, western South America, and, possibly, Japan (Fitzgerald, 2005; Vélez-Juarbe *et al.*, 2015; Whitmore and Kaltenbach, 2008). Kogiids are short-snouted and characterized by the presence of a sagittal crest in the supracranial basin, as well as the loss of both nasals (Figure 4.29). In both of the living and some of the extinct species, the upper teeth are vestigial or absent (Lambert, 2008b). Besides *Kogia*, four fossil genera are currently recognized: *Aprixokogia*, *Nanokogia*, *Praekogia*, and *Thalassocetus*. A fifth, *Scaphokogia* from the Late Miocene of Peru, is classified in its own subfamily, Scaphokogiinae. The status of *Kogiopsis* from the late Middle Miocene of Florida is uncertain. Both morphological and molecular data support a close sister-group relationship of kogiids and physeterids (Figure 4.28).

Physeteridae are represented by the extant, globally distributed giant sperm whale *Physeter*. Like its smaller kogiid relatives, *Physeter* is a deep diver and a suction feeder, preying mainly on squid—including the giant squid *Architeuthis* (Plate 6b) (Werth, 2004). The earliest record of the family dates from the Middle Miocene of California (16–14.5 Ma), with additional material coming from eastern and western North America, Australia, Europe, and, possibly, Japan and South America (Bianucci and Landini, 2006; Fitzgerald, 2004; Oishi and Hasegawa, 1995; Whitmore and Kaltenbach, 2008). However, much of the physeterid record is based on fragmentary material and isolated teeth, which are often difficult to distinguish from those of other physeteroids. Physeterids themselves are poorly defined as a clade, with all phylogenetic hypotheses proposed so far having

Figure 4.29 Skull of a representative member of Kogiidae, *Kogia breviceps*, in (a) dorsal and (b) lateral views. Abbreviations as in Figure 4.4.

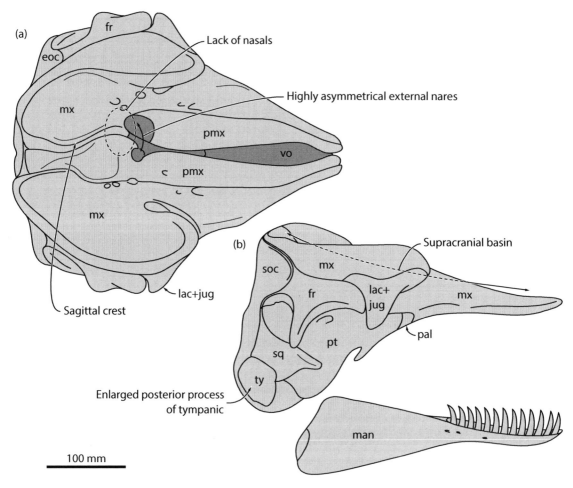

to invoke a considerable degree of evolutionary convergence. Besides *Physeter*, the extinct *Idiophyseter*, *Physeterula*, and, probably, *Aulophyseter* also likely belong to this family, with *Idiorophus*, *Orycterocetus*, *Placoziphius*, and "*Ontocetus*" being additional candidates (Bianucci and Landini, 2006; Lambert *et al.*, 2010; Vélez-Juarbe *et al.*, 2015). Out of these taxa, *Aulophyseter* and *Physeter* are both large-bodied, share the position of the eye socket well ventral to the margin of the rostrum (in lateral view), and, convergently with some kogiids, have reduced the upper dentition (Figure 4.30).

Early phylogenetic analyses disagreed on the position of physeteroids within Cetacea, with some studies advocating a close relationship with beaked whales (discussed further here) or even mysticetes (Geisler and Sanders, 2003; Milinkovitch *et al.*, 1993). By contrast, more recent analyses uniformly support a monophyletic Odontoceti in which physeteroids form the earliest diverging extant lineage. All extant odontocetes except sperm whales share the confluence of the soft tissue nasal passages along most of their lengths (i.e., just distal to the external bony nares), and together they are now known as the **Synrhina**, or "together noses" (Figure 4.28) (Geisler *et al.*, 2011).

The oldest extant lineage within Synrhina are the **Platanistidae**, whose sole modern survivor is

Figure 4.30 Skull of two representative members of Physeteridae, (a, b) *Orycterocetus crocodilinus* and (c) *Physeter macrocephalus* in (a) dorsal and (b, c) lateral views. Abbreviations as in Figure 4.4; adapted in part from Kellogg (1965) and Van Beneden and Gervais (1880).

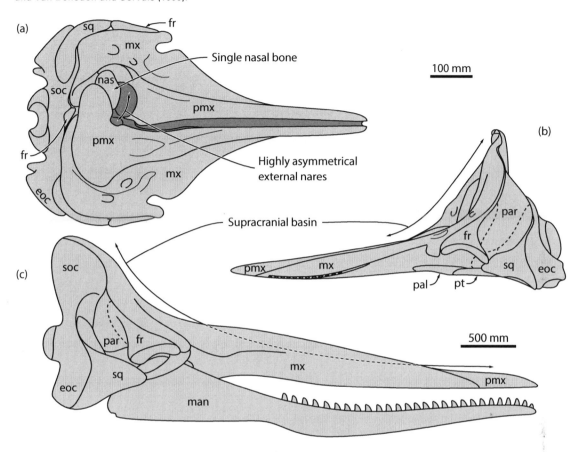

the South Asian river dolphin, *Platanista*—a strictly freshwater, nearly blind animal inhabiting the turbid waters of the Ganges, Indus, and Brahmaputra river systems of the Indian subcontinent (Plate 9). Fossil members of this family have been recovered from Early–Late Miocene marine sediments distributed across Europe, Korea, and North America (Barnes, 2006; Lee *et al.*, 2012). In addition, freshwater occurrences are known from the Middle–Late Miocene of Peruvian Amazonia and the south-eastern United States (Bianucci *et al.*, 2013a; Hulbert and Whitmore, 2006). No undisputed platanistid is known from Pliocene deposits, which suggests that the family may only have persisted in cryptic areas, such as river systems. Platanistids are generally small (2–4 m) and long-snouted, with each

jaw bearing several dozen teeth (up to 87 in some fossil taxa). Other characteristics include a deep groove on the lateral side of the rostrum and mandible, a dorsoventrally expanded zygomatic process (shared with squalodelphinids; discussed further in this section), a hook-like articular process on the periotic, and a conspicuous facial crest formed by the maxilla and/or the frontal (Figure 4.31) (Barnes, 2006; de Muizon, 1987). Fossil platanistids (*Pomatodelphis*, *Prepomatodelphis*, and *Zarhachis*) differ from *Platanista* in having a transversely wider, dorsoventrally flattened rostrum, and hence are included in their own subfamily, Pomatodelphininae (Barnes, 2006). The platanistid affinities of *Araeodelphis* from the Early Miocene of the eastern United States remain to be firmly established (Godfrey *et al.*, 2006).

Figure 4.31 Skull of a representative member of Platanistidae, *Platanista gangetica*, in (a) dorsal and (b) lateral views. Abbreviations as in Figure 4.4; adapted from Van Beneden and Gervais (1880).

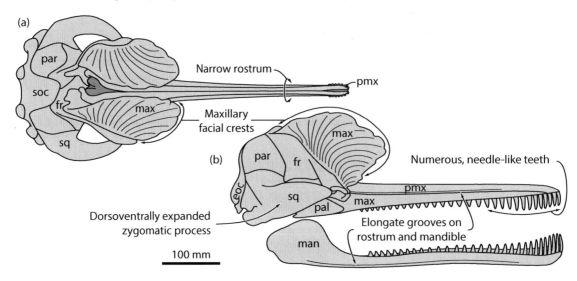

Platanista was originally thought to form a clade (the original superfamily **Platanistoidea**) with the three other extant "river dolphins": the baiji, or Yangtze River dolphin (*Lipotes*), the Amazon River dolphin (*Inia*), and the Franciscana (*Pontoporia*) (section 4.4.4) (Simpson, 1945). However, subsequent analyses convincingly demonstrated that platanistids are only distantly related to other "river dolphins," and that the freshwater habit of these species likely arose as a result of convergence (e.g., de Muizon, 1984). Instead, platanistids appear to be part of a large, now mostly extinct clade also including allodelphinids and squalodelphinids, as well as, possibly, dalpiazinids, squalodontids, waipatiids, and *Prosqualodon* (Figure 4.25) (section 4.4.2). It is in this sense that the term Platanistoidea is now generally used (Barnes, 2006; de Muizon, 1994; Geisler *et al.*, 2011; Tanaka and Fordyce, 2015). Like its scope, the phylogenetic position of Platanistoidea within Odontoceti is controversial. Whereas molecular analyses generally envisage platanistids as an independent lineage and sister to the remainder of Synrhina (McGowen *et al.*, 2009; Steeman *et al.*, 2009), some morphological and total evidence analyses support a close relationship with beaked whales (Ziphiidae) or even interpret platanistoids to be the most basal lineage within crown odontocetes (Geisler *et al.*,

2011, 2014; Murakami *et al.*, 2014; Tanaka and Fordyce, 2015) (Figure 4.28).

The inclusion of the **Squalodelphinidae** within Platanistoidea has been tested and confirmed multiple times (e.g., de Muizon, 1987; Geisler *et al.*, 2014; Murakami *et al.*, 2014; Tanaka and Fordyce, 2015). Occurrences of this family are currently limited to the Early Miocene, but distributed as widely as Europe, the eastern United States, and both the Atlantic and Pacific coasts of South America (Lambert *et al.*, 2014). Squalodelphinids are small to medium-sized odontocetes with a moderately elongated, strongly attenuated rostrum, a markedly V-shaped antorbital notch and anteriorly pointed antorbital process, a dorsoventrally expanded zygomatic process of the squamosal (shared with platanistids), a square-shaped pars cochlearis of the periotic, an elongate median furrow of the tympanic bulla, and, unusually among crown odontocetes, the retention of a slight degree of heterodonty (de Muizon, 1991) (Figure 4.32). There are currently five recognized genera, including *Huaridelphis*, *Medocinia*, *Notocetus*, *Phocageneus*, and *Squalodelphis*. Virtually all phylogenetic studies have found squalodelphinids to be most closely related to platanistids, independent of whether or not they included squalodontids and waipatiids in an

Figure 4.32 Skull of a representative member of Squalodelphinidae, *Huaridelphis raimondii*, in (a) dorsal and (b) lateral views. Abbreviations as in Figure 4.4; adapted from Lambert *et al.* (2014).

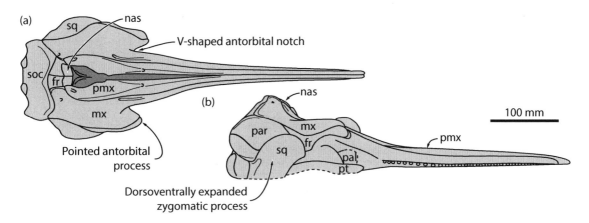

expanded Platanistoidea (Figure 4.25) (e.g., Barnes, 2006; Fordyce, 1994; Geisler *et al.*, 2014).

Allodelphinidae are a small group of extinct odontocetes currently restricted to the Miocene (Aquitanian to early Tortonian) of western North America and Japan (Barnes and Reynolds, 2009). Allodelphinids are small- to medium-sized (4–5 m) and, like their platanistid relatives, long-snouted. Their jaws are dorsoventrally flattened and bear numerous, small, single-rooted teeth. Again as in platanistids, there is a longitudinal groove running along the lateral side of both the rostrum and mandible. The facial region lacks elevated crests, except for a thick prominence near the base of the rostrum in *Zarhinocetus* (Figure 4.33). Long cervical vertebrae and an elongate humerus suggest a flexible neck and a large flipper, respectively (Barnes, 2006; Barnes and Reynolds, 2009). Besides *Zarhinocetus*, so far the only other member of this family is *Allodelphis* itself. Despite at least superficial similarities with platanistids, allodelphinids are regarded as basal platanistoids, even when squalodontids and waipatiids are included (Figure 4.25). However, unlike those of squalodelphinids, the evolutionary relationships and platanistoid affinities of allodelphinids have only rarely been tested (Barnes, 2006; Lambert *et al.*, 2014).

The **Eurhinodelphinidae** are a family of highly distinctive, long-snouted odontocetes from the Early–Middle Miocene of Europe and eastern North America (Lambert, 2005b). Potentially older records have been reported from the Late Oligocene of the Caucasus (*Iniopsis*) and Australia, but are in need of reassessment (Fordyce, 1983; Mchedlidze, 1976). Eurhinodelphinids are small- to medium-sized odontocetes, and primarily stand out for two unique characteristics of their elongate snout: a long, edentulous portion of the premaxilla that may contribute more than half of the total length of the rostrum; and a mandible that is markedly shorter than the upper jaw. Laterally, both the rostrum and mandible bear a lateral groove (Figure 4.34) (de Muizon, 1991; Lambert, 2005b). The specialized snout morphology of eurhinodelphinids is shared with some Mesozoic marine reptiles (ichthyosaurs) as well as some large fish (e.g., swordfish and marlins). Unlike marlins, however, eurhinodelphinids probably were slow swimmers, judging from their relatively long neck vertebrae. The function of the elongated premaxilla remains unknown, but could plausibly have been tactile.

Currently, eurhinodelphinids comprise five genera: *Eurhinodelphis*, *Mycteriacetus*, *Schizodelphis*, *Xiphiacetus*, and *Ziphiodelphis*. *Argyrocetus* and *Macrodelphinus* from the Early Miocene of Argentina and California, United States, were formerly also included in the family, but may instead form a separate clade with the recently described *Chilcacetus* from the Early Miocene of Peru (Lambert *et al.*, 2015). Hypotheses as to the evolutionary relationships of the family are varied, and range from proposed sister-group relationships

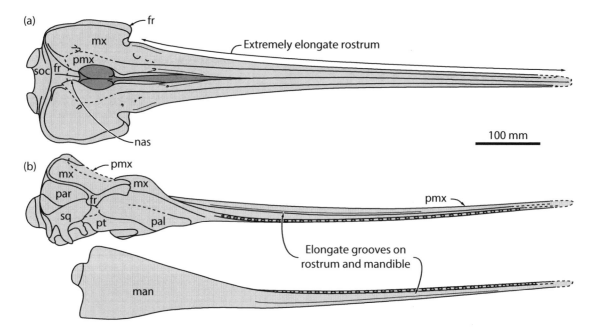

Figure 4.33 Skull of a representative member of Allodelphinidae, *Zarhinocetus errabundus*, in (a) dorsal and (b) lateral views. Abbreviations as in Figure 4.4.

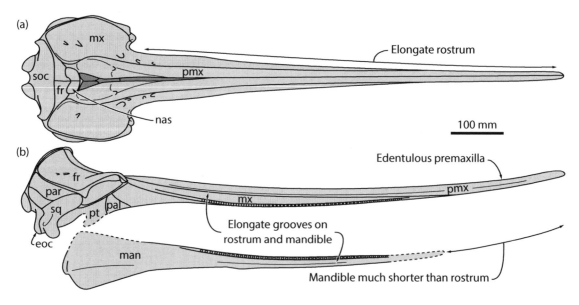

Figure 4.34 Skull of a representative member of Eurhinodelphinidae, *Xiphiacetus bossi*, in (a) dorsal and (b) lateral views. Abbreviations as in Figure 4.4.

with eoplatanistids, platanistoids, delphinidans, and ziphiids, to an independent lineage basal to most or all of Synrhina (Figure 4.28) (de Muizon, 1991; Fordyce, 1994; Geisler *et al.*, 2014; Lambert, 2005a; Lambert *et al.*, 2015; Murakami *et al.*, 2014).

At least superficially similar to eurhinodelphinids are the **Eoplatanistidae**, which are known only from the Early Miocene of north-eastern Italy. Eoplatanistids are relatively small, yet had a long and proportionally robust snout bearing numerous, single-rooted teeth. They are characterized by a flat skull vertex with widely exposed frontals, the presence of a longitudinal groove on the rostrum and mandible, mesial and distal keels on the teeth, and the lack of a median furrow on the tympanic bulla (de Muizon, 1988a; Pilleri, 1985) (Figure 4.35). The family currently includes just a single genus, *Eoplatanista*. Eoplatanistids are sometimes regarded as the sister group of eurhinodelphinids, but a recent analysis instead placed them just outside a clade including eurhinodelphinids, platanistoids, ziphiids, and delphinidans (de Muizon, 1991; Lambert *et al.*, 2015).

Ziphiidae (beaked whales) are the second most diverse family of living cetaceans, comprising 21 extant species (mostly belonging to *Mesoplodon*) spread throughout all of the world's oceans. Modern ziphiids are medium to large-sized, with the largest, Baird's beaked whale (*Berardius bairdii*),

reaching a length of up to 12 m. Several of the extant species are surprisingly poorly known, primarily because of their pelagic, deep-diving lifestyle, which limits opportunities for research. All of them are suction feeders preying mainly on squid, but also fish and other invertebrates (Heyning and Mead, 1996). The earliest fossil ziphiid is *Archaeoziphius* from the Middle Miocene (ca 13–15 Ma) of Europe (Lambert and Louwye, 2006). The ziphiid fossil record is comparatively rich, with specimens known from every continent except Antarctica. Unusually, much of this record has not been gathered from marine rocks exposed on land, but through deep-water trawling—especially off southern Africa and the Iberian Peninsula (Bianucci *et al.*, 2007, 2013b).

The ziphiid skull is characterized by an elevated vertex with conspicuous transverse premaxillary crests, an enlarged hamular process of the pterygoid, and, often, an elongated rostrum (Heyning, 1989; Lambert *et al.*, 2013). The rostral bones are frequently pachyosteosclerotic, with those of the extant Blainville's beaked whale, *Mesoplodon densirostris*, being the densest mammalian bones on record. The thickening of the rostral elements may result in the development of a diverse array of crests and projections, and/or the infilling of the mesorostral groove by the compact vomer (Lambert *et al.*, 2011). Many taxa also

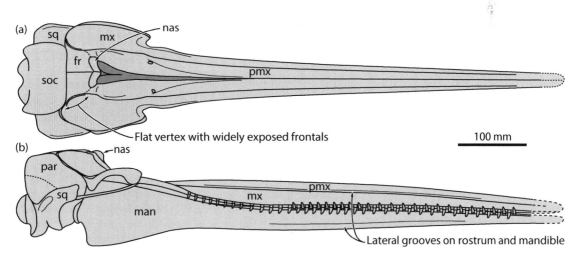

Figure 4.35 Skull of a representative member of Eoplatanistidae, *Eoplatanista gresalensis*, in (a) dorsal and (b) lateral views. Abbreviations as in Figure 4.4.

Figure 4.36 Skull of a representative member of Ziphiidae, *Mesoplodon europaeus*, in (a) dorsal and (b) lateral views. Abbreviations as in Figure 4.4; adapted from Van Beneden and Gervais (1880).

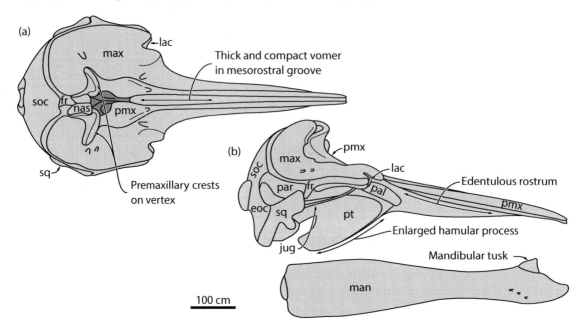

display a drastic reduction of the dentition, leaving little more than one or two pairs of tusk-like mandibular teeth. The latter generally only erupt in males, and may be used as weapons in intraspecific fighting (Figure 4.36) (Heyning, 1984). Both the head and flippers of extant ziphiids are small relative to the rest of the body. Externally, they carry a pair of throat grooves that aid in suction feeding (section 6.1.3).

Besides *Berardius* and *Mesoplodon*, there are a further four extant as well as 21 extinct genera, mostly falling into three subfamilies: (1) Berardiinae (*Archaeoziphius*, the extant *Berardius*, and *Microberardius*); (2) Hyperoodontinae (*Africanacetus, Ihlengesi, Khoikhoicetus*, and the extant *Indopacetus, Mesoplodon*, and *Hyperoodon*); and (3) Ziphiinae (*Choneziphius, Globicetus, Imocetus, Izikoziphius, Tusciziphius*, and the extant *Ziphius*). In addition, the extinct genera *Aporotus, Beneziphius, Notoziphius, Messapicetus*, and *Ziphirostrum* may form a fourth, currently unnamed "*Messapicetus* clade." The subfamilial affinities of *Caviziphius, Nazcacetus, Nenga, Ninoziphius, Pterocetus, Xhosacetus*, and the extant *Tasmacetus* remain uncertain. Nevertheless, recent phylogenetic analyses grouped *Nenga, Pterocetus*, and *Xhosacetus* with Hyperoodontinae, and identified both *Ninoziphius* and the "*Messapicetus* clade" as stem ziphiids (Buono and Cozzuol, 2013; Lambert *et al.*, 2013). Within the context of Odontoceti as a whole, morphologists have sometimes considered ziphiids to be sister to physeteroids (de Muizon, 1991; Fordyce, 1994; Geisler and Sanders, 2003; Murakami *et al.*, 2014). However, other morphological studies and virtually all molecular and total evidence analyses support their inclusion within Synrhina (Figure 4.28) (e.g., Geisler *et al.*, 2014; Heyning, 1989; McGowen *et al.*, 2009; Steeman *et al.*, 2009).

4.4.4 Delphinida

Delphinida is a well-supported, diverse clade comprising (1) the "river dolphins" *Inia, Lipotes*, and *Pontoporia*, each in its own family; (2) the superfamily Delphinoidea, which in turn includes the extant oceanic dolphins, porpoises, and monodontids (section 4.4.5), as well as two related extinct families; and (3) a likely polyphyletic

Figure 4.37 Simplified phylogeny of Delphinida. K and P denote alternative placements of Kentriodontinae and Pithanodelphininae, respectively (see text for details). Life reconstructions © C. Buell.

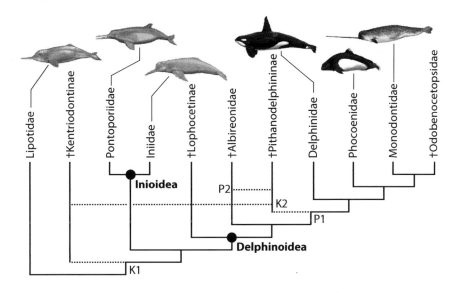

assemblage of fossil taxa commonly referred to as Kentriodontidae (Figure 4.37) (e.g., Cassens *et al.*, 2000; Geisler *et al.*, 2011; Heyning, 1989; McGowen *et al.*, 2009). Morphologically, delphinidans share a reduced posterior extent of the premaxilla, a posteriorly elongated lateral lamina of the palatine, a shortening of the anterior process of the periotic, a depressed dorsal surface of the involucrum of the tympanic bulla, and a muscular process of the malleus that is higher than the manubrium (de Muizon, 1988b; Fordyce, 1994).

Lipotidae are maybe the least known of the "extant" delphinidans. Until the end of the 20th century, the only living representative of this family, the baiji (*Lipotes vexillifer*), inhabited the Yangtze River in China (Plate 9). Since then, the species has likely become extinct—the first cetacean to do so because of human activity (Turvey *et al.*, 2007). Lipotids are small animals measuring no more than 2–3 m in length, and they are characterized by an elongate rostrum bearing numerous teeth (shared with most other "river dolphins"), a posteriorly located constriction of the rostrum, narrow and elevated frontals on the skull vertex, and the retention of an anterior bulla facet on the pointed anterior process of the periotic (Figure 4.38) (Barnes, 1985b; de Muizon, 1988b).

Only one fossil genus, *Parapontoporia* from the latest Miocene–Pliocene of western North America and, possibly, Japan, is currently thought to belong to this family. Note, however, that *Parapontoporia* has formerly also been included in another family of "river dolphins," the Pontoporiidae (see below; Godfrey and Barnes, 2008). The type material of *Prolipotes*, a fragmentary odontocete mandible from the ?Miocene of China, is non-diagnostic. Unlike *Lipotes*, *Parapontoporia* seems to have been mostly marine, although there is at least one freshwater occurrence (Boessenecker and Poust, 2015). Depending on the analysis, lipotids are interpreted either as the earliest-branching lineage within Delphinida, or as part of a "river dolphin" clade also including iniids and pontoporiids (Figures 4.28 and 4.37) (Cassens *et al.*, 2000; Geisler *et al.*, 2011; Hamilton *et al.*, 2001; McGowen *et al.*).

Like *Lipotes*, *Inia*—the only extant member of the **Iniidae**—is a freshwater dolphin (Plate 9). *Inia* is widely distributed in the Amazon, Araguaia–Tocantins and Orinoco river basins of northern South America, where it may have differentiated into as many as three species (Hrbek *et al.*, 2014). The fossil record of the family is restricted to relatively few occurrences from the Late Miocene, Pliocene, and Pleistocene of South America, Panama,

Figure 4.38 Skull of a representative member of Lipotidae, *Parapontoporia sternbergi*, in (a) dorsal and (b) lateral views. Abbreviations as in Figure 4.4; adapted from Barnes (1985b).

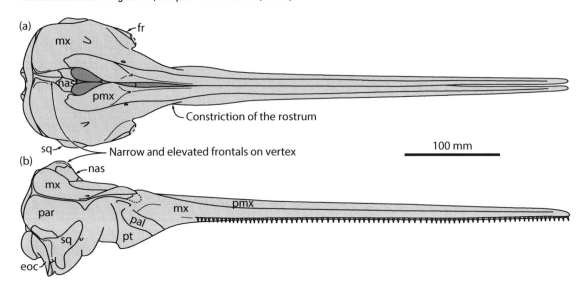

Figure 4.39 Skull of a representative member of Iniidae, *Inia geoffrensis*, in (a) dorsal and (b) lateral views. Abbreviations as in Figure 4.4.

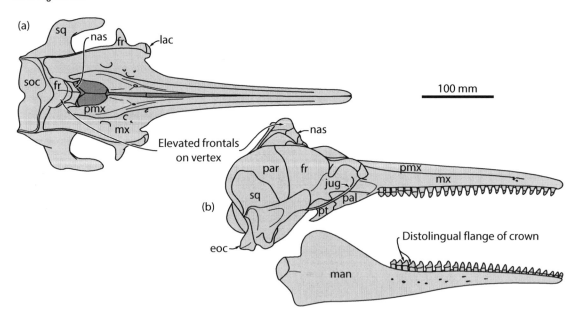

and the eastern United States (Cozzuol, 2010; Geisler *et al.*, 2012; Pyenson *et al.*, 2015). Iniids are relatively small (2–2.5 m in the case of *Inia*) and share elevated frontals forming a boss on the vertex of the skull. In addition, *Inia* is characterized by the drastic reduction of the posterior process of the periotic, a heterodont dentition capable of crushing armored prey, a flexible body with a long neck, and the retention of vibrissae (Figure 4.39). Fossil iniids include *Goniodelphis, Ischyrorhynchus, Isthminia, Meherrinia,* and *Saurocetes.* Of these, *Isthminia* and *Meherrinia* from the Late Miocene of Panama and eastern North America are demonstrably of marine origin, with the remainder coming from fluvial deposits or mixed terrestrial–marine faunas (Geisler *et al.,* 2012; Pyenson *et al.,* 2015). Both molecular and morphological analyses generally group iniids together with pontoporiids (discussed further in this section), with which they form the basal delphinidan superfamily **Inioidea** (Figure 4.37) (Cassens *et al.,* 2000; Geisler *et al.,* 2011; Steeman *et al.,* 2009).

Unlike all other "river dolphins," no member of the **Pontoporiidae** actually occurs in a freshwater environment. This includes the extant Franciscana (*Pontoporia*), which inhabits the coastal waters off Argentina, Brazil, and Uruguay (Plate 9). The oldest pontoporiid is known from the late Middle–early Late Miocene (ca 12 Ma) of Peru or, possibly, the early Middle Miocene (15–16 Ma) of California, United States (Barnes *et al.,* 2015). Beyond these earliest records, fossil pontoporiids are widely distributed, and have been reported from the Late Miocene and Pliocene of Europe, the Atlantic and Pacific coasts of South America, eastern North America, and, possibly, the North Pacific (Lambert and de Muizon, 2013).

At a length of less than 1.8 m, *Pontoporia* is one of the smallest living cetaceans. Fossil members of the family are similarly small and, along with *Pontoporia,* characterized by the presence of premaxillary eminences located in front of the external nares, a low skull vertex with anteroposteriorly long nasals, a lacrimal wrapping itself around the anterior edge of the frontal, and a blade-like posterior extension of the posterior process of the periotic (Figure 4.40) (de Muizon, 1988c; Geisler *et al.,* 2012). While *Pontoporia* has a long rostrum similar to that of other extant "river dolphins," the length of the rostrum in the fossil species is variable. Besides the extant *Pontoporia,* Pontoporiidae currently comprise six fossil genera: *Auroracetus, Brachydelphis, Pliopontos, Pontistes, Protophocaena,* and *Stenasodelphis.* The family is commonly divided into two subfamilies, Brachydelphininae and Pontoporiinae, which are in turn sometimes considered to have familial rank within Inioidea. However, the contents of these subfamilies remain a matter of debate (Cozzuol, 2010; Geisler *et al.,* 2012)

Kentriodontidae is a summary term for a likely polyphyletic assemblage of generally small, Late Oligocene–Late Miocene fossil dolphins not clearly assignable to any of the established delphinidan families. Nevertheless, it is likely that many

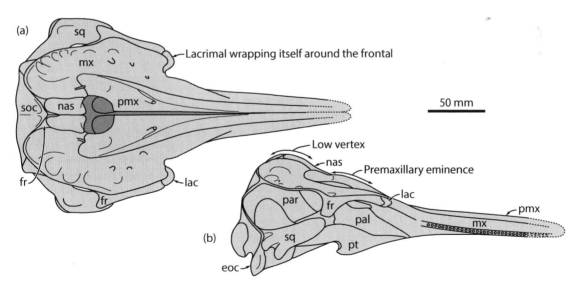

Figure 4.40 Skull of a representative member of Pontoporiidae, *Pliopontos littoralis,* in (a) dorsal and (b) lateral views. Abbreviations as in Figure 4.4; adapted from de Muizon (1984).

kentriodontids are members of the superfamily **Delphinoidea**, along with the extinct albireonids and odobenocetopsids (see section 4.4.5) and extant dolphins, porpoises, and monodontids (section 4.4.5). Features focusing mostly on the narial area, such as the presence of an internasal fossa and a vertical notch on the anterior margin of the nasals, were previously considered to be potential kentriodontid synapomorphies. However, these traits are not present in all of the referred genera, and sometimes also occur in non-kentriodontid taxa (e.g., *Pontoporia*). In the absence of a comprehensive analysis establishing the phylogenetic structure of the group as a whole, the slightly better defined kentriodontid subfamilies Kentriodontinae, Lophocetinae, and Pithanodelphininae are here treated as possibly genuine clades (Figure 4.37). However, it needs to be stressed that there is currently no consensus as to their respective contents.

The highly diverse and widely distributed **Kentriodontinae** are known from the Early to the late Middle or early Late Miocene, with the main records hailing from Europe, Japan, eastern and western North America, and, possibly, New Zealand and Peru (Aguirre-Fernández and Fordyce, 2011; Ichishima *et al.*, 1995; Kazár and Hampe, 2014; Lambert *et al.*, 2005). Kentriodontines are united by a tabular and proportionally wide skull vertex, as well as an often well-developed narial process of the frontal (Figure 4.41a,b). Apart from the procumbent anterior incisors of some species, they have simple, conical teeth, with the posteriormost ones sometimes bearing tiny accessory denticles. As currently understood, the subfamily includes *Delphinodon*, *Kentriodon*, *Macrokentriodon*, *Rudicetus*, and *Tagicetus*, as well as possibly *Belonodelphis*, *Incacetus*, and *Kampholophos*. Like the other kentriodontids, kentriodontines (especially *Kentriodon*) are sometimes regarded as intermediate between inioids and extant delphinoids (discussed further here), but alternatively might be more basal than both (Figure 4.37) (Aguirre-Fernández and Fordyce, 2014; Geisler *et al.*, 2012; Murakami *et al.*, 2014).

Unlike the widely distributed kentriodontines, members of the **Lophocetinae** are currently only known from the Middle–early Late Miocene of North America, as well as possibly Portugal (Estevens and Antunes, 2004; Ichishima *et al.*, 1995). Lophocetines include the largest kentriodontids (bizygomatic width of up to 350 mm) and differ from the other subfamilies in having transversely pinched nasals and frontals exposed on the skull vertex, a comparatively vast temporal fossa, and a highly convex anterodorsal margin of the supraoccipital. The rostrum and teeth are robust, with the premaxilla being laterally swollen slightly anterior to the level of the antorbital notches (Figure 4.41c,d). Of the three described genera (*Hadrodelphis*, *Liolithax*, and *Lophocetus*), only *Hadrodelphis* has been included in a large phylogenetic analysis, in which it occupied a position at the base of Delphinoidea (Figure 4.37) (Murakami *et al.*, 2014).

The geographical distribution of the third kentriodontid subfamily, the **Pithanodelphininae**, is wide but patchy. A limited number of fossils are known from the late Middle and early Late Miocene of Europe, the Caucasus, New Zealand, western North America, and Peru (Ichishima *et al.*, 1995; Kazár and Grigorescu, 2005). Pithanodelphinines are generally small, with a moderately elongated rostrum. Their main defining features are found on the skull vertex, where large, inflated nasals (bearing an internasal fossa) are posteriorly followed by a narrow exposure of the frontals (Figure 4.41e,f). The subfamily includes *Atocetus*, *Pithanodelphis*, and *Sarmatodelphis*, as well as possibly *Leptodelphis* and *Sophianaecetus*. *Atocetus* is the only pithanodelphinine that has been included in phylogenetic analyses, which variably found it to be sister to albireonids, extant delphinoids, or, together with *Kentriodon* (just discussed), extant delphinoids plus albireonids (Figure 4.37) (Aguirre-Fernández and Fordyce, 2014; Geisler *et al.*, 2012; Murakami *et al.*, 2014).

The extinct **Albireonidae** have a relatively poor fossil record, and at present are only known from the Late Miocene (ca 9 Ma) and Pliocene of western North America and Japan (Barnes, 2008). Albireonids are medium-sized compared to other delphinoids, and characterized by a pointed rostrum, proportionally large eye sockets, premaxillary eminences anterior to the external nares, and the presence of a steep posterior wall of the external nares formed by the flattened premaxillae, nasals and frontals (Barnes, 2008) (Figure 4.42). The family is monogeneric and includes only *Albireo*. Despite certain similarities with extant porpoises, such as the presence of premaxillary

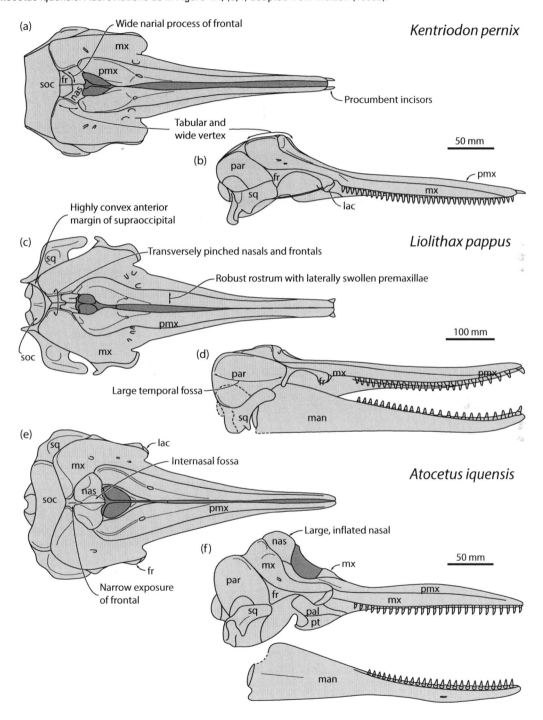

Figure 4.41 Skulls of representative members of the three kentriodontid subfamilies in (a, c, e) dorsal and (b, d, f) lateral views. (a, b) Kentriodontinae, *Kentriodon pernix*; (c, d) Lophocetinae, *Liolithax pappus*; and (e, f) Pithanodelphininae, *Atocetus iquensis*. Abbreviations as in Figure 4.4; (e, f) adapted from Muizon (1988c).

Figure 4.42 Skull of a representative member of Albireonidae, *Albireo whistleri*, in (a) dorsal and (b) lateral views. Abbreviations as in Figure 4.4; adapted from Barnes (2008).

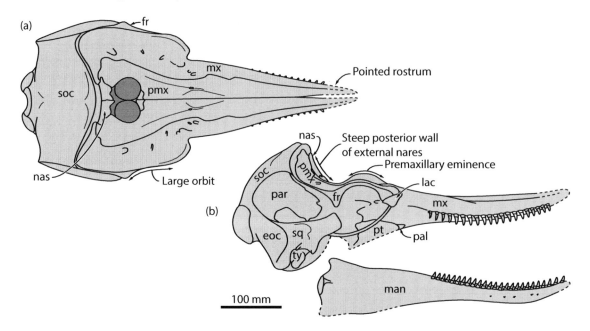

eminences, phylogenetic analyses have tended to recover albireonids as stem delphinoids that diverged from the crown group along with, just before or just after pithanodelphinines (Figure 4.37) (Aguirre-Fernández and Fordyce, 2014; Geisler *et al.*, 2012; Murakami *et al.*, 2014).

4.4.5 Crown Delphinoidea

Unlike those of stem delphinoids, the affinities and higher level relationships of crown delphinoids are largely agreed upon. By definition, crown delphinoids include the extant oceanic dolphins (Delphinidae), the porpoises (Phocoenidae), and the beluga and narwhal (Monodontidae) (e.g., Geisler *et al.*, 2011; Heyning, 1989; McGowen *et al.*, 2009; Murakami *et al.*, 2014). The extinct Odobenocetopsidae are also commonly thought to form part of this group, although their affinities have recently been questioned (de Muizon, 1993; Pyenson *et al.*, 2013).

The **Monodontidae** are the smallest of the modern delphinoid families. Although the two extant species, the beluga (*Delphinapterus leucas*) and the narwhal (*Monodon monoceros*), share an Arctic to sub-Arctic distribution, the whole Neogene record of the family comes from outside this region. The

oldest fossil monodontids were recovered from Late Miocene deposits of western North America, from localities as far south as Mexico. Younger specimens are known from the Early Pliocene of Europe, eastern North America, and, perhaps, Peru (Vélez-Juarbe and Pyenson, 2012). Defining monodontid features include the exposure of the maxilla medial to the premaxilla along the bony nares, a relatively small pterygoid sinus fossa compared to other delphinidans, the loss of the lateral lamina of the pterygoid in the hamular region, and a thickening of the alisphenoid in the area of the foramen ovale (Figure 4.43) (de Muizon, 1988b; Fraser and Purves, 1960; Vélez-Juarbe and Pyenson, 2012). *Monodon* itself is mostly defined by the loss of the entire dentition, except for a pair of anteriorly projected maxillary tusks. The left tusk only erupts in adult males and can reach a length of more than 2.5 m, compared with a body length (without the tusk) of around 4–5 m.

Besides its extant representatives, the family includes two fossil genera: the Late Miocene *Denebola* and the Early Pliocene *Bohaskaia*, with the latter sharing with the beluga and the narwhal the loss of premaxillary teeth and a dorsally convex outline of the rostrum (Vélez-Juarbe and

Figure 4.43 Skull of a representative member of Monodontidae, *Delphinapterus leucas*, in (a) dorsal and (b) lateral views. Abbreviations as in Figure 4.4; adapted from Van Beneden and Gervais (1880).

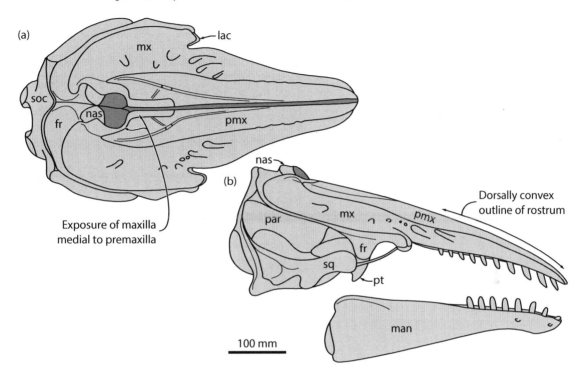

Pyenson, 2012). Molecular and total evidence analyses strongly support a close relationship of monodontids and phocoenids (Geisler *et al.*, 2011; Steeman *et al.*, 2009). In addition, monodontids may be sister to the extinct odobenocetopsids (discussed next) (Figure 4.37).

Odobenocetopsidae are a small, monogeneric family of bizarre odontocetes known only from the Late Miocene and possibly the earliest Pliocene of Peru and Chile (Plate 11) (de Muizon, 1993; Pyenson *et al.*, 2013). The only included genus, *Odobenocetops*, stands out for its highly fore-shortened rostrum, arched palate, anteriorly placed external nares, anteriorly oriented orbits, premaxillary processes housing posteroventrally directed tusks, and well-developed occipital condyles (Figure 4.44). Taken together, these highly unusual modifications cause *Odobenocetops* to appear superficially similar to the walrus, with which it may have shared a similar, suction-based feeding strategy. The dramatic shortening of the rostrum

means that the melon must have been small or even entirely absent. This suggests that, uniquely among odontocetes, *Odobenocetops* may have lost the ability to echolocate, which it likely compensated for by instead relying on binocular vision. The tusks appear to be sexually dimorphic, and may have been used as guides during feeding and in social interactions, such as competition between males (de Muizon *et al.*, 2002). Because of the highly modified skull, the phylogenetic affinities of *Odobenocetops* are difficult to infer. The most widely accepted hypothesis relates odobenocetopsids to monodontids (Figure 4.37), but this notion was recently challenged based on new material from Chile (de Muizon, 1993; Murakami *et al.*, 2012b; Pyenson *et al.*, 2013).

Extant porpoises (**Phocoenidae**) are a moderately diverse family comprising six species in three genera: *Phocoena*, *Phocoenoides*, and *Neophocaena*. *Phocoenoides dalli* (Dall's porpoise) and *Neophocaena phocaenoides* (the finless

Figure 4.44 Skull of a representative member of Odobenocetopsidae, *Odobenocetops peruvianus*, in (a) dorsal and (b) lateral views. Abbreviations as in Figure 4.4; adapted from de Muizon (1993).

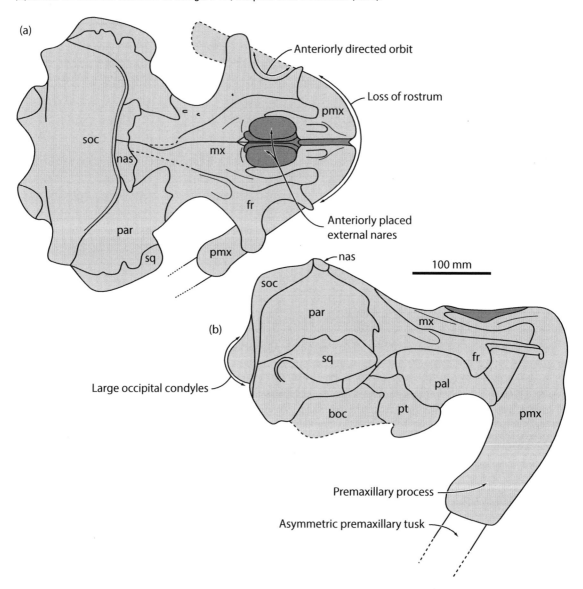

porpoise) are restricted to the North Pacific and the coastal waters off southern and eastern Asia, respectively. By contrast, the four species that make up *Phocoena* between them give this genus a virtually worldwide distribution. *Phocoena sinus*, also known as the vaquita, is both the smallest and, being restricted to the northernmost Gulf of California, the most endangered of the liv-

ing cetaceans (Jaramillo-Legorreta *et al.*, 2007). The phocoenid fossil record goes back to the Late Miocene, with all but two genera (*Brabocetus* and *Septemtriocetus* from the Pliocene of Europe) known only from the Late Miocene and Pliocene of the Pacific realm (Japan, Peru, and western North America) (Barnes, 1985a; Colpaert *et al.*, 2015; Murakami *et al.*, 2012a).

Figure 4.45 Skull of a representative member of Phocoenidae, *Phocoena phocoena*, in (a) dorsal and (b) lateral views. Abbreviations as in Figure 4.4.

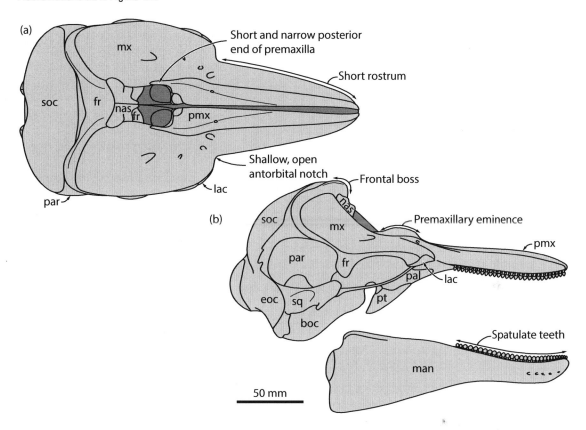

Extant porpoises are small, with only two species exceeding 2 m in length. They are characterized by a short snout, a proportionally large brain case and a comparatively low degree of cranial asymmetry—all of which may be evidence of pedomorphism (section 8.4). In addition, they share the presence of spatulate teeth, a shallow and anterolaterally open antorbital notch, a high premaxillary eminence anterior to the external naris, a narrow posterior termination of the premaxilla, a moderately developed frontal boss on the skull vertex, and a dorsal extension of the air sinus system located between the frontal and the maxilla (Figure 4.45) (Barnes, 1985a; Colpaert *et al.*, 2015). Some of these features are absent in fossil phocoenids, which may be somewhat larger; have a longer snout, deeper antorbital notches, and a lower premaxillary eminence; and, sometimes, lack spatulate teeth.

Besides *Brabocetus*, *Septemtriocetus* and the three extant genera, there are a further 10 extinct taxa, all of them from the Pacific: *Archaeophocaena*, *Australithax*, *Haborophocoena*, *Lomacetus*, *Miophocaena*, *Numataphocoena*, *Piscolithax*, *Pterophocaena*, *Salumiphocaena*, and *Semirostrum*. Originally, some of these genera were thought to fall into two separate subfamilies, Phocoeninae and Phocoenoidinae, each of which was also proposed to contain some of the extant species (Barnes, 1985a). This framework was not, however, corroborated by subsequent phylogenetic analyses, which instead affirmed the monophyly of the phocoenid crown group (Lambert, 2008a; Murakami *et al.*, 2014; Racicot *et al.*, 2014). As a whole, phocoenids are closely related to monodontids (Figure 4.37).

The **Delphinidae** (oceanic dolphins) are the most diverse of all the extant cetacean families, and they are represented today by as many as 17 genera (*Cephalorhynchus, Delphinus, Feresa, Globicephala, Grampus, Lagenodelphis, Lagenorhynchus, Lissodelphis, Orcaella, Orcinus, Peponocephala, Pseudorca, Sotalia, Sousa, Stenella, Steno,* and *Tursiops*) (Plate 10). Between them, modern delphinids not only inhabit all of the world's oceans, but have even ventured into some estuaries and a few river systems (e.g., in South America and Southeast Asia). Surprisingly, the fossil record of the family is still rather sparse. The oldest putative delphinid has been reported from the Late Miocene of California, United States (ca 10–11 Ma), but remains undescribed (Barnes, 1977). A roughly similar age for the origin of the family is indicated by a slightly younger record (7–9 Ma) from Japan. Apart from these Miocene occurrences, all diagnostic delphinid remains described so far date from the Pliocene and Pleistocene. The geographical distribution of fossil delphinids is patchy, but more or less global, with Italy and the eastern United States being among the richest localities (Bianucci, 2013; Boessenecker *et al.*, 2015; Murakami *et al.*, 2014).

Delphinids are disparate both in terms of their size (<2.5–9 m) and in terms of their overall morphology and feeding strategy. Nevertheless, all delphinids share distinctly asymmetrical premaxillae (with the left one being shorter and not making contact with the corresponding nasal), the development of a high mesethmoid plate on the posterior wall of the external nares, anteroposteriorly flattened nasals, and the presence of a keel on the ventral surface of the hamular process of the pterygoid (Figure 4.46). Nine fossil genera have been described to complement the extant taxa: *Arimidelphis, Astadelphis, Australodelphis, Eodelphinus, Etruridelphis, Hemisyntrachelus, Platalearostrum, Protoglobicephala,* and *Septidelphis* (Bianucci, 2013; Murakami *et al.*, 2014). The monophyly of delphinids is widely recognized, as is their sister-group relationship with Monodontidae and Phocoenidae (Figure 4.37) (e.g., Geisler *et al.*, 2011; Murakami *et al.*, 2014; Steeman *et al.*, 2009). Given the large number of genera, it is hardly surprising that as many as six subfamilies have been proposed to contain them. Three of these subfamilies have been supported by a string of molecular analyses: (1) Globicephalinae (*Feresa, Globicephala, Grampus, Peponocephala* and *Pseudorca*, and, possibly, *Orcaella*); (2) Delphininae (*Delphinus, Lagenodelphis, Sousa, Stenella,* and *Tursiops*); and (3) Lissodelphininae (*Cephalor-*

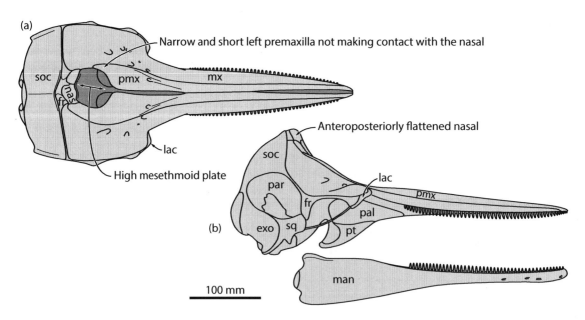

Figure 4.46 Skull of a representative member of Delphinidae, *Delphinus delphis*, in (a) dorsal and (b) right lateral views. Abbreviations as in Figure 4.4.

hynchus, *Lagenorhynchus*, and *Lissodelphis*) (e.g., Caballero *et al.*, 2008; McGowen *et al.*, 2009; Steeman *et al.*, 2009; Vilstrup *et al.*, 2011). Of the fossil genera, *Platalearostrum* and *Protoglobicephala* likely belong to Globicephalinae, (Aguirre-Fernández *et al.*, 2009; Post and Kompanje, 2010), whereas the Italian *Astadelphis*, *Etruridelpis*, and *Septidelphis* are all delphinines (Bianucci, 2013).

4.5 Consensus, conflicts, and diversification dates

Rapidly increasing numbers of molecular and morphological phylogenies provide an invaluable opportunity to cross-check results, and to stimulate discussion on both traditional and entirely novel cladistic hypotheses. The study of cetacean evolution offers a prime example of how molecules and morphology can come together to solve a long-standing enigma, such as the origin of whales from artiodactyls (section 4.1). Congruence of molecular and morphological data has helped to confirm the monophyly of many cetacean families and even superfamilies, but also led to the revision of some traditional concepts we now know to be incorrect—for example, the original definition of Platanistoidea as comprising all extant "river dolphins" (section 4.2.2). There are, however, some persistent points of conflict between the two approaches that have refused to go away over the years, and plague cetacean phylogenetics to this day.

4.5.1 High-level conflicts and possible solutions

In the case of mysticetes, molecular and morphological analyses have remained at loggerheads over the relationships of both the **pygmy right whale** (*Caperea marginata*) and the **gray whale** (*Eschrichtius robustus*): whereas most morphological analyses find *Caperea* to be related to balaenids, and the gray whale to be sister to a monophyletic Balaenopteridae, molecular analyses instead consistently ally *Caperea* with rorquals and gray whales, and nest *Eschrichtius* within modern balaenopterids (Bisconti, 2014; McGowen *et al.*, 2009; Steeman, 2007; Steeman *et al.*, 2009). Recent years have seen some movement on both sides of the argument, with some molecular studies recovering a monophyletic Balaenopteridae to the exclusion of the gray whale (Steeman *et al.*, 2009; Yang, 2009), and some morphological and total evidence analyses placing *Caperea* in the position suggested by the molecular data (Deméré *et al.*, 2008; Fordyce and Marx, 2013; Gol'din and Steeman, 2015; Marx and Fordyce, 2015). Nevertheless, for the most part, these rather major contradictions remain, as do several smaller disagreements regarding the finescale interrelationships of extant balaenopterids.

In the case of odontocetes, some early molecular studies suggested a possible relationship of **physeteroids** with mysticetes (Milinkovitch *et al.*, 1993). Both morphological and subsequent molecular studies were quick to refute this result, and ultimately decided the dispute in favor of the traditional view that classifies sperm whales as *bona fide* odontocetes (Árnason and Gullberg, 1994; Heyning, 1997; Nikaido *et al.*, 2001). Nevertheless, the position of physeteroids is not entirely settled: whereas molecular analyses uniformly recover them as the earliest diverging lineage within crown Odontoceti, some morphological analyses instead ally them with **ziphiids** (Geisler and Sanders, 2003; McGowen *et al.*, 2009; Murakami *et al.*, 2014; Steeman *et al.*, 2009).

Although the non-monophyly of "river dolphins" is widely agreed upon, the position of *Platanista* and *Lipotes* still varies between different analyses, both morphological and molecular. ***Platanista*** is often regarded as the second lineage to diverge from other crown odontocetes, following physeteroids (e.g., McGowen *et al.*, 2009), but it has also been interpreted as the sister group of delphinidans (e.g., Árnason *et al.*, 2004), all other extant odontocetes (Murakami *et al.*, 2014) and even ziphiids (e.g., Cassens *et al.*, 2000; Geisler *et al.*, 2011). Similarly, ***Lipotes*** may be sister either to Inioidea (e.g., Steeman *et al.*, 2009) or to all other delphinidans (Geisler *et al.*, 2012; Hamilton *et al.*, 2001). Potential reasons for this lack of consensus may lie in a phase of rapid diversification early during (crown) odontocete history, as well as the convergent evolution of morphological traits related to a freshwater existence (Cassens *et al.*, 2000; Geisler *et al.*, 2011). Finally, among **delphinoids**, several morphological studies continue to contradict the grouping of monodontids and phocoenids advocated by most molecular analyses, and instead propose a clade comprising delphinids and phocoenids (e.g., Geisler and Sanders, 2003; Murakami *et al.*, 2012b; Steeman *et al.*, 2009).

What is causing such a level of disagreement to persist, especially in the face of ever-more rapidly increasing amounts of available data? Variable methodologies (e.g., Bayesian vs parsimony) may partly be to blame, but there are various other sources of error that may also play a role. On the molecular side, these include: (1) the choice of genes to be analyzed; (2) the reliability of the sequence alignment; (3) the number and scope of the partitions into which the data are divided (e.g., lumping together all mitochondrial and nuclear data vs defining one partition for every gene); and (4) for each partition, the choice of substitution model. In the case of morphology, subjective decisions have to be made on (1) the number and definition of individual characters; (2) the number of states per character and, in the case of multistate characters, the decision of whether to order them; and (3) the actual scoring of specimens, which will inevitably be subject to individual biases. A further problem affecting many morphological datasets is a widespread lack of repeatability, because particular states and scoring decisions are often not explained or illustrated in sufficient detail (Jenner, 2004; Vogt, 2009).

Some approaches designed to tackle these problems have recently become available, and are starting to be employed in the study of cetacean evolution. Among others, these include the use of morphological databases to create interactive, fully illustrated cladistic data matrices (e.g., Ekdale *et al.*, 2011; Marx and Fordyce, 2015), more extensive character discussions (e.g., Ekdale *et al.*, 2011; Geisler and Sanders, 2003) and the introduction of programmes capable of automatically selecting best-fit partitioning schemes (Lanfear *et al.*, 2012). Comparing and combining molecular and morphological data have a long tradition in cetacean paleobiology, and hopefully will further contribute to pinpointing, and ultimately resolving, long-standing phylogenetic conundrums.

4.5.2 Divergence dates

The flourishing of comprehensive cladistic analyses has made it possible to estimate the time of divergence of virtually all extant cetacean species, based on the principle of the molecular clock (section 1.4.2). Molecular divergence dates provide another source of information that can be directly compared with, and indeed partially relies on, data from the fossil record. Most molecular clock studies date the initial divergence of Neoceti to around 34–36 Ma, which is only slightly older than the oldest known fossil neocete, *Llanocetus* (note, however, that *Llanocetus* is frequently used to calibrate this particular node). The same analyses date the split of balaenids from Plicogulae to 28–30 Ma, the separation of *Caperea* from balaenopteroids to about 23–26 Ma, and the initial divergence of crown balaenopteroids to about 14–19 Ma. Within odontocetes, physeteroids diverged from Synrhina around 32–35 Ma; *Platanista* from other synrhinans around 28–33 Ma, closely followed by the split of ziphiids from delphinidans just 1 million years or so afterward; lipotids plus inioids from delphinoids around 23–25 Ma; and delphinids from phocoenids plus monodontids around 14–19 Ma (Figure 4.47) (McGowen *et al.*, 2009; Steeman *et al.*, 2009; Xiong *et al.*, 2009).

Many of these estimates are considerably older than the oldest fossil representatives of the lineages to which they refer. For example, the oldest beaked whale, *Archaeoziphius*, is nearly 15 Ma younger than the putative origin of the ziphiid lineage. Such long ghost lineages imply either that the fossil record has been substantially undersampled or that the ghost lineage is a placeholder for one or more related, entirely extinct clades whose true evolutionary affinities have not yet been understood. In the case of ziphiids, the proposed relationship with eurhinodelphinids could markedly shorten the existing ghost lineage by extending their fossil record back to the Early Miocene. Similarly, the inclusion of *Platanista* in an expanded Platanistoidea (including waipatiids and squalodontids; section 4.2.2) could push back the earliest record of this lineage to the Late Oligocene, thus virtually eliminating its associated ghost lineage altogether. Of course, the simple existence of a ghost lineage by itself does not mean that a related, extinct clade must necessarily exist. Nevertheless, it can help to formulate novel cladistic hypotheses, which can then be tested more rigorously through careful anatomical study and further phylogenetic analyses. Most importantly, the existence of long ghost lineages also highlights

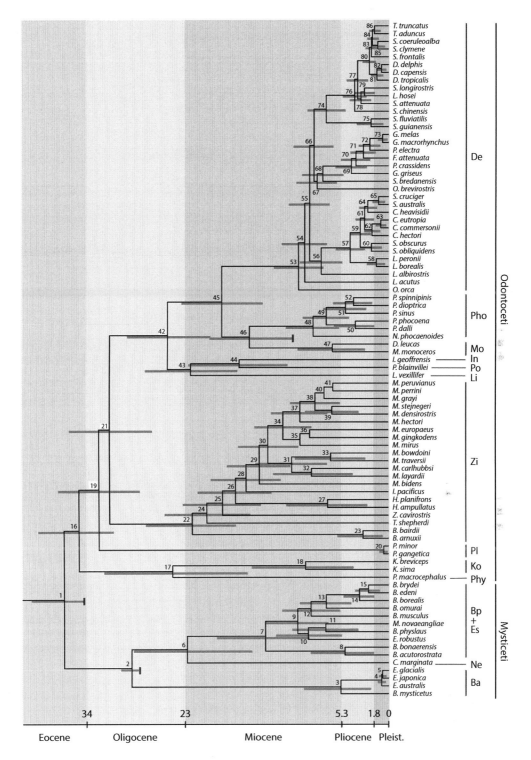

Figure 4.47 Time-calibrated phylogeny of extant cetaceans, based on molecular data. Horizontal gray bars correspond to the 95% highest posterior density for each node. Vertical bars denote calibration. Reproduced from McGowen *et al.* (2009), with permission of Elsevier.

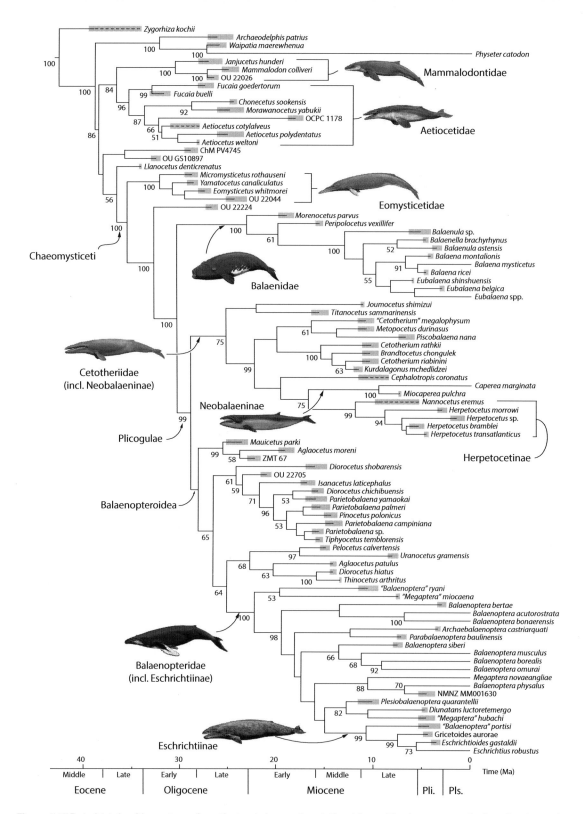

Figure 4.48 Dated total-evidence tree of mysticete phylogenetic relationships, taking into account both molecular and morphological data. All nodes in this tree have been dated using a combined molecular–morphological clock approach. Adapted from Marx and Fordyce (2015) under a Creative Commons Attribution license.

the complex nature of cetacean evolutionary history, which has been shaped by a dynamic of origination and extinction that can only be fully captured through the study of fossils.

Until relatively recently, clock dating relied solely on molecular data, and thus could only be applied to nodes that gave rise to extant descendants. Recent advances have extended the molecular clock methodology to morphological data, thus creating a "morphological clock" that can be applied to both extant and entirely extinct lineages. The details of this method are still being worked on and, so far, it has only been applied to cetacean data once. The result was a fully dated, comprehensive tree of extinct and extant mysticetes that is roughly consistent with previous, purely molecular age estimates, although it predates many of them by about 2–5 Ma (Figure 4.48) (Marx and Fordyce, 2015). Crucially, this tree is the first to provide direct date estimates for nodes with completely extinct descendants, such as the divergence of eomysticetids from balaenomorphs (ca 35 Ma) or the first divergence within Cetotheriidae (ca 25 Ma). Fully dated trees such as this not only provide a much more detailed picture of evolutionary relationships and diversification, but also can be used to infer clade-wide macroevolutionary dynamics, such as rates of diversification and character evolution (section 7.3).

4.6 Suggested readings

de Muizon, C. 1994. Are the squalodontids related to the platanistoids? Proceedings of the San Diego Society of Natural History 29:135–146.

Fordyce, R. E., and C. Muizon, de. 2001. Evolutionary history of cetaceans: a review; pp. 169–233 in J.-M. Mazin and V. de Buffrénil (eds.), Secondary Adaptation of Tetrapods to Life in Water. Verlag Dr. Friedrich Pfeil, München.

Geisler, J. H., M. R. McGowen, G. Yang, and J. Gatesy. 2011. A supermatrix analysis of genomic, morphological, and paleontological data for crown Cetacea. BMC Evolutionary Biology 11:1–22.

Marx, F. G., and Fordyce, R. E. 2015. Baleen boom and bust—a synthesis of mysticete phylogeny, diversity and disparity. Royal Society Open Science 2:140434.

References

Abel, O. 1914. Die Vorfahren der Bartenwale. Denkschriften der Kaiserlichen Akademie der Wissenschaften 90:155–224.

Aguirre-Fernández, G., L. G. Barnes, F. J. Aranda-Manteca, and J. R. Fernández-Rivera. 2009. *Protoglobicephala mexicana*, a new genus and species of Pliocene fossil dolphin (Cetacea; Odontoceti; Delphinidae) from the Gulf of California, Mexico. Boletín de la Sociedad Geológica Mexicana 61:245–265.

Aguirre-Fernández, G., and R. E. Fordyce. 2011. New Zealand fossils reveal early Miocene global distribution for small *Kentriodon* dolphins. Journal of Vertebrate Paleontology:Program and Abstracts, 60.

Aguirre-Fernández, G., and R. E. Fordyce. 2014. *Papahu taitapu*, gen. et sp. nov., an early Miocene stem odontocete (Cetacea) from New Zealand. Journal of Vertebrate Paleontology 34:195–210.

Árnason, Ú., and A. Gullberg. 1994. Relationship of baleen whales established by cytochrome *b* gene sequence comparison. Nature 367:726–728.

Árnason, Ú., A. Gullberg, and A. Janke. 2004. Mitogenomic analyses provide new insights into cetacean origin and evolution. Gene 333:27–34.

Bajpai, S., and P. D. Gingerich. 1998. A new Eocene archaeocete (Mammalia, Cetacea) from India and the time of origin of whales. Proceedings of the National Academy of Sciences USA 95:15464–15468.

Bajpai, S., J. G. M. Thewissen, and R. W. Conley. 2011. Cranial anatomy of Middle Eocene *Remingtonocetus* (Cetacea, Mammalia) from Kutch, India. Journal of Paleontology 85:703–718.

Barnes, L. C., and S. A. McLeod. 1984. The fossil record and phyletic relationships of gray whales; pp. 3–32 in M. L. Jones, L. S. Swartz, and S. Leatherwood (eds.), The Gray Whale: *Eschrichtius robustus*. Academic Press, Orlando, FL.

Barnes, L. G. 1977. Outline of eastern North Pacific fossil cetacean assemblages. Systematic Zoology 25:321–343.

Barnes, L. G. 1985a. Evolution, taxonomy and antitropical distributions of the porpoises (Phocoenidae, Mammalia). Marine Mammal Science 1:149–165.

Barnes, L. G. 1985b. Fossil pontoporiid dolphins (Mammalia: Cetacea) from the Pacific coast of North America. Contributions in Science, Natural History Museum of Los Angeles County 363:1–34.

Barnes, L. G. 2006. A phylogenetic analysis of the superfamily Platanistoidea (Mammalia, Cetacea, Odontoceti). Beiträge zur Paläontologie 30:25–42.

Barnes, L. G. 2008. Miocene and Pliocene Albireonidae (Cetacea, Odontoceti), rare and unusual fossil dolphins from the eastern North Pacific Ocean. Natural History Museum of Los Angeles County Science Series 41:99–152.

Barnes, L. G., J. L. Goedert, and H. Furusawa. 2001. The earliest known echolocating toothed whales (Mammalia; Odontoceti): preliminary observations of fossils from Washington State. Mesa Southwest Museum Bulletin 8:91–100.

Barnes, L. G., M. Kimura, H. Furusawa, and H. Sawamura. 1995. Classification and distribution of Oligocene Aetiocetidae (Mammalia; Cetacea; Mysticeti) from western North America and Japan. The Island Arc 3:392–431.

Barnes, L. G., and R. E. Reynolds. 2009. A new species of early Miocene allodelphinid dolphin (Cetacea, Odontoceti, Platanistoidea) from Cajon Pass, Southern California, U.S.A. Museum of Northern Arizona Bulletin 65:483–507.

Barnes, L. G., L. Tohill, and S. Tohill. 2015. A new pontoporiid dolphin from the Sharktooth Hill Bonebed, central California; the oldest known member of its family. PaleoBios 32 (Suppl.):4.

Bebej, R. M., M. Ul-Haq, I. S. Zalmout, and P. D. Gingerich. 2012a. Morphology and function of the vertebral column in *Remingtonocetus domandaensis* (Mammalia, Cetacea) from the Middle Eocene Domanda Formation of Pakistan. Journal of Mammalian Evolution 19:77–104.

Bebej, R. M., I. S. Zalmout, A. A. Abed El-Aziz, M. S. M. Antar, and P. D. Gingerich. 2012b. First evidence of Remingtonocetidae (Mammalia, Cetacea) outside Indo-Pakistan: new genus from the early Middle Eocene of Egypt. Journal of Vertebrate Paleontology: Program and Abstracts, 62.

Beddard, F. E. 1901. Contribution towards a knowledge of the osteology of the pigmy whale (*Neobalaena marginata*). Transactions of the Zoological Society of London 16:87–115.

Bianucci, G. 2013. *Septidelphis morii*, n. gen. et sp., from the Pliocene of Italy: new evidence of the explosive radiation of true dolphins (Odontoceti, Delphinidae). Journal of Vertebrate Paleontology 33:722–740.

Bianucci, G., M. Gatt, R. Catanzariti, S. Sorbi, C. G. Bonavia, R. Curmi, and A. Varola. 2011. Systematics, biostratigraphy and evolutionary pattern of the Oligo-Miocene marine mammals from the Maltese Islands. Geobios 44:549–585.

Bianucci, G., and P. D. Gingerich. 2011. *Aegyptocetus tarfa*, n. gen. et sp. (Mammalia, Cetacea), from the middle Eocene of Egypt: clinorhynchy, olfaction, and hearing in a protocetid whale. Journal of Vertebrate Paleontology 31:1173–1188.

Bianucci, G., O. Lambert, and K. Post. 2007. A high diversity in fossil beaked whales (Mammalia, Odontoceti, Ziphiidae) recovered by trawling from the sea floor off South Africa. Geodiversitas 29:561–618.

Bianucci, G., O. Lambert, R. Salas-Gismondi, J. Tejada, F. Pujos, M. Urbina, and P.-O. Antoine. 2013a. A Miocene relative of the Ganges River dolphin (Odontoceti, Platanistidae) from the Amazonian Basin. Journal of Vertebrate Paleontology 33:741–745.

Bianucci, G., and W. Landini. 2006. Killer sperm whale: a new basal physeteroid (Mammalia, Cetacea) from the Late Miocene of Italy. Zoological Journal of the Linnean Society 148:103–131.

Bianucci, G., I. Miján, O. Lambert, K. Post, and O. Mateus. 2013b. Bizarre fossil beaked whales (Odontoceti, Ziphiidae) fished from the Atlantic Ocean floor off the Iberian Peninsula. Geodiversitas 35:105–153.

Bisconti, M. 2003. Evolutionary history of Balaenidae. Cranium 20:9–50.

Bisconti, M. 2005. Skull morphology and phylogenetic relationships of a new diminutive balaenid from the lower Pliocene of Belgium. Palaeontology 48:793–816.

Bisconti, M. 2007. A new basal balaenopterid whale from the pliocene of northern Italy. Palaeontology 50:1103–1122.

Bisconti, M. 2008. Morphology and phylogenetic relationships of a new eschrichtiid genus (Cetacea: Mysticeti) from the early Pliocene of northern Italy. Zoological Journal of the Linnean Society 153:161–186.

Bisconti, M. 2010. A new balaenopterid whale from the late Miocene of the Stirone River, northern Italy (Mammalia, Cetacea, Mysticeti). Journal of Vertebrate Paleontology 30:943–958.

Bisconti, M. 2012. Comparative osteology and phylogenetic relationships of *Miocaperea pulchra*, the first fossil pygmy right whale genus and species (Cetacea, Mysticeti, Neobalaenidae). Zoological Journal of the Linnean Society 166:876–911.

Bisconti, M. 2014. Anatomy of a new cetotheriid genus and species from the Miocene of Herentals, Belgium, and the phylogenetic and palaeobiogeographical relationships of Cetotheriidae s.s. (Mammalia, Cetacea, Mysticeti). Journal of Systematic Palaeontology 13:1–19.

Bisconti, M., O. Lambert, and M. Bosselaers. 2013. Taxonomic revision of *Isocetus depauwi* (Mammalia, Cetacea, Mysticeti) and the phylogenetic relationships of archaic "cetothere" mysticetes. Palaeontology 56:95–127.

Bisconti, M., and A. Varola. 2006. The oldest eschrichtiid mysticete and a new morphological diagnosis of Eschrichtiidae (gray whales). Rivista Italiana di Paleontologia e Stratigrafia 112:447–457.

Bloodworth, B., and C. D. Marshall. 2005. Feeding kinematics of *Kogia* and *Tursiops* (Odontoceti: Cetacea): characterization of suction and ram feeding. The Journal of Experimental Biology 208:3721–3730.

Boessenecker, R. W., and R. E. Fordyce. 2014. A new eomysticetid (Mammalia: Cetacea) from the Late Oligocene of New Zealand and a re-evaluation of "*Mauicetus*" *waitakiensis*. Papers in Palaeontology 1:107–140.

Boessenecker, R. W., and R. E. Fordyce. 2015a. Anatomy, feeding ecology, and ontogeny of a transitional baleen whale: a new genus and species of Eomysticetidae (Mammalia: Cetacea) from the Oligocene of New Zealand. PeerJ 3:e1129.

Boessenecker, R. W., and R. E. Fordyce. 2015b. A new genus and species of eomysticetid (Cetacea: Mysticeti) and a reinterpretation of "*Mauicetus*" *lophocephalus* Marples, 1956: transitional baleen whales from the upper Oligocene of New Zealand. Zoological Journal of the Linnean Society 175:607–660.

Boessenecker, R. W., F. A. Perry, and J. H. Geisler. 2015. Globicephaline whales from the Mio-Pliocene Purisima Formation of central California, USA. Acta Palaeontologica Polonica 60:113–122.

Boessenecker, R. W., and A. Poust. 2015. Freshwater occurrence of the extinct dolphin *Parapontoporia* (Cetacea: Lipotidae) from the upper Pliocene nonmarine Tulare Formation of California. Palaeontology 58:489–496.

Bouetel, V., and C. de Muizon. 2006. The anatomy and relationships of *Piscobalaena nana* (Cetacea, Mysticeti), a Cetotheriidae s.s. from the early Pliocene of Peru. Geodiversitas 28:319–396.

Boyden, A., and D. Gemeroy. 1950. The relative position of the Cetacea among the orders of Mammalia as indicated by precipitin tests. Zoologica 35:145–151.

Brandt, J. F. 1873. Untersuchungen über die fossilen und subfossilen Cetaceen Europa's. Mémoires de l'Académie impériale des sciences de St.-Pétersbourg 20:1–372.

Buchholtz, E. 1998. Implications of vertebral morphology for locomotor evolution in early Cetacea; pp. 325–351 in J. G. M. Thewissen (ed.), The Emergence of Whales. Plenum Press, New York.

Buchholtz, E. A. 2011. Vertebral and rib anatomy in *Caperea marginata*: Implications for evolutionary patterning of the mammalian vertebral column. Marine Mammal Science 27:382–397.

Buono, M. R., and M. A. Cozzuol. 2013. A new beaked whale (Cetacea, Odontoceti) from the Late Miocene of Patagonia, Argentina. Journal of Vertebrate Paleontology 33:986–997.

Buono, M. R., M. T. Dozo, F. G. Marx, and R. E. Fordyce. 2014. A Late Miocene potential neobalaenine mandible from Argentina sheds light on the origins of the living pygmy right whale. Acta Palaeontologica Polonica 59:787–793.

Caballero, S., J. Jackson, A. A. Mignucci-Giannoni, H. Barrios-Garrido, S. Beltran-Pedreros, M. A. Montiel-Villalobos, K. M. Robertson, and C. S. Baker. 2008. Molecular systematics of South American dolphins Sotalia: sister taxa determination and phylogenetic relationships, with insights into a multi-locus phylogeny of the Delphinidae. Molecular Phylogenetics and Evolution 46:252–268.

Cabrera, Á. 1926. Cetáceos fósiles del Museo de La Plata. Revista del Museo de La Plata 29:363–411.

Canto, J., J. Yañez, and J. Rovira. 2010. Estado actual del conocimiento do los mamiferos fósiles de Chile. Estudios Geológicos 66:1–37.

Cassens, I., S. Vicario, V. G. Waddell, H. Balchowsky, D. Van Belle, W. Ding, C. Fan, R. S. Lal Mohan, P. C. Simões-Lopes, R. Bastida, A. Meyer, M. J. Stanhope, and M. C. Milinkovitch. 2000. Independent adaptation to riverine habitats allowed survival of ancient cetacean lineages. Proceedings of the National Academy of Sciences USA 97:11343–11347.

Churchill, M., A. Berta, and T. Deméré. 2011. The systematics of right whales (Mysticeti: Balaenidae). Marine Mammal Science 28:497–521.

Clementz, M. T., R. E. Fordyce, S. Peek, L., and D. L. Fox. 2014. Ancient marine isoscapes and isotopic evidence of bulk-feeding by Oligocene cetaceans. Palaeogeography, Palaeoclimatology, Palaeoecology 400:28–40.

Clementz, M. T., A. Goswami, P. D. Gingerich, and P. L. Koch. 2006. Isotopic records from early whales and sea cows: contrasting patterns of ecological transition. Journal of Vertebrate Paleontology 26:355–370.

Colpaert, W., M. Bosselaers, and O. Lambert. 2015. Out of the Pacific: a second fossil porpoise from the Pliocene of the North Sea Basin. Acta Palaeontologica Polonica 60:1–10.

Cozzuol, M. A. 1996. The record of the aquatic mammals in southern South America. Münchner Geowissenschaftliche Abhandlungen, Reihe A 30:321–342.

Cozzuol, M. A. 2010. Fossil record and the evolutionary history of Inioidea; pp. 193–217 in M. Ruiz-García, and J. Shostell (eds.), Biology, Evolution and Conservation of River Dolphins. Nova Publishers, Hauppauge, NY.

Cranford, T. W. 1999. The sperm whale's nose: sexual selection on a grand scale? Marine Mammal Science 15:1133–1157.

Dathe, F. 1983. *Megaptera hubachi* n. sp., ein fossiler Bartenwal aus marinen Sandsteinschichten des tieferen Pliozaens Chiles. Zeitschrift für Geologische Wissenschaften 11:813–848.

de Muizon, C. 1984. Les Vertébrés de la Formation Pisco (Pérou). Deuxième partie: Les Odontocètes (Cetacea, Mammalia) du Pliocène inférieur du Sud-Sacaco. Travaux de l'Institut Français d'Etudes Andines 27:1–188.

de Muizon, C. 1987. The affinities of *Notocetus vanbenedeni*, an Early Miocene platanistoid (Cetacea, Mammalia) from Patagonia, southern Argentina. American Museum Novitates 2904:1–27.

de Muizon, C. 1988a. Le polyphylétisme des Acrodelphidae, Odontocètes longirostres du Miocène européen. Bulletin du Muséum National d'Histoire Naturelle. Section C: Sciences de la Terre, Paléontologie, Géologie, Minéralogie 10:31–88.

de Muizon, C. 1988b. Les relations phylogénétiques des Delphinida (Cetacea, Mammalia). Annales de Paléontologie 74:159–227.

de Muizon, C. 1988c. Les vertébrés fossiles de la Formation Pisco (Pérou). Troisième partie: Les Odontocètes (Cetacea, Mammalia) du Miocène. Travaux de l'Institut Français d'Etudes Andines 42:1–244.

de Muizon, C. 1991. A new Ziphiidae (Cetacea) from the Early Miocene of Washington State (USA) and phylogenetic analysis of the major groups of odontocetes. Bulletin du Muséum National d'Histore Naturelle, Paris 12:279–326.

de Muizon, C. 1993. Walrus-like feeding adaptation in a new cetacean from the Pliocene of Peru. Nature 365:745–748.

de Muizon, C. 1994. Are the squalodonts related to the platanistoids? Proceedings of the San Diego Society of Natural History 29:135–146.

de Muizon, C., D. P. Domning, and D. R. Ketten. 2002. *Odobenocetops peruvianus*, the walrus-convergent delphinoid (Mammalia: Cetacea) from the Early Pliocene of Peru. Smithsonian Contributions to Paleobiology 93:223–261.

Deméré, T. A., and A. Berta. 2008. Skull anatomy of the Oligocene toothed mysticete *Aetiocetus weltoni* (Mammalia; Cetacea): implications for mysticete evolution and functional anatomy. Zoological Journal of the Linnean Society 154:308–352.

Deméré, T. A., A. Berta, and M. R. McGowen. 2005. The taxonomic and evolutionary history of fossil and modern balaenopteroid mysticetes. Journal of Mammalian Evolution 12:99–143.

Deméré, T. A., M. R. McGowen, A. Berta, and J. Gatesy. 2008. Morphological and molecular evidence for a stepwise evolutionary transition from teeth to baleen in mysticete whales. Systematic Biology 57:15–37.

Dooley, A. C., Jr. 2005. A new species of *Squalodon* (Mammalia, Cetacea) from the Middle Miocene of Virginia. Virginia Museum of Natural History Memoirs 8:1–14.

Dubrovo, I. A., and A. E. Sanders. 2000. A new species of *Patriocetus* (Mammalia, Cetacea) from the Late Oligocene of Kazakhstan. Journal of Vertebrate Paleontology 20:577–590.

Ekdale, E.G., A. Berta, and T. A. Deméré. 2011. The comparative osteology of the petrotympanic complex (ear region) of extant baleen whales (Cetacea: Mysticeti). PLoS One 6:e21311.

Ekdale, E.G., and R. A. Racicot. 2015. Anatomical evidence for low frequency sensitivity in an archaeocete whale: comparison of the inner ear of *Zygorhiza kochii* with that of crown Mysticeti. Journal of Anatomy 226:22–39.

El Adli, J. J., T. A. Deméré, and R. W. Boessenecker. 2014. *Herpetocetus morrowi* (Cetacea: Mysticeti), a new species of diminutive baleen whale from the Upper Pliocene (Piacenzian) of California, USA, with observations on the evolution and relationships of the Cetotheriidae. Zoological Journal of the Linnean Society 170:400–466.

Emlong, D. R. 1966. A new archaic cetacean from the Oligocene of northwest Oregon. Bulletin of the Oregon University Museum of Natural History 3:1–51.

Estevens, M., and M. T. Antunes. 2004. Fragmentary remains of odontocetes (Cetacea, Mammalia) from the Miocene of the Lower Tagus Basin (Portugal). Revista Española de Paleontología 19:93–108.

Fahlke, J. M., K. A. Bastl, G. M. Semprebon, and P. D. Gingerich. 2013. Paleoecology of archaeocete whales throughout the Eocene: dietary adaptations revealed by microwear analysis. Palaeogeography, Palaeoclimatology, Palaeoecology 386:690–701.

Fitzgerald, E. M. G. 2004. A review of the Tertiary fossil Cetacea (Mammalia) localities in Australia. Memoirs of Museum Victoria 61:183–208.

Fitzgerald, E. M. G. 2005. Pliocene marine mammals from the Whalers Bluff Formation of Portland, Victoria, Australia. Memoirs of Museum Victoria 62:67–89.

Fitzgerald, E. M. G. 2006. A bizarre new toothed mysticete (Cetacea) from Australia and the early evolution of baleen whales. Proceedings of the Royal Society B 273:2955–2963.

Fitzgerald, E. M. G. 2010. The morphology and systematics of *Mammalodon colliveri* (Cetacea: Mysticeti), a toothed mysticete from the Oligocene of Australia. Zoological Journal of the Linnean Society 158:367–476.

Fitzgerald, E. M. G. 2011. A fossil sperm whale (Cetacea, Physeteroidea) from the Pleistocene of Nauru, Equatorial Southwest Pacific. Journal of Vertebrate Paleontology 31:929–931.

Fitzgerald, E. M. G. 2012. Possible neobalaenid from the Miocene of Australia implies a long evolutionary history for the pygmy right whale *Caperea marginata* (Cetacea, Mysticeti). Journal of Vertebrate Paleontology 32:976–980.

Flower, W. H. 1883. On whales, past and present, and their probable origin. Nature 28:199–202; 226–230.

Flynn, T. T. 1948. Description of *Prosqualodon davidi* Flynn, a fossil cetacean from Tasmania. Transactions of the Zoological Society of London 26:153–195.

Fordyce, R. E. 1981. Systematics of the odontocete whale *Agorophius pygmaeus* and the family Agorophidae (Mammalia: Cetacea). Journal of Paleontology 55:1028–1045.

Fordyce, R. E. 1983. Rhabdosteid dolphins (Mammalia: Cetacea) from the Middle Miocene, Lake Frome area, South Australia. Alcheringa 7:27–40.

Fordyce, R. E. 1994. *Waipatia maerewhenua*, new genus and new species (Waipatiidae, new family), an archaic Late Oligocene dolphin (Cetacea: Odontoceti: Platanistoidea) from New Zealand. Proceedings of the San Diego Society of Natural History 29:147–176.

Fordyce, R. E. 2002. *Simocetus rayi* (Odontoceti: Simocetidae, new family): a bizarre new archaic Oligocene dolphin from the eastern North Pacific. Smithsonian Contributions to Paleobiology 93:185–222.

Fordyce, R. E. 2003. Early crown-group Cetacea in the southern ocean: the toothed archaic mysticete *Llanocetus*. Journal of Vertebrate Paleontology 23 (Suppl. to 3):50A.

Fordyce, R. E., and F. G. Marx. 2013. The pygmy right whale *Caperea marginata*: the last of the cetotheres. Proceedings of the Royal Society B 280:20122645.

Fraser, F. C., and P. E. Purves. 1960. Hearing in cetaceans. Bulletin of the British Museum (Natural History) 7:1–140.

Gatesy, J. 1998. Molecular evidence for the phylogenetic affinities of Cetacea; pp. 63–112 in J. G. M. Thewissen (ed.), The Emergence of Whales. Plenum Press, New York.

Geisler, J. H., M. W. Colbert, and J. L. Carew. 2014. A new fossil species supports an early origin for toothed whale echolocation. Nature 508:383–386.

Geisler, J. H., S. J. Godfrey, and O. Lambert. 2012. A new genus and species of late Miocene inioid (Cetacea, Odontoceti) form the Meherrin River, North Carolina, USA. Journal of Vertebrate Paleontology 32:198–211.

Geisler, J. H., and Z.-X. Luo. 1998. Relationships of Cetacea to terrestrial ungulates and the evolution of cranial vasculature in Cete; pp. 163–212 in J. G. M. Thewissen (ed.), The Emergence of Whales. Plenum Press, New York.

Geisler, J. H., M. R. McGowen, G. Yang, and J. Gatesy. 2011. A supermatrix analysis of genomic, morphological, and paleontological data from crown Cetacea. BMC Evolutionary Biology 11:1–33.

Geisler, J. H., and A. E. Sanders. 2003. Morphological evidence for the phylogeny of Cetacea. Journal of Mammalian Evolution 10:23–129.

Geisler, J. H., and J. M. Theodor. 2009. *Hippopotamus* and whale phylogeny. Nature 458:E1–E4.

Geisler, J. H., and M. D. Uhen. 2003. Morphological support for a close relationship between hippos and whales. Journal of Vertebrate Paleontology 23:991–996.

Geisler, J. H., and M. D. Uhen. 2005. Phylogenetic relationships of extinct Cetartiodactyls: results of simultaneous analyses of molecular, morphological, and stratigraphic data. Journal of Mammalian Evolution 12:145–160.

Gingerich, P. D., M. S. M. Antar, and I. Zalmout. 2014. Skeleton of new protocetid (Cetacea, Archaeoceti) from the lower Gehannam Formation of Wadi al Hitan in Egypt: survival of a protocetid into the Priabonian Late Eocene. Journal of Vertebrate Paleontology: Program and Abstracts, 138.

Gingerich, P. D., M. u. Haq, I. S. Zalmout, I. H. Khan, and M. S. Malakani. 2001. Origin of whales from early artiodactyls: hands and feet of Eocene Protocetidae from Pakistan. Science 293:2239–2242.

Gingerich, P. D., B. H. Smith, and E. L. Simons. 1990. Hind limbs of Eocene *Basilosaurus*: evidence of feet in whales. Science 249:154–157.

Gingerich, P. D., M. Ul-Haq, W. von Koenigswald, W. J. Sanders, B. H. Smith, and I. S. Zalmout. 2009. New protocetid whale from the Middle Eocene of Pakistan: birth on land, precocial development, and sexual dimorphism. PLoS One 4:1–20.

Gingerich, P. D., N. A. Wells, D. E. Russell, and S. M. I. Shah. 1983. Origin of whales in epicontinental remnant seas: new evidence from the early Eocene of Pakistan. Science 220:403–406.

Godfrey, S. J., and L. G. Barnes. 2008. A new genus and species of late Miocene pontoporiid dolphin (Cetacea: Odontoceti) from the St. Marys Formation in Maryland. Journal of Vertebrate Paleontology 28:520–528.

Godfrey, S. J., L. G. Barnes, and D. Bohaska. 2006. *Araeodelphis natator* Kellogg, 1957, the most primitive known member of the Platanistidae (Odontoceti, Cetacea), and relationships to other clades within the Platanistoidea. Journal of Vertebrate Paleontology 26 (Suppl. 3):68A.

Godfrey, S. J., M. D. Uhen, and J. E. Osborne. in press. A new specimen of *Agorophius pygmaeus* (Agorophiidae, Odontoceti, Cetacea) from the early Oligocene Ashley Formation of South Carolina, U.S.A. Journal of Paleontology.

Goedert, J. L., R. L. Squires, and L. G. Barnes. 1995. Paleoecology of whale-fall habitats from deep-water Oligocene rocks, Olympic Peninsula, Washington State. Palaeogeography, Palaeoclimatology, Palaeoecology 118:151–158.

Gol'din, P., D. Startsev, and T. Krakhmalnaya. 2014. The anatomy of the Late Miocene baleen whale *Cetotherium riabinini* from Ukraine. Acta Palaeontologica Polonica 59:795–814.

Gol'din, P., and E. Zvonok. 2013. *Basilotritus uheni*, a new cetacean (Cetacea, Basilosauridae) from the late Middle Eocene of Eastern Europe. Journal of Paleontology 87:254–268.

Gol'din, P., and M. E. Steeman. 2015. From problem taxa to problem solver: a new Miocene family, Tranatocetidae, brings perspective on baleen whale evolution. PLoS One 10:e0135500.

Hamilton, H., S. Caballero, A. G. Collins, and R. L. Brownell, Jr. 2001. Evolution of river dolphins. Proceedings of the Royal Society B 268:549–556.

Hanna, G. D., and M. McLellan. 1924. A new species of whale from the type locality of the Monterey Group. Proceedings of the California Academy of Sciences 13:237–241.

Hassanin, A., F. Delsuc, A. Ropiquet, C. Hammer, B. J. van Vurren, C. A. Matthee, M. Ruiz-Garcia, F. M. Catzeflis, V. Areskoug, T. T. Nguyen, and A. Couloux. 2012. Pattern and timing of diversification of Cetartiodactyla (Mammalia, Laurasiatheria), as revealed by a comprehensive analysis of mitochondrial genomes. Comptes Rendus Biologies:32–50.

Heyning, J. E. 1984. Functional morphology involved in intraspecific fighting of the beaked whale, *Mesoplodon carlhubbsi*. Canadian Journal of Zoology/Revue Canadienne de Zoologie 62:1645–1654.

Heyning, J. E. 1989. Comparative facial anatomy of beaked whales (Ziphiidae) and a systematic revision among the families of extant Odontoceti. Contributions in Science, Natural History Museum of Los Angeles County 405:1–64.

Heyning, J. E. 1997. Sperm whale phylogeny revisited: analysis of the morphological evidence. Marine Mammal Science 13:596–613.

Heyning, J. E., and J. G. Mead. 1996. Suction feeding in beaked whales: morphological and observational evidence. Contributions in Science, Natural History Museum of Los Angeles County 464:1–12.

Hrbek, T., V. M. F. da Silva, N. Dutra, W. Gravena, A. R. Martin, and I. P. Farias. 2014. A new species of river dolphin from Brazil or: how little do we know our biodiversity. PLoS One 9:e83623.

Hulbert, R. C., and F. C. Whitmore. 2006. Late Miocene mammals from the Mauvilla local fauna, Alabama. Bulletin of the Florida Museum of Natural History 46:1–28.

Hunter, J. 1787. Observations on the structure and oeconomy of whales. Philosophical Transactions of the Royal Society of London B 77:371–450.

Ichishima, H., L. G. Barnes, R. E. Fordyce, M. Kimura, and D. J. Bohaska. 1995. A review of kentriodontine dolphins (Cetacea; Delphinoidea; Kentriodontidae): systematics and biogeography. The Island Arc 3:486–492.

Irwin, D., and Ú. Árnason. 1994. Cytochrome *b* gene of marine mammals: phylogeny and evolution. Journal of Mammalian Evolution 2:37–55.

Jaramillo-Legorreta, A., L. Rojas-Bracho, R. L. Brownell, A. J. Read, R. R. Reeves, K. Ralls, and B. L. Taylor. 2007. Saving the vaquita: immediate action, not more data. Conservation Biology 21:1653–1655.

Jenner, R. A. 2004. The scientific status of metazoan cladistics: why current research practice must change. Zoologica Scripta 33:293–310.

Kazár, E., and D. Grigorescu. 2005. Revision of *Sarmatodelphis moldavicus* Kirpichnikov, 1954 (Cetacea: Delphinoidea), from the Miocene of Kishinev, Republic of Moldavia. Journal of Vertebrate Paleontology 25:929–935.

Kazár, E., and O. Hampe. 2014. A new species of Kentriodon (Mammalia, Odontoceti, Delphinoidea) from the middle/late Miocene of Groß Pampau (Schleswig-Holstein, North Germany). Journal of Vertebrate Paleontology 34:1216–1230.

Kellogg, R. 1936. A review of the Archaeoceti. Carnegie Institution of Washington Publication 482:1–366.

Kellogg, R. 1965. The Miocene Calvert sperm whale *Orycterocetus*. United States National Museum Bulletin 247:47–63.

Kemper, C. M. 2009. Pygmy right whale *Caperea marginata*; pp. 939–941 in W. F. Perrin, B. Würsig, and J. G. M. Thewissen (eds.), Encyclopedia of Marine Mammals. Academic Press, Burlington, MA.

Kimura, T. 2009. Review of the fossil balaenids from Japan with a re-description of *Eubalaena shinshuensis* (Mammalia, Cetacea, Mysticeti). Quaderni del Museo di Storia Naturale di Livorno 22:3–21.

Kohno, N., H. Koike, and K. Narita. 2007. Outline of fossil marine mammals from the Middle Miocene Bessho and Aoki Formations, Nagano Prefecture, Japan. Research Report of the Shinshushinmachi Fossil Museum 10:1–45.

Lambert, O. 2005a. Phylogenetic affinities of the long-snouted dolphin *Eurhinodelphis* (Cetacea, Odontoceti) from the Micoene of Antwerp, Belgium. Palaeontology 48:653–679.

Lambert, O. 2005b. Review of the Miocene long-snouted dolphin *Priscodelphinus cristatus* du Bus, 1872 (Cetacea, Odontoceti) and phylogeny among eurhinodelphinids. Bulletin de l'Institut Royal des Sciences Naturelles de Belgique: Sciences de la Terre 75:211–235.

Lambert, O. 2008a. A new porpoise (Cetacea, Odontoceti, Phocoenidae) from the Pliocene of the North Sea. Journal of Vertebrate Paleontology 28:863–872.

Lambert, O. 2008b. Sperm whales from the Miocene of the North Sea: a re-appraisal. Bulletin de l'Institut Royal des Sciences Naturelles de Belgique: Sciences de la Terre 78:277–216.

Lambert, O., G. Bianucci, K. Post, C. de Muizon, R. Salas-Gismondi, M. Urbina, and J. Reumer. 2010. The giant bite of a new raptorial sperm whale from the Miocene epoch of Peru. Nature 466:105–108.

Lambert, O., G. Bianucci, and M. Urbina. 2014. *Huaridelphis raimondii*, a new early Miocene Squalodelphinidae (Cetacea, Odontoceti) from the Chilcatay Formation, Peru. Journal of Vertebrate Paleontology 34:987–1004.

Lambert, O., V. de Buffrénil, and C. de Muizon. 2011. Rostral densification in beaked whales: diverse processes for a similar pattern. Comptes Rendus Palevol 10:453–468.

Lambert, O., and C. de Muizon. 2013. A new long-snouted species of the Miocene pontoporiid dolphin *Brachydelphis* and a review of the Mio-Pliocene marine mammal levels in the Sacaco Basin, Peru. Journal of Vertebrate Paleontology 33:709–721.

Lambert, O., C. de Muizon, and G. Bianucci. 2013. The most basal beaked whale *Ninoziphius platyrostris* Muizon, 1983: clues on the evolutionary history of the family

Ziphiidae (Cetacea: Odontoceti). Zoological Journal of the Linnean Society 167:569–598.

Lambert, O., C. de Muizon, and G. Bianucci. 2015. A new archaic homodont toothed cetacean (Mammalia, Cetacea, Odontoceti) from the early Miocene of Peru. Geodiversitas 37:79–108.

Lambert, O., M. Estevens, and R. Smith. 2005. A new kentriodontine dolphin from the middle Miocene of Portugal. Acata Palaeontologica Polonica 50:239–248.

Lambert, O., and S. Louwye. 2006. *Archaeoziphius microglenoideus*, a new primitive beaked whale (Mammalia, Cetacea, odontoceti) from the Middle Miocene of Belgium. Journal of Vertebrate Paleontology 26:182–191.

Lanfear, R., B. Calcott, S. Y. W. Ho, and S. Guindon. 2012. PartitionFinder: combined selection of partitioning schemes and substitution models for phylogenetic analyses. Molecular Biology and Evolution 29:1695–1701.

Lee, Y.-N., H. Ichishima, and D. K. Choi. 2012. First record of a platanistoid cetacean from the Middle Miocene of South Korea. Journal of Vertebrate Paleontology 32:231–234.

Loch, C., J. A. Kieser, and R. E. Fordyce. 2015. Enamel ultrastructure in fossil cetaceans (Cetacea: Archaeoceti and Odontoceti). PLoS One 10:e0116557.

Lydekker, R. 1894. Cetacean skulls from Patagonia. Annales del Museo de la Plata 2:1–13.

Madar, S. I. 2007. The postcranial skeleton of Early Eocene pakicetid cetaceans. Journal of Paleontology 81:176–200.

Madar, S. I., J. G. M. Thewissen, and S. T. Hussain. 2002. Additional holotype remains of *Ambulocetus natans* (Cetacea, Ambulocetidae), and their implications for locomotion in early whales. Journal of Vertebrate Paleontology 22:405–422.

Martínez Cáceres, M., and C. de Muizon. 2011. A new basilosaurid (Cetacea, Pelagiceti) from the Late Eocene to Early Oligocene Otuma Formation of Peru. Comptes Rendus Palevol 10:517–526.

Martínez Cáceres, M., C. de Muizon, O. Lambert, G. Bianucci, R. Salas-Gismondi, and M.

Urbina Schmidt. 2011. A toothed mysticete from the Middle Eocene to Lower Oligocene of the Pisco Basin, Peru: new data on the origin and feeding evolution of Mysticeti. Sixth Triennial Conference on Secondary Adaptation of Tetrapods to Life in Water, San Diego, 56–57.

Marx, F. G. 2011. The more the merrier? A large cladistic analysis of mysticetes, and comments on the transition from teeth to baleen. Journal of Mammalian Evolution 18:77–100.

Marx, F. G., and R. E. Fordyce. 2015. Baleen boom and bust: a synthesis of mysticete phylogeny, diversity and disparity. Royal Society Open Science 2:140434.

Marx, F. G., Tsai, C.-H., and Fordyce R. E. 2015. A new Early Oligocene toothed "baleen" whale (Mysticeti: Aetiocetidae) from western North America – one of the oldest and the smallest. Royal Society Open Science 2:150476.

McGowen, M. R., M. Spaulding, and J. Gatesy. 2009. Divergence date estimation and a comprehensive molecular tree of extant cetaceans. Molecular Phylogenetics and Evolution 53:891–906.

Mchedlidze, G. A. 1976. Osnovnye Cherty Paleobiologischeskoi Istorii Kitoobraznykh. Metsnierba Publishers, Tbilisi, Georgia.

Mead, J. G., and E. D. Mitchell. 1984. Atlantic gray whales; pp. 33–53 in M. L. Jones, S. L. Swartz, and S. Leatherwood (eds.); The Gray Whale, *Eschrichtius robustus*. Academic Press, Orlando, FL.

Milinkovitch, M. C., G. Orti, and A. Meyer. 1993. Revised phylogeny of whales suggested by mitochondrial ribosomal DNA sequences. Nature 361:346–348.

Mitchell, E. D. 1989. A new cetacean from the late Eocene La Meseta Formation, Seymour Island, Antarctic Peninsula. Canadian Journal of Fisheries and Aquatic Science 46:2219–2235.

Montgelard, C., F. M. Catzeflis, and E. Douzery. 1997. Phylogenetic relationships of artiodactyls and cetaceans as deduced from the comparison of cytochrome b and 12S rRNA mitochondrial sequences. Molecular Biology and Evolution 14:550–559.

Murakami, M., C. Shimada, Y. Hikida, and H. Hirano. 2012a. A new basal porpoise, *Pterophocaena nishinoi* (Cetacea, Odontoceti, Delphinoidea), from the upper Miocene of Japan and its phylogenetic relationships. Journal of Vertebrate Paleontology 32:1157–1171.

Murakami, M., C. Shimada, Y. Hikida, and H. Hirano. 2012b. Two new extinct basal phocoenids (Cetacea, Odontoceti, Delphinoidea), from the upper Miocene Koetoi Formation of Japan and their phylogenetic significance. Journal of Vertebrate Paleontology 32:1172–1185.

Murakami, M., C. Shimada, Y. Hikida, Y. Soeda, and H. Hirano. 2014. *Eodelphis kabatensis*, a new name for the oldest true dolphin *Stenella kabatensis* Horikawa, 1977 (Cetacea, Odontoceti, Delphinidae), from the upper Miocene of Japan, and the phylogeny and paleobiogeography of Delphinoidea. Journal of Vertebrate Paleontology 34:491–511.

Nikaido, M., F. Matsuno, H. Hamilton, R. L. Brownell, Y. Cao, W. Ding, Z. Zuoyan, A. M. Shedlock, R. E. Fordyce, M. Hasegawa, and N. Okada. 2001. Retroposon analysis of major cetacean lineages: the monophyly of toothed whales and the paraphyly of river dolphins. Proceedings of the National Academy of Sciences 98:7384–7389.

Nummela, S., S. T. Hussain, and J. G. M. Thewissen. 2006. Cranial anatomy of Pakicetidae (Cetacea, Mammalia). Journal of Vertebrate Paleontology 26:746–759.

O'Leary, M. A., and J. Gatesy. 2008. Impact of increased character sampling on the phylogeny of Cetartiodactyla (Mammalia): combined analysis including fossils. Cladistics 24:397–442.

O'Leary, M. A., and M. D. Uhen. 1999. The time of origin of whales and the role of behavioral changes in the terrestrial-aquatic transition. Paleobiology 25:534–556.

Oishi, M., and Y. Hasegawa. 1995. A list of fossil cetaceans in Japan. The Island Arc 3:493–505.

Okazaki, Y. 1987. Additional materials of *Metasqualodon symmetricus* (Cetacea: Mammalia) from the Oligocene Ashiya Group,

Japan. Bulletin of the Kitakyushu Museum of Natural History 7:133–138.

Okazaki, Y. 2012. A new mysticete from the upper Oligocene Ashiya Group, Kyushu, Japan and its significance to mysticete evolution. Bulletin of the Kitakyushu Museum of Natural History and Human History, Series A (Natural History) 10:129–152.

Pilleri, G. 1985. The Miocene Cetacea of the Belluno sandstones (Eastern Southern Alps). Memorie di Scienze Geologische 37:1–250.

Pivorunas, A. 1979. The feeding mechanisms of baleen whales. American Scientist 67:432–440.

Pledge, N. S. 2005. A new species of early Oligocene Cetacean from Port Willunga, South Australia. Memoirs of the Queensland Museum 51:121–133.

Post, K., and E. J. O. Kompanje. 2010. A new dolphin (Cetacea, Delphinidae) from the Plio-Pleistocene of the North Sea. Deinsea 14:1–12.

Pyenson, N. D., C. S. Gutstein, M. A. Cozzuol, J. Velez-Juarbe, and M. Suárez. 2013. New material of *Odobenocetops* form the late Miocene of Chile clarifies the systematics and paleobiology of walrus-convergent odontocetes. Journal of Vertebrate Paleontology: Program and Abstracts, 195.

Pyenson, N. D., J. Vélez-Juarbe, C. S. Gutstein, H. Little, D. Vigil, and A. O'Dea. 2015. *Isthminia panamensis*, a new fossil inioid (Mammalia, Cetacea) from the Chagres Formation of Panama and the evolution of "river dolphins" in the Americas. PeerJ 3:e1227.

Racicot, R. A., T. A. Deméré, B. L. Beatty, and R. W. Boessenecker. 2014. Unique feeding morphology in a new prognathous extinct porpoise from the Pliocene of California. Current Biology 24:774–779.

Reguero, M. A., S. Marenssi, A., and S. N. Santillana. 2012. Weddellian marine/coastal vertebrates diversity from a basal horizon (Ypresian, Eocene) of the Cucullaea I Allomember, La Meseta formation, Seymour (Marambio) Island, Antarctica. Revista Peruana de Biología 19:275–284.

Rivin, M. A. 2010. Early Miocene cetacean diversity in the Vaqueros Formation, Laguna

Canyon, Orange County, California. MSc Thesis, California State University, Fullerton.

Rothausen, K. 1968. Die systematische Stellung der europäischen Squalodontidae (Odontoceti, Mamm.). Paläontologische Zeitschrift 42:83–104.

Sanders, A. E., and L. G. Barnes. 2002. Paleontology of the Late Oligocene Ashley and Chandler Bridge Formations of South Carolina, 3: Eomysticetidae, a new family of primitive mysticetes (Mammalia: Cetacea). Smithsonian Contributions to Paleobiology 93:313–356.

Sanders, A. E., and J. H. Geisler. 2015. A new basal odontocete from the upper Rupelian of South Carolina, U.S.A., with contributions to the systematics of *Xenorophus* and *Mirocetus* (Mammalia, Cetacea). Journal of Vertebrate Paleontology 35:e890107.

Sawamura, H., M. Kimura, L. G. Barnes, and H. Furusawa. 1996. Late Oligocene Cetacea from Ashoro-cho, Hokkaido, Japan; The fauna and its ecologic imiplications. The Paleontological Society Special Publication 8:342.

Scheinin, A. P., D. Kerem, C. D. MacLeod, M. Gazo, C. A. Chicote, and M. Castellote. 2011. Gray whale (*Eschrichtius robustus*) in the Mediterranean Sea: anomalous event or early sign of climate-driven distribution change? Marine Biodiversity Records 4: doi:10.1017/S1755267211000042.

Sears, R., and W. F. Perrin. 2009. Blue Whale *Balaenoptera musculus*; pp. in W. F. Perrin, B. Würsig, and J. G. M. Thewissen (eds.), Encyclopedia of Marine Mammals. Academic Press, Burlington, MA.

Sekiguchi, K., P. B. Best, and B. Z. Kaczmaruk. 1992. New information on the feeding habits and baleen morphology of the pygmy right whale *Caperea marginata*. Marine Mammal Science 8:288–293.

Simpson, G. G. 1945. The principles of classification and a classification of mammals. Bulletin of the American Museum of Natural History 85:1–350.

Spaulding, M., M. A. O'Leary, and J. Gatesy. 2009. Relationships of Cetacea (Artiodactyla) among mammals: increased taxon sampling alters interpretations of key fossils and character evolution. PLoS One 4:e7062.

Steeman, M. E. 2007. Cladistic analysis and a revised classification of fossil and recent mysticetes. Zoological Journal of the Linnean Society 150:875–894.

Steeman, M. E., M. B. Hebsgaard, R. E. Fordyce, S. Y. W. Ho, D. L. Rabosky, R. Nielsen, C. Rhabek, H. Glenner, M. V. Sørensen, and E. Willerslev. 2009. Radiation of extant cetaceans driven by restructuring of the oceans. Systematic Biology 58:573–585.

Tanaka, Y., and R. E. Fordyce. 2014. Fossil dolphin *Otekaikea marplesi* (Latest Oligocene, New Zealand) expands the morphological and taxonomic diversity of Oligocene cetaceans. PLoS One 9:e107972.

Tanaka, Y., and R. E. Fordyce. 2015. Historically significant late Oligocene dolphin *Microcetus hectori* Benham 1935: a new species of *Waipatia* (Platanistoidea). Journal of the Royal Society of New Zealand 45:135–150.

Tarasenko, K. K., and A. V. Lopatin. 2012. New baleen whale genera (Cetacea, Mammalia) from the Miocene of the northern Caucasus and Ciscaucasia: 2. *Vampalus* gen. nov. from the Middle–Late Miocene of Chechnya and Krasnodar Region. Paleontological Journal 46:620–629.

Thewissen, J. G. M., and S. Bajpai. 2001. Dental morphology of Remingtonocetidae (Cetacea, Mammalia). Journal of Paleontology 75:463–465.

Thewissen, J. G. M., and S. Bajpai. 2009. New skeletal material of *Andrewsiphius* and *Kutchicetus*, two Eocene cetaceans from India. Journal of Paleontology 83:635–663.

Thewissen, J. G. M., S. T. Hussain, and M. Arif. 1994. Fossil evidence for the origin of aquatic locomotion in archaeocete whales. Science 263:210–212.

Thewissen, J. G. M., S. I. Madar, and S. T. Hussain. 1996. *Ambulocetus natans*, an Eocene cetacean (Mammalia) from Pakistan. Courier Forschungsinstitut Senckenberg 191:1–86.

Thewissen, J. G. M., E. M. Williams, L. J. Roe, and S. T. Hussain. 2001. Skeletons of terrestrial cetaceans and the relationship of whales to artiodactyls. Nature 413:277–281.

Tsai, C.-H., and T. Ando. 2015. Niche partitioning in Oligocene toothed mysticetes

(Mysticeti: Aetiocetidae). Journal of Mammalian Evolution 22:1–9.

Turvey, S. T., R. L. Pitman, B. L. Taylor, J. Barlow, T. Akamatsu, L. A. Barrett, X. Zhao, R. R. Reeves, B. S. Stewart, K. Wang, Z. Wei, X. Zhang, L. T. Pusser, M. Richlen, J. R. Brandon, and D. Wang. 2007. First human-caused extinction of a cetacean species? Biology Letters 3:537–540.

Uhen, M. D. 1998. Middle to Late Eocene basilosaurines and dorudontines; pp. 29–61 in J. G. M. Thewissen (ed.), The Emergence of Whales. Plenum Press, New York.

Uhen, M. D. 2004. Form, function, and anatomy of *Dorudon atrox* (Mammalia: Cetacea): An archaeocete from the Middle to Late Eocene of Egypt. University of Michigan Papers on Paleontology 34:1–222.

Uhen, M. D. 2007. Evolution of marine mammals: back to the sea after 300 million years. The Anatomical Record 290:514–522.

Uhen, M. D. 2008a. New protocetid whales from Alabama and Mississippi, and a new cetacean clade, Pelagiceti. Journal of Vertebrate Paleontology 28:589–593.

Uhen, M. D. 2008b. A new *Xenorophus*-like odontocete cetacean from the Oligocene of North Carolina and a discussion of the basal odontocete radiation. Journal of Systematic Palaeontology 6:433–452.

Uhen, M. D. 2010. The origin(s) of whales. Annual Review of Earth and Planetary Sciences 38:189–219.

Uhen, M. D. 2014. New material of *Natchitochia jonesi* and a comparison of the innominata and locomotor capabilities of Protocetidae. Marine Mammal Science 30:1029–1066.

Uhen, M. D. 2015. Cetacea: Paleobiology Database Data Archive 9. https://paleobiodb.org

Uhen, M. D., N. D. Pyenson, T. J. Devries, M. Urbina, and P. R. Renne. 2011. New Middle Eocene whales from the Pisco Basin of Peru. Journal of Paleontology 85:955–969.

Van Beneden, P.-J., and P. Gervais. 1880. Ostéographie des cétacés vivants et fossiles. Bertrand, Paris.

Van Valen, L. M. 1966. Deltatheridia, a new order of mammals. Bulletin of the American Museum of Natural History 132:1–126.

Vélez-Juarbe, J., and N. D. Pyenson. 2012. *Bohaskaia monodontoides*, a new monodontid (Cetacea, Odontoceti, Delphinoidea) from the Pliocene of the western North Atlantic Ocean. Journal of Vertebrate Paleontology 32:476–484.

Vélez-Juarbe, J., A. R. Wood, C. De Gracia, and A. J. W. Hendy. 2015. Evolutionary patterns among living and fossil kogiid sperm whales: evidence from the Neogene of Central America. PLoS One 10:e0123909.

Vilstrup, J. T., S. Y. Ho, A. D. Foote, P. A. Morin, D. Kreb, M. Krutzen, G. J. Parra, K. M. Robertson, R. de Stephanis, P. Verborgh, E. Willerslev, L. Orlando, and M. T. Gilbert. 2011. Mitogenomic phylogenetic analyses of the Delphinidae with an emphasis on the Globicephalinae. BMC Evolutionary Biology 11:65.

Vogt, L. 2009. The future role of bio-ontologies for developing a general data standard in biology: chance and challenge for zoo-morphology. Zoomorphology 128:201–217.

Waddell, P. J., N. Okada, and M. Hasegawa. 1999. Towards resolving the interordinal relationships of placental mammals. Systematic Biology 48:1–5.

Werth, A. J. 2000. Feeding in marine mammals; pp. 487–526 in K. Schwenk (ed.), Feeding: Form, Function and Evolution in Tetrapods. Academic Press, San Diego.

Werth, A. J. 2004. Functional morphology of the sperm whale (*Physeter macrocephalus*) tongue, with reference to suction feeding. Aquatic Mammals 30:405–418.

Whitmore, F. C., Jr., and L. G. Barnes. 2008. The Herpetocetinae, a new subfamily of extinct baleen whales (Mammalia, Cetacea, Cetotheriidae). Virginia Museum of Natural History Special Publication 14:141–180.

Whitmore, F. C., Jr., and J. A. Kaltenbach. 2008. Neogene Cetacea of the Lee Creek Phosphate Mine, North Carolina. Virginia Museum of Natural History Special Publication 14:181–269.

Xiong, Y., M. Brandley, S. Xu, K. Zhou, and G. Yang. 2009. Seven new dolphin mitochondrial genomes and a time-calibrated phylogeny of whales. BMC Evolutionary Biology 9:20.

Yang, X.-G. 2009. Bayesian inference of cetacean phylogeny based on mitochondrial genomes. Biologia 64:811–818.

5 Major Steps in the Evolution of Cetaceans

5.1 From land to sea: the last steps

5.1.1 Initial forays into the water

The fossil record of early cetaceans provides a remarkably complete picture of how the terrestrial ancestors of modern whales turned into the highly specialized marine mammals we know today (Figure 5.1). The origin of cetaceans was originally thought to be defined by their adoption of an aquatic lifestyle, and the morphological and physiological changes that such a step necessitates. However, this idea was challenged twice: first, by the discovery that the postcranial morphology of the most archaic whales, the **pakicetids**, looks fundamentally terrestrial (Madar, 2007; Thewissen *et al.*, 2001); and, second, by the realization that at least some of the closest relatives of cetaceans, the **raoellids**, already spent much of their time in rivers or lakes—possibly to escape predation (Thewissen *et al.*, 2007). The emergence of unambiguous aquatic adaptations thus only occurred some time after the ancestors of cetaceans had first stepped into the water (Clementz *et al.*, 2006; Thewissen *et al.*, 2007).

Terrestrial artiodactyls are **cursorial**, or specialized runners (Rose, 1985). Evidence of such a lifestyle persists in raoellids and pakicetids, both of which retain a suite of features typical of running animals, including: (1) long and slender, **digitigrade** (digit-supported) fore- and hind limbs; (2) an elongate tibia; (3) an elbow that is specialized for extensive flexion and prevents supination; (4) revolute zygapophyses in the lumbar vertebrae, which largely prevent spinal torsion; and (5) an astragalus that is shaped like a double pulley (Figure 5.2) (Cooper *et al.*, 2012; Madar, 2007). Based on these features, pakicetids were originally considered to have been poor swimmers, comparable to extant tapirs (Thewissen *et al.*, 2001). However, besides their superficially cursorial looks, pakicetids are also characterized by elongate digits, strong epaxial muscles, and a robust tail, all of which are more typical of later archaeocetes than of archaic artiodactyls (Madar, 2007). The ilium is relatively short, as in many specialized aquatic mammals (Gingerich, 2003), and the phalanges bear muscular crests that once provided support for strong digital abductors. The latter may be indicative of webbed feet, which would have assisted in swimming and/or facilitated walking on wet substrates (Gingerich *et al.*, 2001; Madar, 2007) (Plate 1A). In addition, the presence of numerous foramina near the tip of the rostrum suggests that pakicetids may have possessed well-developed **vibrissae** (whiskers), similar to those of other marine mammals (Thewissen and Nummela, 2008).

Raoellids lack any externally visible aquatic adaptations, but at least one of them (*Indohyus*) shares with pakicetids the presence of pronounced **osteosclerosis** (increased

Cetacean Paleobiology, First Edition. Felix G. Marx, Olivier Lambert, and Mark D. Uhen.
© 2016 John Wiley & Sons, Ltd. Published 2016 by John Wiley & Sons, Ltd.

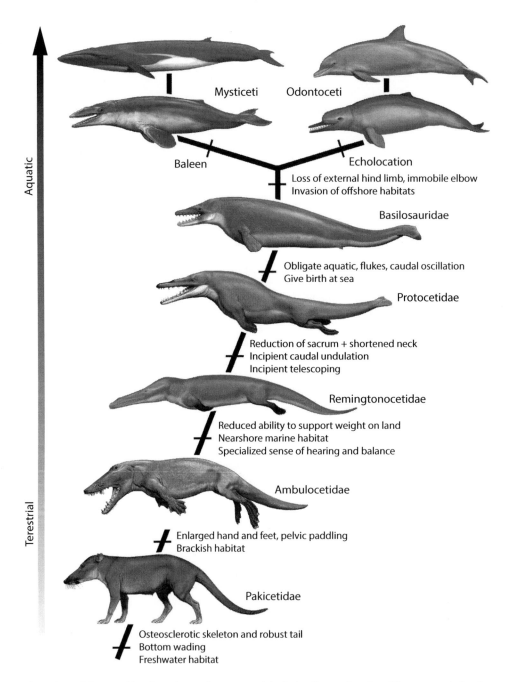

Figure 5.1 Overview of the transition from the earliest terrestrial whales (bottom) to the obligate aquatic (top) taxa of today. Annotations show major changes in morphology, habitat, and lifestyle. Life reconstructions © C. Buell.

Figure 5.2 Cursorial and aquatic (in boldface) adaptations of *Pakicetus* (a). Below are alternative reconstructions depicting (b) the raoellid artiodactyl *Indohyus* and (c) *Pakicetus* as either more terrestrially (top) or more aquatically adapted. Drawing of *Pakicetus* skeleton reproduced from Thewissen *et al.* (2001), with permission of Macmillan Publishers. Life reconstructions © C. Buell.

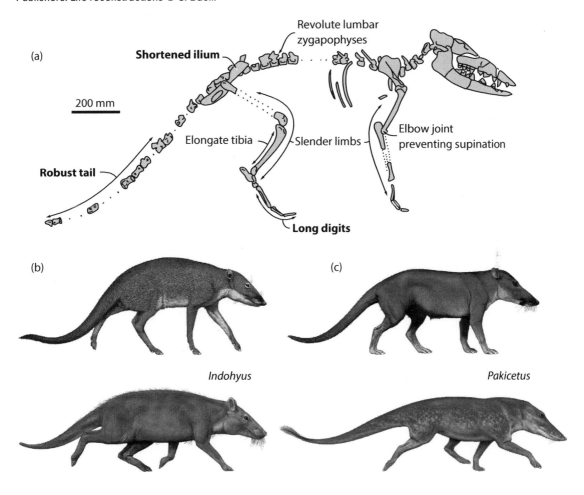

(a)

Revolute lumbar zygapophyses

Shortened ilium

200 mm

Elongate tibia — Slender limbs — Elbow joint preventing supination

Robust tail

Long digits

(b)

Indohyus

(c)

Pakicetus

bone density) (Cooper *et al.*, 2012). Although both certainly were able to move on land, the sheer weight of their skeleton likely prevented sustained running, and would have made terrestrial locomotion in general energetically expensive. In addition, hypermineralization of the bones means that they are brittle and thus relatively easy to break (Gray *et al.*, 2007; Madar, 2007; Thewissen *et al.*, 2007). Together, these features suggest a largely aquatic mode of locomotion, such as bottom walking on river beds (Cooper *et al.*, 2012; Madar, 2007), during which the dense skeleton would have served as ballast (Wall, 1983). In addition, pakicetids may

have derived a limited amount of propulsion from their well-developed tail, while at the same time paddling through the water with their webbed feet (Madar, 2007).

Direct evidence for the transition from land to water comes from the analysis of carbon and oxygen **stable isotope ratios** (section 1.4.3). *Indohyus* and pakicetids are characterized by relatively stable, low oxygen isotope values, which are typical of freshwater environments (Figure 5.3) (Clementz *et al.*, 2006; Thewissen *et al.*, 2007). This is in agreement with the terrestrial nature of the deposits—the Kuldana and Subathu formations

Figure 5.3 Bivariate plot of oxygen (δ¹⁸O) and carbon (δ¹³C) stable isotope ratios of selected Eocene and Oligocene cetaceans, with ranges indicating particular environmental settings labeled on the bottom and to the right of the graph (see section 1.4.3 for a detailed explanation of stable isotope values). Note that the oxygen values of the Eocene taxa (Pakicetidae, *Ambulocetus*, *Dalanistes*, Protocetidae, and Basilosauridae) are enriched by circa 1‰ relative to the younger Eocene taxa, owing to intense evaporation in the tropical Tethys Sea (Clementz *et al.*, 2006). Data are from Clementz *et al.* (2006, 2014) and Thewissen *et al.* (2007, 2011b). Life reconstructions © C. Buell.

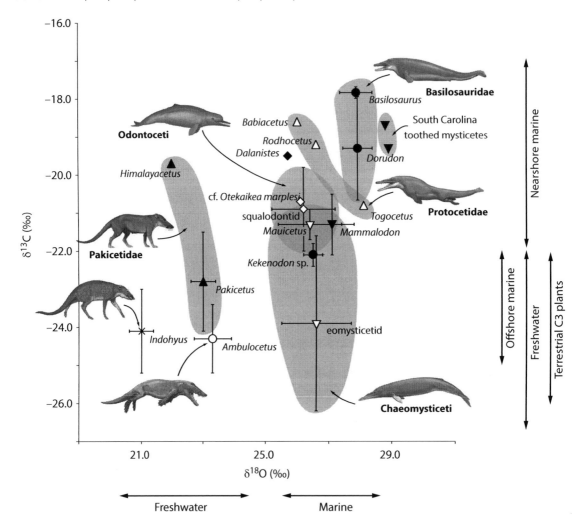

of northern India and Pakistan—in which these taxa are typically found. Interestingly, the oxygen isotope values for *Khirtharia*, a close relative of *Indohyus*, are much more variable, implying that not all raoellids regularly spent much time in the water (Thewissen *et al.*, 2007). The carbon isotope values of raoellids and pakicetids are also generally low, and indicative of either a terrestrial (raoellids) or a freshwater (pakicetids) diet (Thewissen *et al.*, 2011b). The only exception to this pattern is the oldest known cetacean, *Himalayacetus*, which was recovered from marine strata and presents an unusual combination of high carbon and low oxygen isotope values (Bajpai and Gingerich, 1998; Clementz *et al.*, 2006). Foraging along the coast could explain the high carbon values, which makes *Himalayacetus* the oldest known cetacean that ventured out to sea. Nevertheless, it still had to return to land or a river in order to drink (Clementz *et al.*, 2006).

The conservative morphology of pakicetids and their ecological resemblance to raoellids raise an important question: what factors eventually set early cetaceans on a path to becoming fully aquatic? The answer to this may lie in the morphology of the skull and teeth. Pakicetids differ from raoellids and other artiodactyls in having closely spaced eyes and a narrow, elongated temporal region for the attachment of the jaw muscles. In addition, molar crushing basins, which are large in raoellids and other artiodactyls, are greatly reduced or absent in cetaceans, thus hinting at a drastic change in diet (likely to fish; section 6.1.1). Together, these developments describe a major reorganization of the skull architecture, which must have affected the workings of most sense organs and the way early whales caught and processed food. In terms of cetacean origins, specializing on aquatic food sources may thus have been as important as the ability to move and remain in water for long periods of time (Thewissen *et al.*, 2007).

The emergence of pakicetids was followed by that of the larger **ambulocetids**, such as *Ambulocetus*, around 49 Ma. Like pakicetids, ambulocetids foraged in an environment dominated by freshwater (Figure 5.3). Unlike them, however, ambulocetids occurred in coastal deposits, and thus possibly lived in or near river mouths (Roe *et al.*, 1998; Thewissen *et al.*, 1994).

A well-developed sacrum and robust hind limbs demonstrate that *Ambulocetus* was able to move on land (Madar *et al.*, 2002). Terrestrial locomotion would furthermore have been supported by the presence of distinct, load-bearing columns of spongy (**trabecular**) bone in the limb elements (Madar, 1998). Compared to most mammals, however, the stance of *Ambulocetus* was sprawling, likely **plantigrade** (with the entire hand or foot flat on the ground) at least in the hind limbs, and relatively unsuited to prolonged or fast terrestrial locomotion (Thewissen *et al.*, 1994). This is in agreement with the persistence of osteosclerosis, which indicates that ambulocetids spent much of their time in water (Gray *et al.*, 2007; Madar, 1998).

Unlike those of pakicetids, the aquatic adaptations of ambulocetids are easy to see (Figure 5.4). Besides a robust tail, they include a reduced femur, greatly enlarged hands and feet, well-developed axial muscles, and an elongated pre-sacral spine (Madar *et al.*, 2002) (Plate 2), although the last of these features is equivocal (Bebej *et al.*, 2012a). Ambulocetids thus represent the first step on a path that transformed the still largely terrestrial locomotory pattern of pakicetids into the highly specialized swimming style of modern cetaceans. Based on comparisons with other aquatic and semiaquatic mammals, this path comprised up to five stages (Figure 5.5): (1) **quadrupedal paddling**: alternating strokes of the four limbs; this swimming style is typical of terrestrial mammals and, alongside bottom walking, likely was used by the earliest cetaceans; (2) **pelvic paddling**: alternate or simultaneous strokes of the enlarged hind limbs

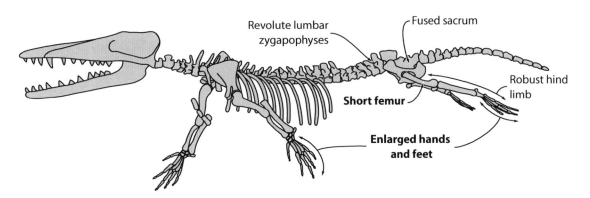

Figure 5.4 Terrestrial and aquatic (in boldface) locomotory adaptations of *Ambulocetus*.

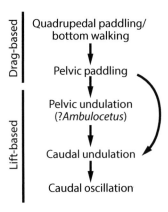

Figure 5.5 Hypothetical sequence of swimming styles adopted by early cetaceans during their transition from land to water (Fish, 1996; Thewissen and Fish, 1997).

only; (3) **pelvic undulation**: wave-like motions of the vertebral column swinging the enlarged (usually webbed) feet, which act as a **hydrofoil**; (4) **caudal undulation**, in which wave-like motions of the lumbar and caudal portions of the vertebral column swing a hydrofoil located at the end of the tail; the waves traveling along the vertebral column are sinusoidal, with different parts of the body corresponding to different phases of the sinusoid; and (5) **caudal oscillation**, which resembles caudal undulation, except that in oscillation the entire body of the swimmer remains in the same phase (Fish, 1996; Thewissen and Fish, 1997). Quadrupedal and pelvic paddling rely on **drag** forces and are powered by the limb muscles, whereas all of the undulatory swimming styles work by generating **lift** and mainly involve the axial musculature (Thewissen and Fish, 1997).

Within this model of locomotor evolution, the style employed by ambulocetids remains contentious. Undulations of the robust tail probably provided some lift, but were not the primary means of generating propulsion (Madar *et al.*, 2002; Thewissen and Fish, 1997). Instead, it seems that *Ambulocetus* mostly relied on pelvic paddling with its enlarged feet, which resemble those of river otters in their general outline and proportions (Thewissen and Fish, 1997). Alternatively, the main propulsive force may have come from pelvic undulation, with the enlarged feet acting as hydrofoils (Madar *et al.*, 2002; Thewissen *et al.*, 1994). The

argument about the locomotory style largely depends on the flexibility of the vertebral column. If reconstructed correctly, the elongated presacral spine of *Ambulocetus* would theoretically allow more dorsoventral movement, which would in turn be consistent with pelvic undulation (Madar *et al.*, 2002). Against this stands the observation that the lumbar vertebrae are firmly held in place by revolute zygapophyses and anteriorly pointing neural spines, which points to a relatively stiff back supporting foot-powered locomotion (Bebej *et al.*, 2012a). Likewise, well-developed epaxial musculature may have served to extend the column, but could equally well have helped to stabilize it, as in modern cetaceans (Bebej *et al.*, 2012a).

5.1.2 Transition to marine environments

Apart from pakicetids and ambulocetids, virtually all archaeocetes are characterized by relatively high carbon (δ^{13}C) and oxygen (δ^{18}O) isotope values (Figure 5.3) (section 1.4.3). Together with sedimentological evidence, this indicates a definite shift not only to marine habitats but also to nearshore marine food sources maybe as early as 48 Ma (Clementz *et al.*, 2006). The oldest of these seagoing cetaceans include the **remingtonocetids** and the **protocetids**. Protocetids are a diverse and morphologically disparate group which primarily lived in relatively clear and open waters. By contrast, at least some remingtonocetids seem to have kept somewhat closer to the coast, where they inhabited marshes and swamps (Thewissen and Bajpai, 2009).

The move into marine waters posed a particular challenge for these early cetaceans, which now had to cope with an excessive environmental salt load. Like other mammals, cetaceans can only tolerate minor variations in the electrolyte (ion) concentration and volume of their body fluids. Because their internal electrolyte concentration is much lower than (i.e., **hypotonic** relative to) that of the surrounding seawater, they actively need to regulate water and electrolyte intake and loss in order to preserve the status quo (**homeostasis**). This essential balance between what goes in and out of the body is called **osmoregulation** (Costa, 2009). To maintain homeostasis, modern cetaceans mostly rely on their intake of dietary and metabolic water, as well as minimizing the amount of water lost through respiration and lactation. Ingestion of seawater, although seemingly not uncommon, apparently plays no role in replacing lost water. It may, however, facilitate the excretion of the large amounts of urea that result from the protein-rich, fish- and invertebrate-based cetacean diet (Costa, 2009; Hui, 1981).

The fossil record reveals little about the osmoregulatory strategies of early cetaceans, but it does provide evidence about the timing of their emergence (Thewissen *et al.*, 1996). Thus, the remingtonocetid *Attockicetus* displays a carbon isotope value indicative of nearshore marine foraging, but an oxygen isotope value more closely resembling that of pakicetids and ambulocetids. This indicates that at least some remingtonocetids hunted in the ocean, but had to spend significant amounts of time in freshwater—presumably, to deal with the excess salt load of their diet or ingested seawater (Roe *et al.*, 1998). The oxygen isotope values of later remingtonocetids and protocetids are closer to those of the fully marine basilosaurids and neocetes, but still somewhat intermediate (Figure 5.3) (Clementz *et al.*, 2006). Fully fledged marine osmoregulation thus slightly lagged behind the adoption of a marine diet. Nevertheless, it seems that even a semideveloped ability to osmoregulate in seawater may have been enough to enable these archaeocetes to disperse through the Tethys to modern-day Egypt and, ultimately, across the Atlantic Ocean (section 7.6.1) (Bebej *et al.*, 2012b; Clementz *et al.*, 2006).

Like that of ambulocetids, the postcranial morphology of remingtonocetids and protocetids still shows clear signs of their terrestrial heritage. Both have well-developed hind limbs and pelves fused to the sacral region of the vertebral column (Bebej *et al.*, 2012a; Uhen, 2014) (Plates 1b and 3a). The neural spines of the anterior thoracics are tall, as in artiodactyls, and likely helped to support the weight of the head outside the water (Gingerich *et al.*, 1994). Furthermore, the presence of trabecular columns indicates that the limb bones of both groups were load-bearing, and thus involved in locomotion on land—although a shared trend toward high levels of bone compactness in the femur and ribs would have made the latter extremely inefficient (Houssaye *et al.*, 2015; Madar, 1998). There are, however, differences in

degree: whereas remingtonocetids retain the ancestral count of four sacral vertebrae (Gingerich *et al.*, 1995), protocetids gradually reduce the length of the sacral region—from four vertebrae in *Maiacetus* and *Rodhocetus*, to three in *Qaisracetus*, two in *Natchitochia*, and eventually none in *Georgiacetus* and an undescribed protocetid from the Late Eocene of Egypt (Gingerich *et al.*, 2014; Uhen, 2014). Likewise, all sacral vertebrae tend to be wholly or partially fused in remingtonocetids, but to varying degrees are free in protocetids (Gingerich *et al.*, 1994, 2009). Finally, trabecular columns, while present, are more weakly expressed in protocetids than in remingtonocetids (Madar, 1998).

The hind limb of remingtonocetids is poorly known, but the relative length of the femur is shorter than in *Ambulocetus* (Madar *et al.*, 2002). The caudal vertebrae are robust and, at least in some cases, seemingly flattened, which suggests a broad tail involved in creating propulsion (Thewissen and Bajpai, 2009). The presence of a fused sacrum, however, excludes caudal undulation of the entire body as a potential swimming style (Gingerich *et al.*, 1995). The lumbar vertebrae of remingtonocetids are comparatively long, but restricted in their mobility by curved zygapophyses and broad neural spines. This presumably resulted in a relatively stiff back, which was further stabilized by the well-developed epaxial muscles and incapable of marked dorsoventral movements. Like ambulocetids, remingtonocetids thus likely employed pelvic paddling in combination with an undulating tail (Bebej *et al.*, 2012a). Somewhat curiously, the neck of remingtonocetids is comparatively elongate, which at first sight might seem detrimental to the creation of a hydrodynamic body outline and the stabilization of the head during swimming. However, the cervical vertebrae allow only limited neck mobility and provide attachment sites for powerful muscles capable of holding the head firmly in position while underwater (Bebej *et al.*, 2012a).

Compared to ambulocetids and remingtonocetids, protocetids are characterized by a shorter femur and even more enlarged feet, which consequently would have formed efficient paddles or hydrofoils (Figure 5.6) (Gingerich *et al.*, 1994, 2001). The neck of protocetids is considerably

Figure 5.6 Terrestrial and aquatic (in boldface) locomotory adaptations of the protocetid *Maiacetus*. Drawing of skeleton adapted from Gingerich *et al.* (2009), under a Creative Commons Attribution license.

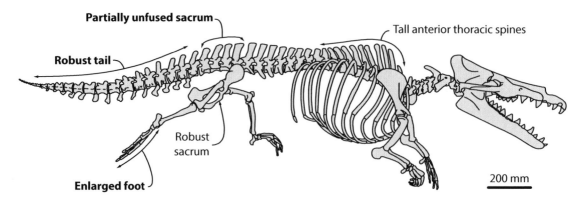

Figure 5.7 Range of relative centrum length (ratio of centrum length to anterior centrum height) of the post-axis cervical vertebra, shown for a range or archaeocetes and some extant neocetes. Data are from Bebej *et al.* (2012a).

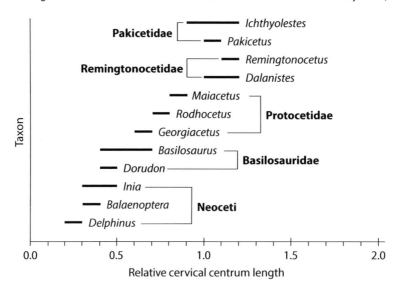

shortened, and hence comparatively immobile, as is typical of aquatic mammals with a streamlined body (Figure 5.7) (Buchholtz, 2001). The lumbar region is relatively flexible and capable of dorsoventral undulation. In some taxa (e.g., *Protocetus* and *Georgiacetus*), the posteriormost thoracics assume the shape of lumbar vertebrae. At the same time, the gradual reduction of the sacrum blurs the boundaries between the lumbar, sacral, and anterior caudal regions. This weakening of traditional vertebral series boundaries results in the creation of two novel functional units: the **chest**, which comprises the anterior thoracics and supports the rib cage; and the **torso**, consisting of the posterior thoracics, the lumbars, and the anterior caudals (Buchholtz, 1998, 2001). The unity of the torso is further enhanced by a gradual widening of the transverse processes of the sacral vertebrae, for the attachment of well-developed axial muscles. Following its emergence within Protocetidae, the functional torso became a key characteristic of all later archaeocetes and modern cetaceans (discussed further in this chapter and in section 8.2).

As a single unit, the torso supports continuous, wave-like dorsoventral movements along the central and posterior spine, which likely enabled protocetids to swim via a combination of pelvic paddling and caudal undulation (Uhen, 2014). At least some of the most crownward protocetids (e.g., *Georgiacetus* and *Natchitochia*) may also have used a form of pelvic undulation, with the axial undulatory wave acting directly on the enlarged feet. This scenario is particularly likely for *Georgiacetus* and the undescribed Late Eocene specimen from Egypt, which are the only protocetids known to have completely lost the bony connection between the pelvis and the vertebral column (Gingerich *et al.*, 2014; Hulbert, 1998). Without this structural support, the hind limb muscles alone would not have been able to power the stroke of the foot (Hulbert *et al.*, 1998). The latter could, however, still have functioned as a hydrofoil, especially given that there is currently no evidence for the presence of tail flukes in any protocetid (Uhen, 2008a, 2014).

Besides osmoregulation and locomotion, the shift to a more permanently aquatic existence also transformed many of the sense organs of the earliest marine cetaceans, including those related to **vision**, **mechanoreception**, **balance**, and **hearing** (Figure 5.8). The most obvious feature related to vision is the size of the eye itself, which can often be gauged from that of the orbit. Cetacean eyes are large in both absolute and, at least in some lineages (e.g., archaic mysticetes), relative terms, which demonstrates the continued importance of vision in water (Marx, 2011; Thewissen and Nummela, 2008). The only exception to this pattern are species inhabiting turbid environments, such as *Remingtonocetus* and the essentially blind South Asian river dolphin, *Platanista* (Bajpai *et al.*, 2011; Reeves and Martin, 2009).

Molecular data indicate that archaeocetes retained the three visual pigments typical of most mammals, expressed in a single type of retinal **rod** and two types of **cone** cells, respectively (Meredith *et al.*, 2013). Like their terrestrial ancestors, early cetaceans were thus likely capable of dichromatic color vision. By contrast, all living cetaceans have convergently lost one or both of the cone-based pigments, and thus have become either cone or rod monochromats (section 6.6). Despite their retention of ancestral visual capabilities, early

cetaceans nevertheless adapted to life in the water through a pronounced **blue shift** in the rod pigment (rod opsin). This condition is advantageous in the open ocean, because the spectrum of visible light shifts toward shorter wavelengths, and thus becomes bluer, as it travels through water (Meredith *et al.*, 2013).

As in *Pakicetus*, the tip of the rostrum of at least some remingtonocetids bears numerous small foramina, which may be indicative of well-developed vibrissae (Thewissen and Bajpai, 2009). No such structures have been identified in protocetids or any more crownward taxa, which suggests that the importance of vibrissae declined as cetaceans moved into clearer, more open waters. Nevertheless, short vibrissae still occur in extant mysticetes and some odontocetes, although most of the latter lose them either before or shortly after birth (Czech-Damal *et al.*, 2012; Slijper, 1962). In the Guiana dolphin, *Sotalia guianensis*, the crypts that originally housed the vibrissae have been transformed into **electroreceptors** capable of detecting the bioelectric fields generated by (benthic) prey. There is currently no evidence for a more widespread occurrence of electroreception among odontocetes, but the presence of well-developed vibrissal crypts in other species (e.g., *Kogia*) suggests that this ability may often simply have gone unnoticed (Czech-Damal *et al.*, 2012).

In terrestrial mammals, the semicircular canals record rotational movements and help to stabilize the head relative to the body via the **vestibulocollic** reflex (Wilson *et al.*, 1995). The size of the canals correlates both with their own sensitivity and the agility of their owner, with large canals being typical of highly maneuverable taxa, such as primates (Spoor, 2003; Spoor *et al.*, 2007). Their three-dimensional habitat makes cetaceans particularly agile, and should thus have led to the evolution of a heightened sense of balance. Yet, surprisingly, the average radius of their semicircular canals is actually very small: in remingtonocetids and virtually all later cetaceans, it is approximately one-third of what might be expected in a mammal of comparable size (Figure 5.9) (Spoor *et al.*, 2002). A similar, albeit less marked, reduction can be seen in the cross section (**lumen**) of the canals, which is even more directly linked to their sensitivity (Spoor and Thewissen, 2008).

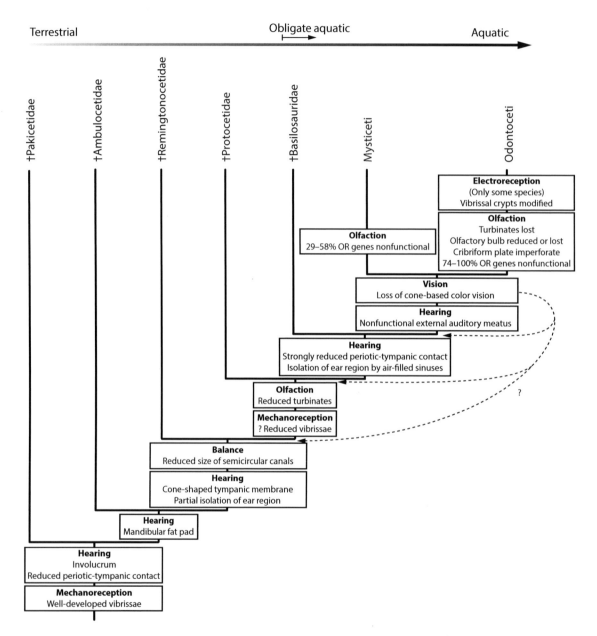

Figure 5.8 Overview of the changes that affected the major sense organs during the terrestrial–aquatic transition of cetaceans. Data are from Berta *et al.* (2014), Nummela *et al.* (2004b), Spoor *et al.* (2002), and Levenson and Dizon (2003). Note that the loss of color vision is based purely on molecular evidence, and could potentially have occurred along several branches leading up to Neoceti, as well as independently within odontocetes and mysticetes.

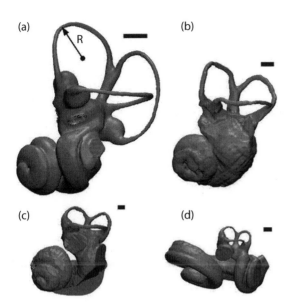

Figure 5.9 Lateral view of the left bony labyrinth of (a) the agile non-cetacean mammal *Galago moholi*, (b) the pakicetid *Ichthyolestes pinfoldi*, (c) the protocetid *Indocetus ramani*, and (d) the extant bottlenose dolphin *Tursiops truncatus*. Scale bars are 1 mm. R denotes the radius of curvature of the anterior semicircular canal. Reproduced from Spoor *et al.* (2002), with permission of Macmillan Publishers.

The reason for this apparent contradiction may lie in the way the canals operate on land. In most mammals, head stabilization via the vestibulo-collic reflex forms part of a feedback loop, in which the stabilization process itself prevents the canals from being constantly overstimulated. In cetaceans, however, immobilization of the neck rendered the vestibulocollic reflex largely nonfunctional, thus breaking the feedback loop on which the semicircular canal system usually depends. The resulting increase in sensory input made it necessary to lower the sensitivity of the canals, and therefore probably led to their reduction. This may also explain why most of the modern odontocetes have somewhat smaller canals than mysticetes, even though they are usually more agile (Spoor *et al.*, 2002; Spoor and Thewissen, 2008). Nevertheless, the cetacean semicircular canal system remains fully functional despite its small size, and thus should not

be interpreted as vestigial (Spoor *et al.*, 2002). For remingtonocetids and protocetids, the reduction of the vestibular apparatus likely made terrestrial locomotion extremely difficult, and marked a profound shift toward an almost entirely aquatic existence.

In terrestrial mammals, airborne sounds are channeled through the external acoustic meatus and picked up by the tympanic membrane. From there, vibrations are passed on to the auditory ossicles and, ultimately, the inner ear. When swimming, however, water building up in the external acoustic meatus creates a pressure differential, which prevents the tympanic membrane from working properly. Instead, sound is conducted to the ear via the tissues of the head (bone conduction), but without the animal being able to gauge its direction. All cetaceans, including pakicetids, have adapted to this situation via the development of a thickened involucrum and the (partial) detachment of the tympanic bulla from the skull (Nummela *et al.*, 2004b). Being loosely suspended from the periotic, the involucrum is able to vibrate independently, and thus provides a crude replacement for the nonfunctional tympanic membrane.

Except for the presence and partial isolation of the involucrum, pakicetids largely retain the ear morphology of terrestrial mammals. By contrast, remingtonocetids, protocetids, and all later cetaceans greatly refine underwater hearing through (1) the further reduction of the contact between the periotic and the tympanic bulla; (2) an increase in the mass of the auditory ossicles; and (3) the development of a mandibular fat pad for the reception of waterborne sounds, as indicated by their enlarged mandibular fossa (Figure 5.10) (Nummela *et al.*, 2004b; Thewissen and Hussain, 1993). The fat pad directly transmits waterborne sounds to the outer lip of the tympanic bulla (also known as the **tympanic plate**), which thus assumes the role of the tympanic membrane as the main sound input area (Nummela *et al.*, 2004b). A fat pad was likely also present in ambulocetids, but it seemingly did not function in the same way as in later archaeocetes (Nummela *et al.*, 2007). Originally, sounds were thought to enter the fat pad mostly through the pan bone (Norris, 1968; Nummela *et al.*, 2004a), but studies of extant species have

Figure 5.10 Schematic representation of the sound transmission mechanism in (a) a generalized terrestrial mammal, (b) a pakicetid, (c) a remingtonocetid/protocetid, and (d) a modern odontocete. Coc, cochlea; Dom, dome-shaped depression for periotic; EAM, external acoustic meatus; FaPa, fat pad; Inc, incus; Inv, involucrum; Mal, malleus; Man, mandible; MeTy, medial synostosis between periotic and tympanic bone (absent in cetaceans); OvW, oval window; Per, periotic bone; PeTy, joint between periotic and tympanic; Sin, air-filled sinuses; Sk, skull; Sta, stapes; TyBo, tympanic bone; TyMe, tympanic membrane; TyPl, tympanic plate. Reproduced from Nummela *et al.* (2004b), with permission of Macmillan Publishers.

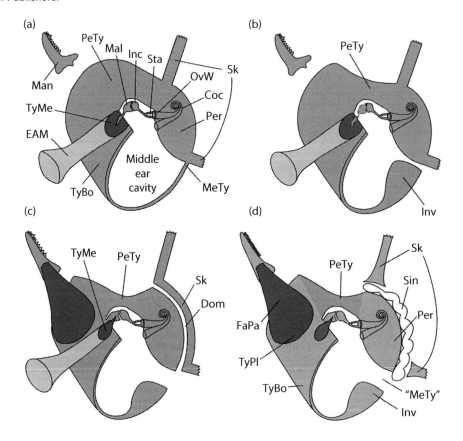

revealed the presence of an alternative, **gular pathway** along the soft tissues of the throat region (Cranford *et al.*, 2008). Pending the publication of further data on this issue, it is thus possible that the exact sound path may differ depending on the species.

From the tympanic plate, vibrations are passed on to the malleus and, ultimately, the fenestra ovalis of the periotic, via the chain of auditory ossicles (Cranford *et al.*, 2010; Hemilä *et al.*, 2010). Based on the shape of its attachment site, the tympanic membrane of remingtonocetids and protocetids probably had a conical shape somewhat reminiscent

of that of modern mysticetes (section 3.4.5) (Nummela *et al.*, 2004b). Airborne sounds could likely still be received via the large external acoustic meatus, but within a lower frequency range than in pakicetids and terrestrial mammals, owing to the heavy auditory ossicles. Despite these modifications, directional hearing underwater was likely still relatively poor in remingtonocetids and many protocetids because of the lingering contact of the tympanoperiotic with the basicranium (Figure 5.10) (Nummela *et al.*, 2004b). In neocetes, the ear bones are isolated from the surrounding skull by the accessory air sinus system (section 3.4.3).

The latter is absent in remingtonocetids and most protocetids, but incipiently developed in some of the more crownward species (Geisler *et al.*, 2005). These earliest traces of the sinus system mark the beginning of a trend, which led to the conspicuous peribullary and pterygoid sinuses of basilosaurids and mysticetes, and culminated in the greatly enlarged sinus system dominating the basicranium of modern odontocetes (Fraser and Purves, 1960).

5.1.3 Divorce from land

The protocetid *Georgiacetus* is the most archaic cetacean to show signs of an obligate aquatic existence, since the detachment of its pelvis from the vertebral column no longer allowed it to support its own weight on land (Hulbert *et al.*, 1998). After their emergence around 40 Ma, the slightly more crownward **basilosaurids** inherited this feature, but not the large hind limbs which still characterized late protocetids. Instead, their legs and feet have shrunk to little more than stumps (Figure 5.11 and Plate 3b) (Gingerich *et al.*, 1990). This reduction is even more pronounced in **neocetes**, which have lost any trace of an external hind limb. Concurrent morphological changes in the forelimb are less drastic, but still involve (1) dorsoventral flattening; (2) greatly reduced mobility of the joints within the hand; (3) a switch from an osteosclerotic to a largely trabecular bone structure in the humerus (except in *Basilosaurus*); and (4) in neocetes, the complete loss of mobility of the elbow joint (section 8.1.1) (Houssaye *et al.*, 2015; Uhen, 2004). The result of these adaptations is the transformation of the

mobile, weight-bearing forelimb of earlier archaeocetes into the flipper of modern cetaceans, which is primarily used for steering, starting and stopping (Benke, 1993). While *Georgiacetus* may still have been able to pull out on to land, the reduction of the hind limb and transformation of the forelimb almost certainly prevented basilosaurids from doing the same (Uhen, 2004). By inference, this means that they had overcome the final hurdle on the road to becoming fully aquatic: unlike protocetids, such as *Maiacetus*, basilosaurids must have given birth at sea (Gingerich *et al.*, 2009).

The obligate aquatic nature of certain protocetids and all basilosaurids is confirmed by their oxygen isotope values, which resemble those of most living cetaceans and point to a fully marine existence (Figure 5.3) (Clementz *et al.*, 2006). They thus represent the final outcome of a land-to-sea transition that took little more than 10 million years, and created the basis for the subsequent radiation of mysticetes and odontocetes. Their aquatic existence allowed basilosaurids to spread around the globe, including the Tethys, Atlantic, and Pacific oceans (Gol'din and Zvonok, 2013; Köhler and Fordyce, 1997; Martínez Cáceres and de Muizon, 2011; Uhen, 2004). Surprisingly, however, high carbon isotope values indicate that basilosaurids did not venture far from the coast to forage (Figure 5.3). It therefore seems that the pelagic lifestyle characteristic of many living cetaceans only arose with the emergence of neocetes, at least 5–6 million years after cetaceans first severed the last ties that had bound them to land (Clementz *et al.*, 2006).

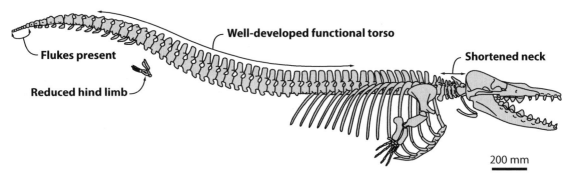

Figure 5.11 Aquatic locomotory adaptations of the basilosaurid *Dorudon*. Drawing of skeleton adapted from Gingerich *et al.* (2009), under a Creative Commons Attribution license.

The reduction of the hind limb in basilosaurids and neocetes marks a definite switch from foot-based to tail-powered locomotion, supported by the presence of a well-developed functional torso and the earliest evidence of tail flukes (Buchholtz, 1998; Uhen, 2008a). In basilosaurids and mysticetes, the vertebrae of the torso are nearly uniform in shape, and thus capable of undulation. The only exception to this pattern is a region of foreshortened and slightly transversely compressed vertebrae immediately anterior to the flukes, which is known as the **caudal tail stock**. The shortening of the vertebrae makes the tail stock more rigid, and thus provides a firm base against with the flukes can oscillate. At the same time, the transverse compression of the tail stock facilitates dorsoventral movements of the flukes by minimizing resistance (Buchholtz, 1998, 2001). In odontocetes, the development of the caudal tail stock is even more marked, and has led to the formation of an elongate, stabilized **peduncle**. In many species, vertebral shortening extends into the anterior torso and is accompanied by corresponding increases in vertebral count. The torso of odontocetes is thus comparatively rigid, which restricts or even prevents undulation and instead causes the peduncle to oscillate (Buchholtz, 2001).

Commensurate with their obligate aquatic lifestyle, advanced protocetids, basilosaurids, and neocetes show a marked posterior shift in the location of the external bony nares (Figure 5.12) (Hulbert et al., 1998; Uhen, 2004). The backward migration of the nares in protocetids and basilosaurids is a precursor of the pronounced facial telescoping characteristic of modern cetaceans, and likely occurred in response to the need to breathe while being submerged in water (Heyning and Mead, 1990). Along with the reorganization of the nasal cavity came a reduction of the turbinates (maxillo- and ethmoturbinates), which opened up the nasal cavity and allowed rapid, unobstructed air exchange (Reidenberg and Laitman, 2008). It is unclear to what degree this reduction of the turbinates impacted olfaction. The cetacean sense of smell is commonly thought to be vestigial or altogether absent, likely as a result of their move into the water (Pihlström, 2008). Extant odontocetes indeed seem to have entirely lost the ability to smell, as shown by the reduction of the olfactory nerve and the virtual absence of ethmoturbinates, which in terrestrial mammals carry the olfactory epithelium (Berta et al., 2014). This loss can also be detected genetically, with nearly all genes involved in olfactory reception (OR) being nonfunctional in crown odontocetes (Figure 5.8) (McGowen et al., 2008).

By contrast, mysticetes retain a better developed, though still reduced, olfactory apparatus, as well as a considerably higher number of functional OR genes (Berta et al., 2014; Thewissen et al., 2011a). The mysticete olfactory system closely resembles that of archaeocetes in comprising separate respiratory and olfactory airflows, as well as a blind chamber (the **olfactory recess**) containing the ethmoturbinates (Godfrey et al., 2013). As a result, the airflow over the nasal epithelium is largely unidirectional, and inhaled air becomes stagnant within the recess. The latter, in turn, facilitates odorant separation and absorption (Craven et al., 2010). A similar arrangement is commonly found in macrosmatic (keen-scented) terrestrial mammals, including artiodactyls, which suggests a relatively well-developed sense of smell functioning in mate and prey detection (Godfrey et al., 2013; Thewissen et al., 2011a).

There are currently two hypotheses as to what may have caused the loss of olfaction in odontocetes, but not mysticetes. The first of these suggests that the emergence of an active biosonar in odontocetes not only replaced olfaction as a tool for prey and congener detection, but also caused a profound reorganization of the nasal apparatus in the process. Ultimately, this reorganization led to the juxtaposition of the nasal passages and the anterior brain case, and thus obliterated the anatomical structures involved in olfaction (Cave, 1988; Godfrey et al., 2013). Alternatively, it is possible that the sense of smell initially declined in cetaceans as a whole, as indicated by the reduction of the associated morphological structures relative to artiodactyls. This decline continued in odontocetes, but stalled and maybe even partially reversed in mysticetes owing to a newfound need for olfaction to detect the particular odors (dimethylsulfide and pyradines) given off by krill (Kishida and Thewissen, 2012).

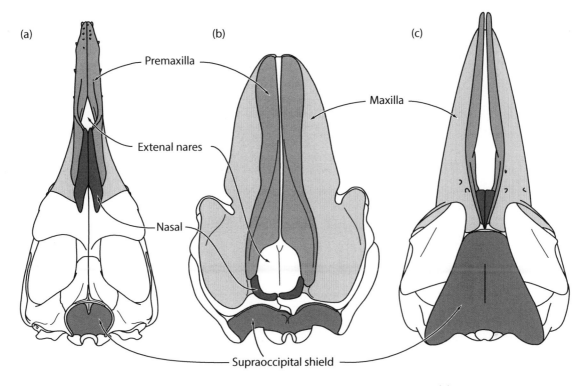

Figure 5.12 Backward migration of the external nares and associated facial telescoping in (a) the basilosaurid *Dorudon*, (b) the delphinid *Orcinus*, and (c) *Balaenoptera*.

5.2 Key innovations: baleen and echolocation

5.2.1 Baleen

Baleen, also known as whalebone, is a summary term for the unique filter-feeding apparatus that is the defining feature of all living mysticetes. Baleen is epidermal in origin and occupies roughly the same position as teeth would in most other mammals. Nevertheless, the two structures are not homologous and are expressed alongside each other in some modern mysticete fetuses (Deméré *et al.*, 2008), even though the development of the baleen apparatus curiously appears to have co-opted some of the molecular signaling usually involved in tooth development (Thewissen *et al.*, 2014). Ultimately, the origin of baleen may lie in the transverse palatal ridges present in many terrestrial mammals, including artiodactyls (Werth, 2000).

The baleen apparatus is organized into a series of thin, continuously growing keratinous **plates**, which are lined up into a **rack** and suspended from the margin of the upper jaw. The number of plates varies by species and can reach more than 400 in some balaenopterids. Within the rack, plates are spaced 1–2 cm apart and vary in size, with the anteriormost and the posteriormost ones always being the shortest (Rice, 2009). The plates are held in position by a rubbery layer of gingival tissue (**Zwischensubstanz**), which may also act as a shock absorber during feeding (Pinto and Shadwick, 2013). In ventral view, the plates are oriented transversely and slightly curved to facilitate the flow of water out of the mouth cavity. In anterior view, each plate is triangular, with a more or less vertically oriented lateral edge and an oblique medial margin.

In terms of its development, baleen is similar to other keratinous structures, such as hooves or hair (Pinto and Shadwick, 2013). Baleen plates originate from a **basal plate** of connective tissue (**corium**), which sends forth a series of thin papillae into gaps within the Zwischensubstanz (Figure 5.13). Here, the **papillae** are enveloped in keratinous tubules, which are held together by an equally keratinous matrix (Fudge *et al.*, 2009; Pinto and Shadwick, 2013). Together, the tubules and the matrix form the medulla of the plate, which is in turn surrounded by a smooth, highly structured cortical layer. By the time the plate

emerges from the Zwischensubstanz, it is fully formed and mostly consists of dead tissue.

Along the ventral and medial margins of the plate, feeding-related wear constantly breaks down the cortical layer and the medullary matrix. This process causes the ends of the more resistant tubules to become exposed in the form of numerous **bristles**, which gives the medial margin of the plate a somewhat comb-like appearance. The bristles of adjacent plates overlap to form a dense

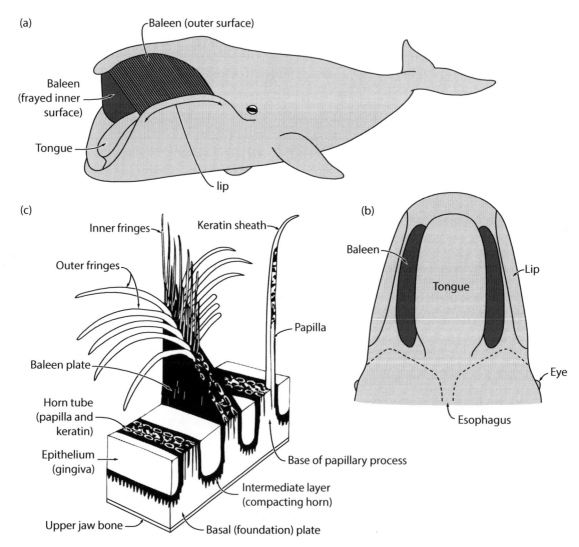

Figure 5.13 Location of the baleen apparatus in (a) anterolateral and (b) dorsal views; and (c) internal structure of a baleen plate and the tissues from which it grows. (a, b) adapted from Werth (2012) under a Creative Commons Attribution license; (c) reproduced from Werth (2000; after Pivorunas, 1979), with permission of Elsevier.

mat on the inside of the baleen apparatus. During feeding, prey-laden water is channeled through the baleen plates, with the bristle mat effectively acting as a sieve (Pivorunas, 1979; Rice, 2009). Any prey items trapped on the inside of the plates are subsequently washed out, shaken out, or licked off by the large tongue, before ultimately being swallowed (Werth, 2001).

The evolution of baleen allowed mysticetes to target vast quantities of minute prey, and thus to forage lower in the food web than their toothed ancestors. This, in turn, led to increased mysticete feeding efficiency, which laid the foundations for their subsequent diversification and notable gigantism (Berger, 2007; Werth, 2000). Many of the features that characterize modern mysticetes—unfused, laterally bowed mandibles, the absence of teeth, broad, thin, kinetic rostra, and pronounced cranial telescoping—are directly linked to baleen-assisted filter feeding (Bouetel, 2005; Deméré et al., 2008). This marked impact on diversity and morphology makes baleen a defining feature, as important to

mysticete evolution as bipedalism is to that of humans. In addition, their dietary specialization and large body size make mysticetes an important mediator of oceanic nutrient cycles, on which nearly all marine life ultimately depends (Nicol et al., 2010; Roman and McCarthy, 2010).

Baleen only fossilizes under exceptional conditions (Plate 15b), but its existence can be inferred from the presence of palatal foramina and sulci for the transmission of the superior alveolar artery, which nourishes the corium (section 3.4.2) (Deméré et al., 2008; Ekdale et al., 2015). Foramina clearly comparable to those of the living species occur in virtually all chaeomysticetes, but are absent in archaeocetes and many archaic mysticetes, such as mammalodontids. The morphological gap between toothed and toothless mysticetes is closed by the aetiocetids, which possess several small, radially arranged palatal foramina and sulci in addition to a fully functional dentition (Figure 5.14 and Plate 5a) (Deméré et al., 2008; Sawamura et al., 2006). Eomysticetids also seem to have

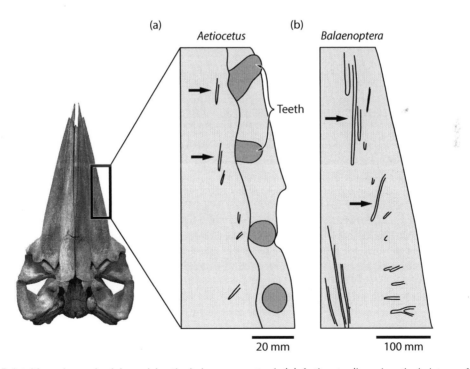

Figure 5.14 Palatal foramina and sulci supplying the baleen apparatus in (a) *Aetiocetus* (based on the holotype of *Aetiocetus weltoni*) and (b) extant *Balaenoptera*. Photo shows *Balaenoptera acutorostrata* (National Museum of Nature and Science of Japan specimen M42450).

retained shallow alveoli and a restricted number of teeth, although the latter were likely nonfunctional in this case (Boessenecker and Fordyce, 2014). The sulci of aetiocetids are considerably smaller than those of the living species, and their homology with the sulci carrying the superior alveolar artery in chaeomysticetes has been questioned on both morphological and phylogenetic grounds (Ichishima et al., 2008; Marx, 2011); however, it has never been disproved (Ekdale et al., 2015). Fine palatal grooves surrounding the tooth alveoli have also been reported in the oldest known fossil mysticete, *Llanocetus*, but it is presently unclear whether these are homologous with the sulci seen in aetiocetids (Fordyce, 2003).

Based on the presence of palatal foramina in aetiocetids, baleen—or its precursor—evolved shortly after the origin of mysticetes as a whole, and may have been present in incipient form as early as 33 Ma. This roughly coincides with the Eocene–Oligocene boundary, and thus the final separation of South America from Antarctica—a major ocean-restructuring event that paved the way for the modern Antarctic Circumpolar Current (ACC). The onset of the ACC is thought to have resulted in heightened levels of marine primary productivity, which may ultimately have been responsible for the emergence of modern whales and dolphins (section 7.1.1) (Fordyce, 1980; Steeman et al., 2009).

5.2.2 Echolocation

Echolocation, also known as biosonar, is the ability to "see" with sound by first producing high-frequency noises, and then listening to their echoes as they bounce off nearby objects. The use of sound to navigate and find prey offers an intuitive alternative to vision during the night or in naturally dark or turbid environments, such as caves, muddy waters, or the deep sea. Because of this, echolocation independently evolved in a variety of rather disparate mammals, including shrews, golden hamsters, and flying lemurs—but never to the degree seen in microchiropteran bats and odontocetes.

Cetacean echolocation capabilities are difficult to test in the wild and have only been experimentally demonstrated in about a dozen species. Nevertheless, most of the living odontocetes are known to produce repeated click-like sounds, and all of them possess the anatomical structures involved in the generation, propagation, and reception of high-frequency sounds (Cranford et al., 1996). Echolocation clicks are generated in the nasal apparatus, where pressurized air is forced past two pairs of **phonic lips**, causing them to slap against each other in the process (Figure 5.15). The sounds are then reflected forward by the concave facial area and a series of nasal air sacs (section 3.4.3), and eventually focused into a beam and sent toward the target by the fatty melon (Cranford et al., 2013; McKenna et al., 2012). When they hit an object, the clicks are again reflected and return to the sender, where they are received via the mandibular fat pad (section 3.4.3). Living mysticetes lack all of the facial structures involved in odontocete echolocation, as well as any other evidence for a functional biosonar. However, a small mass of adipose tissue anterior to the nasal passages may be homologous with the odontocete melon (Heyning and Mead, 1990), which raises the question whether echolocation may have been present in the last common ancestor of neocetes (Milinkovitch, 1995).

None of the soft tissues involved in echolocation have ever been known to fossilize, but the presence of some of them can nonetheless be read from the bones (Fordyce, 2002; Geisler et al., 2014; Heyning, 1989; Mead and Fordyce, 2009). Pronounced facial asymmetry, which in extant odontocetes generally follows that of the overlying soft tissues (with some exceptions, such as phocoenids), has long been interpreted as a major correlate of sound production (Cranford et al., 1996). However, two alternative explanations for the evolution of asymmetry have been put forward, namely, directional hearing and the need to swallow large prey whole (Fahlke et al., 2011; MacLeod et al., 2007). The latter has been argued against based on the concurrent presence of a limited degree of asymmetry and shearing teeth in archaeocetes, including both protocetids and basilosaurids (Fahlke et al., 2011). However, archaeocete asymmetry is largely restricted to the rostrum, rather than the nasal area or the ears, and thus also fits neither sound production nor directional hearing (Fahlke and Hampe, 2015; Fahlke et al., 2011). Possibly, asymmetry in archaeocetes is unrelated to that in odontocetes, and, in any case, it is still in need of explanation. In odontocetes, none of the proposed functions of asymmetry can thus

Figure 5.15 Schematic right lateral view of the head of *Delphinus*, illustrating the main soft tissue elements involved in the production (nasal apparatus, phonic lips), reflection (air sacs), propagation (melon), and reception (fat pad) of high-frequency echolocation sounds.

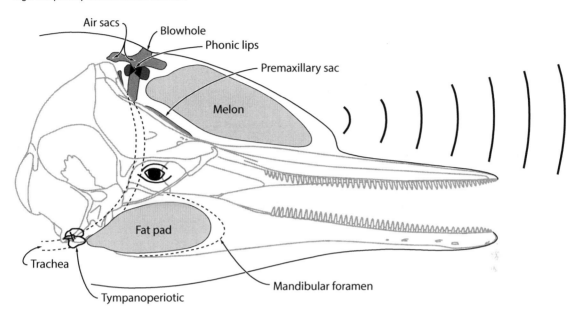

currently be ruled out with complete certainty, although sound production is often regarded as the best fit (Fahlke and Hampe, 2015).

Besides facial asymmetry, further, less controversial indicators of sound production include (1) the broad, concave facial area, which is covered almost entirely by the posteriorly telescoped maxilla and houses hypertrophied nasofacial muscles acting on the blowhole, air sacs, and the melon; (2) the premaxillary sac fossa; (3) the premaxillary foramen, which transmits branches of the maxillary artery and nerve; and (4), in platanistids, the pronounced facial crest, which is likely involved in sound reflection and the formation of the sonar beam. In addition, the inner ear may preserve several characteristic signs of high-frequency hearing, such as a narrow gap between the primary and secondary laminae, numerous foramina for the ganglion cells of the auditory nerve, a marked size difference between the scala tympani and the scala vestibuli, and few, widely separated cochlear turns (Fleischer, 1976; Luo and Eastman, 1995).

Rudimentary versions of at least some of the above features already occur in the earliest diverging stem odontocetes, including xenorophids (Geisler *et al.*, 2014; Uhen, 2008b), *Simocetus*

(Fordyce, 2002), and two as-yet-undescribed Early Oligocene taxa (Barnes *et al.*, 2001). Some form of echolocation was thus most likely present in the last common ancestor of all known odontocetes, but subsequently went through several steps of refinement. In particular, xenorophids such as *Cotylocara* seem to have evolved specialized structures related to ultrasonic sound production, including facial asymmetry and more pronounced facial telescoping, in parallel with other odontocetes (Geisler *et al.*, 2014). By contrast, there is no evidence for an enlarged melon or other specializations of the nasal apparatus in fossil mysticetes or archaeocetes (Fitzgerald, 2006). This suggests that the melon of early neocetes was small and likely served a different function, such as to facilitate the movement of the nasal plugs, before becoming hypertrophied and involved in echolocation in odontocetes (Heyning, 1997; Heyning and Mead, 1990).

Relatively little is known about the inner ear anatomy of fossil odontocetes, but morphologies indicative of high-frequency hearing are present in Late Oligocene "squalodontoids," as well as a range of Miocene and Pliocene taxa (Fleischer, 1976; Luo and Eastman, 1995; Luo and Marsh, 1996).

By contrast, the cochleae of basilosaurids and mysticetes are both adapted to the perception of low-frequency sounds, which likely represents the ancestral condition for all neocetes (Ekdale and Racicot, 2015; Geisler and Luo, 1996). The morphological evidence for the perception of high-frequency sounds is thus in line with that for its production: both are unique to odontocetes, and only emerged after the latter had diverged from their baleen-bearing cousins. This pattern is independently confirmed by molecular data on the evolution of the protein Prestin, which functions in ultrasonic hearing by determining the motility of hair cells within the organ of Corti (section 3.4.5). Prestin underwent two episodes of rapid (ca 5 Ma) sequence changes during the course of cetacean evolution: one along the lineage uniting all extant odontocetes, and a second along the lineage uniting ziphiids and delphinoids (Liu *et al.*, 2010). Like associated morphological adaptations, molecular changes related to high-frequency hearing thus seem to be restricted to odontocetes.

The early evolution and pervasive distribution of echolocation make it a defining feature of odontocetes—although it was subsequently lost in at least one species of the walrus-like delphinoid *Odobenocetops* (section 7.7). Their acquisition of biosonar not only made odontocetes efficient hunters but also gave them access to otherwise inaccessible environments, such as turbid rivers and the deep sea. The need to process the acoustic information provided by echolocation may also at least partly be responsible for the enlargement of the odontocete brain (Marino *et al.*, 2004), although complex brains are also present in non-echolocating mysticetes (section 7.5) (Montgomery *et al.*, 2013).

Echolocating odontocetes appeared roughly at the same time as the first baleen-bearing mysticetes, and hence plausibly in response to a common driving mechanism (Fordyce, 1980). One potential candidate for such a driver may be enhanced marine productivity associated with the onset of the ACC (as discussed in this chapter and in section 7.1.1), which promoted the formation of acoustically detectable, dense agglomerations of vertically migrating prey (deep scattering layers) (Berger, 2007). Initially, echolocation may have facilitated the detection of prey (especially cephalopods) at shallow depths during the night, before becoming more sophisticated and able to track migrating animals at depth (Lindberg and Pyenson, 2007). The use of echolocation to navigate muddy river waters likely followed later, when some odontocete lineages secondarily reinvaded freshwater habitats.

5.3 Invasion of freshwater habitats

After their return to the sea around 48 Ma, cetaceans largely remained marine mammals occupying both coastal and offshore habitats. More than once, however, an opportunity arose for some of the smaller odontocetes to expand their range into nearby river systems. Living examples of this process are the Irrawaddy dolphin, *Orcaella brevirostris*, and the finless porpoise, *Neophocaena phocaenoides*, both of which include populations inhabiting freshwater environments. If successful, speciation may ultimately separate the freshwater populations from their marine relatives, as exemplified by the closely related species pair constituting the genus *Sotalia*: the tucuxi, *Sotalia fluviatilis*, which inhabits the Amazon River, and the Guiana dolphin, *Sotalia guianensis*, which lives along the eastern and northern coasts of South America (Caballero *et al.*, 2007). Fossils of freshwater-adapted cetaceans are relatively rare, even though such animals seem to have existed since at least the Middle Miocene, and possibly even the Late Oligocene (Bianucci *et al.*, 2013; Clementz *et al.*, 2014; Fordyce, 1983). Considering the relatively patchy record of fluvial cetacean-bearing deposits, and the relative ease with which living cetaceans seem to move into rivers, it is likely that the total diversity of freshwater species has been severely underestimated.

Remarkably, three of the 10 extant odontocete families are entirely restricted to freshwater, and commonly known as "river dolphins." They include the **Platanistidae**, represented by the South Asian river dolphin, *Platanista*, in the Ganges, Indus, and Brahmaputra river basins; the **Lipotidae**, represented by the now seemingly extinct Yangtze River dolphin, *Lipotes*; and the **Iniidae**, represented by the Amazon River dolphin, *Inia*, in the

Amazon, Araguaia–Tocantins, and Orinoco river basins (Figure 5.16 and Plate 9). A fourth family of "river" dolphins, the **Pontoporiidae** (represented by the Franciscana, *Pontoporia*), is closely related to *Inia*, but somewhat confusingly only inhabits the coastal waters off eastern South America. All river dolphins were originally thought to form part of a single clade, but both molecular and morphological evidence now point to two or three separate lineages (Figure 5.17) (Cassens *et al.*, 2000; de Muizon, 1988; Geisler *et al.*, 2011).

The living "river dolphins" are the last survivors of their respective families or lineages, all of which were once more diverse and geographically widespread (Boessenecker, 2013; de Muizon, 1991; Geisler *et al.*, 2012; Hamilton *et al.*, 2001). Crucially, all of these families also used to be largely marine, which means that their living representatives must have adapted to freshwater environments independently of each other (Geisler *et al.*, 2012; Pyenson *et al.*, 2015). Sedimentological and morphological evidence demonstrates that platanistids and iniids have been inhabiting nonmarine environments at least since the late Middle and Late Miocene, respectively (Figure 5.16) (Bianucci *et al.*, 2013; Cozzuol, 2010; Gutstein *et al.*, 2014a). Surprisingly, the only record of a fossil freshwater platanistid comes not from the Indian subcontinent, but from

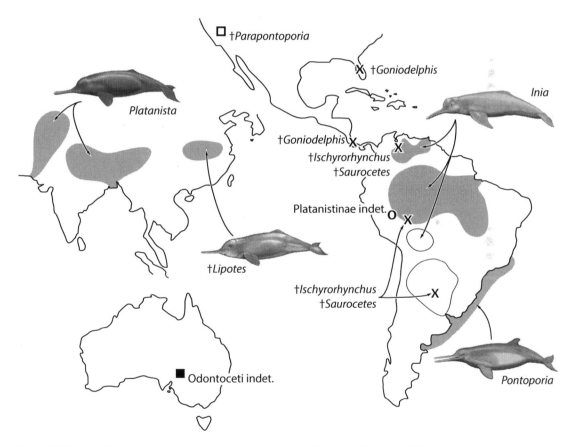

Figure 5.16 Geographic distribution of the extant freshwater dolphins *Inia*, *Lipotes*, and *Platanista* (dark gray areas) and location of the major localities that have yielded fossil odontocetes from nonmarine sediments. X, fossil iniids; open circle, fossil platanistid; open square, fossil lipotid; filled square, Odontoceti indet. Pleistocene records of *Inia* are not shown. Data are from Hamilton *et al.* (2001), Cozzuol (2006, 2010), Boessenecker and Poust (2015), and Valerio and Laurito (2012). Life reconstructions © C. Buell.

Figure 5.17 (a) Phylogenetic position and (b) typical features of the freshwater dolphins *Inia*, *Lipotes*, and *Platanista*, and the coastal–estuarine dolphin *Pontoporia*. There is a lack of consensus regarding the relationships of *Platanista* and *Lipotes*, as illustrated by stippled lines indicating alternative topologies. Based on, among others, de Muizon (1988), Cassens *et al.* (2000), Hamilton *et al.* (2001), McGowen *et al.* (2009), Steeman *et al.* (2009), and Geisler *et al.* (2011). Life reconstructions © C. Buell.

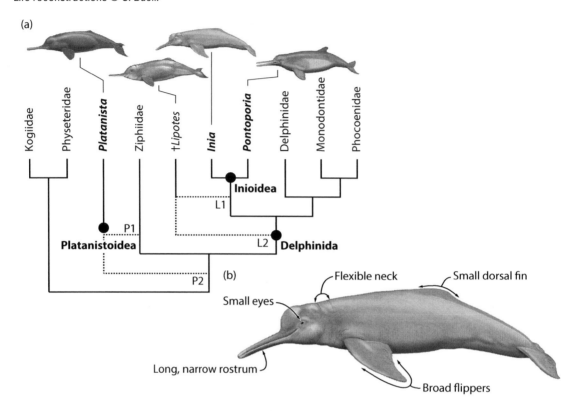

the western part of the Amazon Basin of South America, which implies that this family colonized rivers at least twice (Bianucci *et al.*, 2013). There are no fossil lipotids known exclusively from freshwater deposits, but a nonmarine, Late Pliocene specimen of *Parapontoporia* was recently reported from western North America (Boessenecker and Poust, 2015).

The origin of the modern riverine dolphins may be related to global sea-level change. During the Middle Miocene, elevated sea levels caused shallow epicontinental seas to form in the vast basins of the Indus, Ganges, Amazon, Paraná, and Yangtze rivers. When the oceans finally retreated and those seas were drained, some of their inhabitants were left stranded in the remaining rivers and lakes (Hamilton *et al.*, 2001). Ironically, being marooned in this way ultimately may have

ensured the dolphins' survival. Over the second half of the Miocene, marine platanistids in particular started to decline, possibly in response to large-scale environmental change or the concurrent radiation of delphinoids (Cassens *et al.*, 2000). Both of these notions have been challenged, based on (1) the higher environmental variability of rivers compared to the sea, which means that freshwater environments could likely not have functioned as a refuge from potentially detrimental changes in marine temperature or ocean circulation; and (2) the seemingly unproblematic coexistence of living "river dolphins" and delphinoids (e.g., *Inia* with *Sotalia*) in the Yangtze and Amazon river basins (Geisler *et al.*, 2011). Nevertheless, it remains possible that their restricted contact with delphinoids shielded "river dolphins" from the kind of intense competitive pressure that may have led to the demise of their marine cousins.

Moving into freshwater poses several challenges to an animal adapted to life in the sea, including osmoregulation, increased habitat heterogeneity, and reduced visibility. In addition, differences in temperature and turbidity can affect the acoustic properties of the surrounding water, and thus also echolocation. Despite being only distantly related, all "river dolphins" share a suite of morphological features that may represent a convergent response to freshwater-related selective pressures (Figure 5.17). These features include: (1) a marked shift in echolocation frequencies, along with corresponding changes in the shape of the periotic (Figure 5.18) (Gutstein *et al.*, 2014b); (2) a long and narrow rostrum bearing numerous teeth, accompanied by an equally elongate mandibular symphysis; (3) small eyes and correspondingly reduced orbits; (4) an elongated neck; (5) broad flippers; (6) a poorly developed dorsal fin; and (7) an enlarged zygomatic process (Cassens *et al.*, 2000; Geisler *et al.*, 2011; Gutstein *et al.*, 2014a). The size and shape of the rostrum, teeth, neck, and zygomatic process likely represent

adaptations to raptorial snap feeding (section 6.1.3), whereas broad flippers correlate with increased maneuvrability (Geisler *et al.*, 2011; Werth, 2000). More unique features include the pronounced facial crests of *Platanista* and the presence of a secondary joint between the humerus and the sternum in iniids, which presumably function in echolocation and facilitate maneuvring in complex river environments, respectively (Gutstein *et al.*, 2014a; Klima, 1980).

Based on their phylogenetic distribution, many of these supposed riverine adaptations have been reinterpreted as inherited ancestral conditions (symplesiomorphies), rather than genuine convergences. Notably, raptorial snap feeding also characterizes many of the fossil marine relatives of the living "river dolphins," and hence does not clearly correlate with freshwater habitats (Geisler *et al.*, 2011). Nevertheless, the shared morphology of "river dolphins" could still reflect a case of exaptation (i.e., the acquisition of a selective advantage via a preexisting trait). Thus, snap feeding may have allowed their ancestors to colonize freshwater systems more easily than other, suction-feeding odontocetes (Bianucci *et al.*, 2013), which might explain the geographically

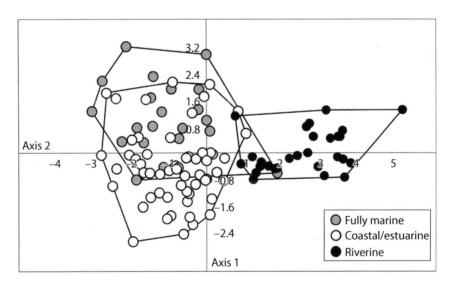

Figure 5.18 Scatterplot showing the results of a canonical variate analysis of the periotic shape of extinct and extant odontocetes. Individual specimens are color-coded by habitat or (in the case of fossils) depositional environment. Note the clear distinction between freshwater and coastal–fully marine taxa. Reproduced from Gutstein *et al.* (2014b), with permission of Elsevier.

disparate distribution of extinct and extant freshwater platanistids and lipotids (Bianucci *et al.*, 2013; Boessenecker and Poust, 2015). For exaptation to work, snap feeding would have to be more effective than suction feeding within the context of a riverine habitat. Further research into the precise function of feeding-related adaptations in the living "river dolphins" is necessary to determine whether this is indeed the case (Geisler *et al.*, 2011).

5.4 Key fossils

As in all sciences, paleontologists strive to maximize sample sizes in order to characterize morphological, geographical, and temporal variation within both species and higher taxonomic categories. Nevertheless, often the discovery of a single specimen, or a single species, is enough to challenge widely held opinions and transform our understanding of evolutionary history. This is especially true for extinct cetaceans, for which the available material is often rather limited (section 2.2). In this section, we describe and illustrate some of the key fossils that

had a dramatic impact on our telling of the story of whale evolution. Our list must necessarily be incomplete, but it should provide at least a flavour of the major discoveries that have shaped the study of fossil cetaceans.

5.4.1 Archaeocetes

Pakicetus Gingerich and Russell 1981. The original description of *Pakicetus inachus* (Figure 5.19 and Plate 1a) provided the first unequivocal evidence that whales had originated in and around the Eocene Tethys, in what is now Indo-Pakistan (Gingerich and Russell, 1981). Its preservation in fluvial deposits indicated that whales had passed through a freshwater stage before becoming fully marine (Gingerich *et al.*, 1983), as was later confirmed by isotopic evidence (section 5.1.1) (Clementz *et al.*, 2006; Roe *et al.*, 1998). In addition, the loss of crushing basins from the teeth suggested that changes in the feeding apparatus, along with the development of the involucrum of the tympanic bulla, were some of the first features to set cetaceans apart from artiodactyls (Gingerich *et al.*, 1983). This was again later confirmed by the description of the essentially

Figure 5.19 Cast of reconstructed composite skeleton of *Pakicetus* (National Museum of Nature and Science of Japan specimen PV 20725) in anterolateral view. Courtesy of National Museum of Nature and Science, Japan. Photo by F. Marx.

terrestrial-looking postcranial skeleton, which resembles that of the closely related artiodactyl *Indohyus* (Thewissen *et al.*, 2001, 2007). Finally, *Pakicetus* helped to confirm that cetaceans are related to artiodactyls, rather than mesonychians, through the discovery of its double-pulley astragalus (Thewissen *et al.*, 2001).

Ambulocetus Thewissen *et al.* 1994. When it was first described, *Ambulocetus natans* from the early Middle Eocene of Pakistan was the earliest cetacean to preserve a relatively complete skeleton (Figure 5.20 and Plate 2). Most importantly, the latter included well-developed hind limbs that showed that *Ambulocetus* could both move on land *and* swim, earning it the name "the walking, swimming whale" (section 5.1.1) (Thewissen *et al.*, 1994).

Rodhocetus Gingerich *et al.* 1994. At the time of its description, the protocetid *Rodhocetus kasranii* from the early Middle Eocene of Pakistan was the oldest whale recovered from open marine shelf deposits, and the first to preserve a genuine mix of postcranial characters indicating both terrestrial and advanced aquatic competence (section 5.1.2) (Gingerich *et al.*, 1994). A few years later, virtually complete fore- and hind limbs were described as part of the holotype specimen of *Rodhocetus balochistanensis* (Gingerich *et al.*, 2001). The latter included a double-pulley astragalus closely resembling that

of artiodactyls, which helped to cement the previously controversial placement of Cetacea within Artiodactyla (section 4.1) (Geisler and Uhen, 2005; Gingerich *et al.*, 2001).

Maiacetus Gingerich *et al.* 2009. The protocetid *Maiacetus inuus* from the early Middle Eocene of Pakistan is known from a virtually complete male skeleton and a partial female skeleton bearing a fetus (Figure 5.21 and Plate 15a) (Gingerich *et al.*, 2009), although the identity of the latter is disputed (Thewissen and McLellan, 2009). *Maiacetus* provides the earliest evidence of sexual dimorphism in archaeocetes, with the male being slightly larger and having better developed canines. Even more importantly, it demonstrates that the near-term fetus of protocetids was positioned with the head pointing caudally (section 6.3). This resembles the situation in terrestrial artiodactyls, but is the exact opposite of the strategy followed by modern cetaceans: giving birth tail first to minimize the risk of drowning. The position of the fetus therefore suggests that protocetids like *Maiacetus* still gave birth on land (Gingerich *et al.*, 2009).

Basilosaurus Harlan 1834 and **Dorudon** Gibbes 1845. *Basilosaurus* (Figure 5.22) was the first archaeocete to be described, albeit wrongly as a marine reptile—which explains its rather odd name, "king lizard" (Harlan, 1834). Realizing the mistake, Richard Owen (1839) later assigned

Figure 5.20 Cast of reconstructed skeleton of the holotype of *Ambulocetus natans* (National Museum of Nature and Science of Japan specimen PV 20730). Courtesy of National Museum of Nature and Science, Japan. Photo by N. Kohno.

Figure 5.21 Holotype (female) skull and anterior skeleton of *Maiacetus inuus* (Geological Survey of Pakistan–University of Michigan collection specimen 3475a). Reproduced from Gingerich *et al.* (2009) under a Creative Commons Attribution license.

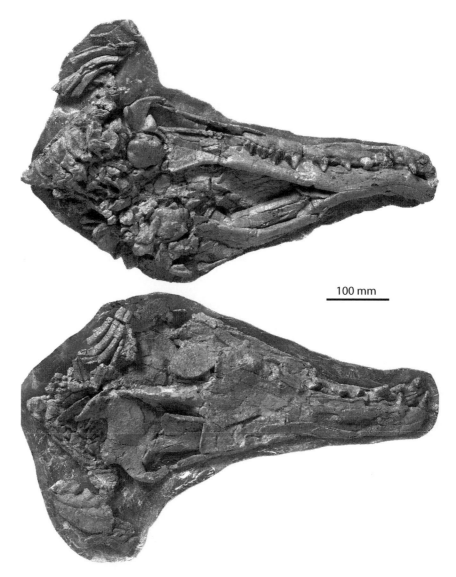

100 mm

Basilosaurus to Cetacea and tried to rename the genus to *Zeuglodon*, but, according to the international rules of zoological nomenclature, the original name stuck. *Basilosaurus* is by far the largest known archaeocete, and it is characterized by unusually elongate trunk vertebrae (Gingerich *et al.*, 1990; Kellogg, 1936). It was the first cetacean ever discovered to preserve hind limb elements (Lucas, 1901) and feet (Gingerich *et al.*, 1990). Both are clearly recognizable, but also greatly reduced and no longer weight-bearing, thus illustrating the transition from four-legged terrestrial mammals to modern cetaceans (sections 5.1.3 and 8.1). *Dorudon* was originally named based on the partial skull and skeleton of a juvenile individual from the Late Eocene of South Carolina. Subsequent discoveries led to a great number of additional specimens covering virtually every component of the skeleton, which arguably makes this animal the best known of all archaeocetes (Figure 5.23 and Plate 3b). Phylogenetically, *Dorudon* is thought to be close to the origin of Neoceti, and thus provides crucial

information on cetacean ecology, distribution, morphology, and even soft tissue anatomy around the time of origin of modern whales and dolphins (Uhen, 2004).

5.4.2 Mysticeti

Llanocetus Mitchell, 1989. *Llanocetus denticrenatus* from the Late Eocene of Antarctica is the oldest neocete discovered to date. Only a small portion of the holotype has been formally described (Mitchell, 1989), but enough is known of the remainder of the specimen to place it securely within Mysticeti (Fitzgerald, 2010; Fordyce, 2003). Because of its age, *Llanocetus* has been widely used to calibrate molecular clock analyses (section 4.5). Unlike archaeocetes, *Llanocetus* has widely spaced teeth surrounded by numerous palatal grooves (Fordyce, 2003). This morphology may be indicative of some incipient form of filter feeding, which could thus have arisen as early as 34 Ma.

Janjucetus Fitzgerald, 2006 and **Mammalodon** Pritchard, 1939. *Janjucetus hunderi* (Figure 5.24 and Plate 4a) and *Mammalodon colliveri* (Figure 5.25 and Plate 4b) constitute a family of small, highly unusual toothed mysticetes known primarily from the Late Oligocene of Australia (Fitzgerald, 2006, 2010). Neither species shows any sign of baleen, but their bizarre morphologies reveal an unexpected diversity of early mysticete feeding strategies (section 6.1.2), as well as

Figure 5.22 Cast of reconstructed skull and anterior skeleton of *Basilosaurus* (National Museum of Nature and Science of Japan specimens PV 20728 and 20748). Courtesy of National Museum of Nature and Science, Japan. Photo by N. Kohno.

Figure 5.23 Cast of reconstructed skull and skeleton of *Dorudon* (National Museum of Nature and Science of Japan specimen PV 20729). Courtesy of National Museum of Nature and Science, Japan. Photo by N. Kohno.

Figure 5.24 Holotype skull of *Janjucetus hunderi* (Museum Victoria, Melbourne specimen P216929) in anterolateral view. Courtesy of E. Fitzgerald and Museum Victoria.

Figure 5.25 Holotype skull of *Mammalodon colliveri* (Museum Victoria, Melbourne specimen P199986) in anterolateral view. Courtesy of E. Fitzgerald and Museum Victoria.

potential evidence of pedomorphism (section 8.4). Despite their relatively archaic morphology (e.g., the presence of well-developed, heterodont teeth and a fused mandibular symphysis), mammalodontids occurred long after the origin of mysticetes, alongside both aetiocetids and a range of chaeomysticetes. Their phylogenetic position is controversial. Overall, *Janjucetus* and *Mammalodon* clearly demonstrate that early mysticete evolution

was both complex and taxonomically, morphologically, and ecologically diverse.

Aetiocetus Emlong, 1966. *Aetiocetus cotylalveus* from the early Late Oligocene of Oregon, United States (Figure 5.26 and Plate 5a), was the first ever toothed mysticete to be described (Emlong, 1966), although it was not immediately recognized as such (Van Valen, 1968). Its discovery began to fill the large morphological gap that separates modern, baleen-bearing mysticetes from their archaeocete ancestors, and provided some of the first insights into the evolution of early neocetes. Later work showed that aetiocetids, including *A. cotylalveus*, may have possessed baleen and teeth at the same time (Deméré *et al.*, 2008), thus turning them into transitional fossils somewhat akin to the famous *Archaeopteryx* (section 5.2.1).

Yamatocetus Okazaki, 2012. *Yamatocetus canaliculatus* from the Early Oligocene of Japan (Figure 5.27) is a member of the most archaic family of chaeomysticetes, the Eomysticetidae (Okazaki, 2012). It was neither the first eomysticetid to be described (Marples, 1956) nor the name-giving taxon of the family (Sanders and Barnes, 2002), but it is by far the most complete. *Yamatocetus* preserves residual alveoli but no functional dentition. It thus not only provides a link between the transitional morphotype represented by aetiocetids and later chaeomysticetes, but also is the oldest clear example of baleen-assisted bulk feeding (section 5.2.1).

Piscobalaena Pilleri and Siber, 1989. The family Cetotheriidae was originally erected to include its name-giving taxon, *Cetotherium rathkii* from the Late Miocene of Russia. Subsequently, however, the term quickly became used as a wastebasket taxon for all fossil chaeomysticetes that did not clearly belong to any of the extant families. This situation lasted for more than 100 years, and only changed when a series of recent papers made a convincing case for a monophyletic, but also much less inclusive, clade centered on *C. rathkii*—the Cetotheriidae *sensu stricto* (Bouetel and de Muizon, 2006; Steeman, 2007; Whitmore and Barnes, 2008). *Piscobalaena nana* from the Late Miocene of Peru (Figure 5.28) is by far the best known member of this family, and played a crucial role in its establishment (section 4.3.2). The redefinition of Cetotheriidae radically altered ideas about mysticete taxonomy

Figure 5.26 Holotype cranium of *Aetiocetus cotylalveus* (United States National Museum of Natural History specimen 25210) in dorsal view. Courtesy of Smithsonian Institution. Photo by F. Marx.

Figure 5.27 Holotype cranium of *Yamatocetus canaliculatus* (Kitakyushu Museum Kitakyushu of Natural History and Human History, Japan, specimen VP000,017) in dorsal view.

Figure 5.28 Skull of *Piscobalaena nana* (Muséum national d'Histoire naturelle, Paris, France, specimen SAS 892) in dorsolateral view. Courtesy of C. de Muizon. Photo by D. Serrette.

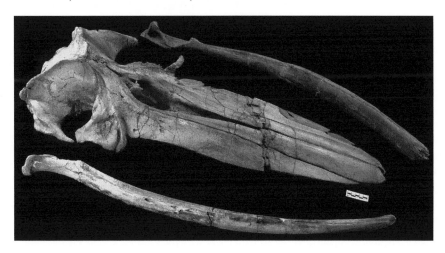

Figure 5.29 Holotype cranium of *Simocetus rayi* (United States National Museum of Natural History specimen 256517) in anterolateral view. Courtesy of Smithsonian Institution. Photo by O. Lambert.

Figure 5.30 Holotype skull of *Cotylocara macei* (College of Charleston Natural History Museum, United States, specimen 101) in lateral view. Courtesy of J. Geisler.

and evolution, with profound implications for at least one extant species: the extant pygmy right whale, *Caperea marginata*, is now considered to be a cetotheriid by some workers, thus effectively resurrecting the family from the dead (section 4.5) (Fordyce and Marx, 2013).

5.4.3 Odontoceti

Simocetus Fordyce, 2002. *Simocetus rayi* (Figure 5.29) from the Early Oligocene of Oregon, United States, is the oldest described fossil odontocete, and one of the most archaic (Fordyce, 2002). *Simocetus* preserves some of the earliest evidence for echolocation,

and provides a glimpse of the diversity and disparity of stem odontocetes. Its downturned, proportionally short rostrum ending in a toothless premaxilla is unique among ancient odontocetes, and suggests a specialized, maybe benthic, feeding strategy.

Cotylocara Geisler *et al.*, 2014. *Cotylocara macei* (Figure 5.30) from the Late Oligocene of South Carolina, United States, is the best known representative of the archaic xenorophids (Geisler *et al.*, 2014). This family has so far only been recorded from the Oligocene of eastern North America, and may represent a geographically limited, early odontocete radiation. The facial

Figure 5.31 Holotype cranium of *Squalodon bariensis* in dorsal (top) and lateral (bottom) views. Courtesy of G. Bianucci, photos taken with support of the National Science Foundation (DEB 1025260 to J. Geisler).

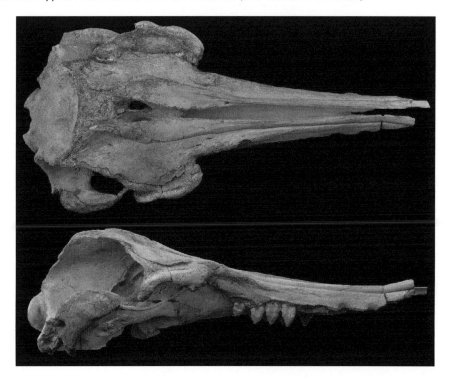

region of *C. macei* preserves some of the best evidence for the early evolution of the odontocete biosonar. Surprisingly, the latter seems to have followed two separate, albeit convergent, paths. Thus, osteological evidence for the presence of echolocation is more pronounced in *Cotylocara* than in the somewhat older *Simocetus*, even though the latter is more closely related to extant odontocetes (section 5.2.2) (Geisler *et al.*, 2014).

Squalodon Grateloup, 1840 and ***Waipatia*** Fordyce, 1994. Squalodontids are an iconic family of fossil "shark-toothed dolphins," known for their spectacular, highly ornamented dentition. They are also rather abundant, and are often seen as a morphological intermediate linking archaic Oligocene odontocetes with their modern descendants. One of the first and most informative species was *Squalodon bariensis* (Figure 5.31) from the Early Miocene of southern France (Jourdan, 1861). The wealth of information provided by the highly complete type specimen formed the basis for subsequent

discussions of the phylogenetic affinities of squalodontids (de Muizon, 1991, 1994). From these arose the concept of the Platanistoidea, a diverse superfamily comprising squalodontids, platanistids, and squalodelphinids, whose only living member is the South Asian river dolphin, *Platanista* (section 4.4).

Shortly afterward, Platanistoidea was further expanded by the addition of the new family Waipatiidae, which is mostly known from the exquisitely preserved type specimen of *Waipatia maerewhenua* (Figure 5.32) from the Late Oligocene of New Zealand (Fordyce, 1994). The interrelationships of squalodontids and waipatiids remain controversial, as does the definition of Platanistoidea as a whole (Geisler *et al.*, 2011; Tanaka and Fordyce, 2014). Nevertheless, the coexistence of squalodontids, waipatiids, and other putative platanistoids highlights the considerable diversity of Late Oligocene–Early Miocene heterodont odontocetes, and hints at the existence of a still not fully appreciated radiation event.

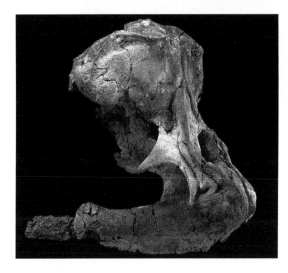

Odobenocetops Muizon, 1993. The discovery of *Odobenocetops peruvianus* (Figure 5.33 and Plate 11) from the latest Miocene to Early Pliocene of Peru had paleontologists puzzled for some time. The mix of dolphin and walrus-like characters in this strange animal finally led to its interpretation as a delphinoid odontocete, and thus the result of an astonishing degree of convergent evolution (section 7.7) (de Muizon, 1993). The extreme reduction of the rostrum and the development of posteroventrally directed premaxillary tusks suggest that *Odobenocetops* was a highly specialized benthic feeder which, uniquely among odontocetes, had lost the ability to echolocate (de Muizon and Domning, 2002; de Muizon *et al.*, 2002). In addition, *Odobenocetops* is one of the only fossil cetaceans to show clear evidence of sexual dimorphism—in this case, in the form of differently sized tusks (section 6.5).

Livyatan Lambert *et al.*, 2010. Large animals and large teeth often attract the attention of a vast audience. With its 3 m-long skull and massive teeth (some of them longer than 36 cm), the extinct sperm whale *Livyatan melvillei* from the Middle–Late Miocene of Peru matches both criteria (Figure 5.34 and Plate 7) (Lambert *et al.*, 2010). The size and morphology of *L. melvillei* suggest that it was capable of preying upon other large animals, including baleen whales (section 6.1.3). Along with other raptorial Miocene physeteroids, this gigantic macropredator, one of the largest of all time, undoubtedly occupied a position at the very top of the food web—in striking contrast to its living, suction-feeding relatives.

Figure 5.34 Holotype skull of *Livyatan melvillei* (Museo de Historia Natural, Lima, Peru, specimen 1676) in right lateral view, with a few detached lower teeth at the same scale. Courtesy of G. Bianucci.

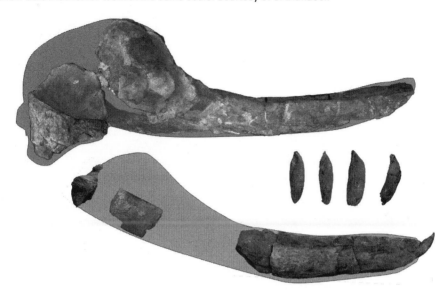

5.5 Suggested readings

Deméré, T., McGowen, M., Berta, A., and Gatesy, J. 2008. Morphological and molecular evidence for a stepwise evolutionary transition from teeth to baleen in mysticete whales. Systematic Biology 57:15–37.

Geisler, J. H., M. W. Colbert, and J. L. Carew. 2014. A new fossil species supports an early origin for toothed whale echolocation. Nature 508:383–386.

McGowen, M. R., Gatesy, J., and Wildman, D. E. 2014. Molecular evolution tracks macroevolutionary transitions in Cetacea. Trends in Ecology & Evolution 29: 336–346.

Thewissen, J. G. M., and S. Nummela. 2008. Sensory Evolution on the Threshold: Adaptations in Secondarily Aquatic Vertebrates. University of California Press, Berkeley.

Uhen, M. D. 2010. The origin(s) of whales. Annual Review of Earth and Planetary Sciences 38:189–219.

References

Bajpai, S., and P. D. Gingerich. 1998. A new Eocene archaeocete (Mammalia, Cetacea) from India and the time of origin of whales. Proceedings of the National Academy of Sciences USA 95:15464–15468.

Bajpai, S., J. G. M. Thewissen, and R. W. Conley. 2011. Cranial anatomy of Middle Eocene *Remingtonocetus* (Cetacea, Mammalia) from Kutch, India. Journal of Paleontology 85:703–718.

Barnes, L. G., J. L. Goedert, and H. Furusawa. 2001. The earliest known echolocating toothed whales (Mammalia; Odontoceti): preliminary observations of fossils from Washington State. Mesa Southwest Museum Bulletin 8:91–100.

Bebej, R. M., M. Ul-Haq, I. S. Zalmout, and P. D. Gingerich. 2012a. Morphology and function of the vertebral column in *Remingtonocetus domandaensis* (Mammalia, Cetacea) from the Middle Eocene Domanda Formation of Pakistan. Journal of Mammalian Evolution 19:77–104.

Bebej, R. M., I. S. Zalmout, A. A. Abed El-Aziz, M. S. M. Antar, and P. D. Gingerich. 2012b. First evidence of Remingtonocetidae (Mammalia, Cetacea) outside Indo-Pakistan: new genus from the early Middle Eocene of Egypt. Journal of Vertebrate Paleontology: Program and Abstracts, 62.

Benke, H. 1993. Investigations on the osteology and functional morphology of the flipper of whales and dolphins. Investigations on Cetacea 24:9–252.

Berger, W. H. 2007. Cenozoic cooling, Antarctic nutrient pump, and the evolution of whales. Deep Sea Research Part II: Topical Studies in Oceanography 54:2399–2421.

Berta, A., E. G. Ekdale, and T. W. Cranford. 2014. Review of the cetacean nose: form, function, and evolution. The Anatomical Record 297:2205–2215.

Bianucci, G., O. Lambert, R. Salas-Gismondi, J. Tejada, F. Pujos, M. Urbina, and P.-O. Antoine. 2013. A Miocene relative of the Ganges River dolphin (Odontoceti, Platanistidae) from the Amazonian Basin. Journal of Vertebrate Paleontology 33:741–745.

Boessenecker, R. W. 2013. A new marine vertebrate assemblage from the Late Neogene Purisima Formation in Central California, part II: Pinnipeds and Cetaceans. Geodiversitas 35:815–940.

Boessenecker, R. W., and R. E. Fordyce. 2014. A new eomysticetid (Mammalia: Cetacea) from the Late Oligocene of New Zealand and a re-evaluation of "*Mauicetus*" *waitakiensis*. Papers in Palaeontology 58:489–496.

Boessenecker, R. W., and A. Poust. 2015. Freshwater occurrence of the extinct dolphin *Parapontoporia* (Cetacea: Lipotidae) from the upper Pliocene nonmarine Tulare Formation of California. Palaeontology 1:107–140.

Bouetel, V. 2005. Phylogenetic implications of skull structure and feeding behavior in balaenopterids (Cetacea, Mysticeti). Journal of Mammalogy 86:139–146.

Bouetel, V., and C. de Muizon. 2006. The anatomy and relationships of *Piscobalaena nana* (Cetacea, Mysticeti), a Cetotheriidae s.s. from the early Pliocene of Peru. Geodiversitas 28:319–396.

Buchholtz, E. 1998. Implications of vertebral morphology for locomotor evolution in early Cetacea; pp. 325–351 in J. G. M. Thewissen (ed.), The Emergence of Whales. Plenum Press, New York.

Buchholtz, E. A. 2001. Vertebral osteology and swimming style in living and fossil whales (Order: Cetacea). Journal of Zoology 253:175–190.

Caballero, S., F. Trujillo, J. A. Vianna, H. Barrios-Garrido, M. G. Montiel, S. Beltrán-Pedreros, M. Marmontel, M. C. Santos, M. Rossi-Santos, F. R. Santos, and C. S. Baker. 2007. Taxonomic status of the genus *Sotalia*: species level ranking for "tucuxi" (*Sotalia fluviatilis*) and "costero" (*Sotalia guianensis*) dolphins. Marine Mammal Science 23:358–386.

Cassens, I., S. Vicario, V. G. Waddell, H. Balchowsky, D. Van Belle, W. Ding, C. Fan, R. S. Lal Mohan, P. C. Simões-Lopes, R. Bastida, A. Meyer, M. J. Stanhope, and M. C. Milinkovitch. 2000. Independent adaptation to riverine habitats allowed survival of ancient cetacean lineages. Proceedings of the National Academy of Sciences USA 97:11343–11347.

Cave, A. J. E. 1988. Note on olfactory activity in mysticetes. Journal of Zoology 214:307–311.

Clementz, M. T., R. E. Fordyce, S. Peek, L., and D. L. Fox. 2014. Ancient marine isoscapes and isotopic evidence of bulk-feeding by Oligocene cetaceans. Palaeogeography, Palaeoclimatology, Palaeoecology 400:28–40.

Clementz, M. T., A. Goswami, P. D. Gingerich, and P. L. Koch. 2006. Isotopic records from early whales and sea cows: contrasting patterns of ecological transition. Journal of Vertebrate Paleontology 26:355–370.

Cooper, L. N., J. G. M. Thewissen, S. Bajpai, and B. N. Tiwari. 2012. Postcranial morphology and locomotion of the Eocene raoellid *Indohyus* (Artiodactyla: Mammalia). Historical Biology 24:279–310.

Costa, D. P. 2009. Osmoregulation; pp. 801–806 in W. F. Perrin, B. Würsig, and J. G. M. Thewissen (eds.), Encyclopedia of Marine Mammals. Academic Press, Burlington, MA.

Cozzuol, M. A. 2006. The Acre vertebrate fauna: age, diversity, and geography. Journal of South American Earth Sciences 21:185–203.

Cozzuol, M. A. 2010. Fossil record and the evolutionary history of Inioidea; pp. 193–217 in M. Ruiz-García, and J. Shostell (eds.), Biology, Evolution and Conservation of River Dolphins. Nova Publishers, Hauppauge, NY.

Cranford, T. W., M. Amundin, and K. S. Norris. 1996. Functional morphology and homology in the odontocete nasal complex: implications for sound generation. Journal of Morphology 228:223–285.

Cranford, T. W., P. Krysl, and M. Amundin. 2010. A new acoustic portal into the odontocete ear and vibrational analysis of the tympanoperiotic complex. PLoS One 5:e11927.

Cranford, T. W., P. Krysl, and J. A. Hildebrand. 2008. Acoustic pathways revealed: simulated sound transmission and reception in Cuvier's beaked whale (*Ziphius cavirostris*). Bioinspiration & Biomimetics 3:016001.

Cranford, T. W., V. Trijoulet, C. R. Smith, and P. Krysl. 2013. Validation of a vibroacoustic finite element model using bottlenose dolphin simulations: the dolphin biosonar beam is focused in stages. Bioacoustics 23:161–194.

Craven, B. A., E. G. Paterson, and G. S. Settles. 2010. The fluid dynamics of canine olfaction: unique nasal airflow patterns as an explanation of macrosmia. Journal of The Royal Society Interface 7:933–943.

Czech-Damal, N. U., A. Liebschner, L. Miersch, G. Klauer, F. D. Hanke, C. Marshall, G. Dehnhardt, and W. Hanke. 2012. Electroreception in the Guiana dolphin (*Sotalia guianensis*). Proceedings of the Royal Society of London B 279:663–668.

de Muizon, C. 1988. Les relations phylogénétiques des Delphinida (Cetacea, Mammalia). Annales de Paléontologie 74:159–227.

de Muizon, C. 1991. A new Ziphiidae (Cetacea) from the Early Miocene of Washington State (USA) and phylogenetic analysis of the major groups of odontocetes. Bulletin du Muséum National d'Histore Naturelle, Paris 12:279–326.

de Muizon, C. 1993. Walrus-like feeding adaptation in a new cetacean from the Pliocene of Peru. Nature 365:745–748.

de Muizon, C. 1994. Are the squalodonts related to the platanistoids? Proceedings of the San Diego Society of Natural History 29:135–146.

de Muizon, C., and D. P. Domning. 2002. The anatomy of *Odobenocetops* (Delphinoidea, Mammalia), the walrus-like dolphin from the Pliocene of Peru and its palaeobiological implications. Zoological Journal of the Linnean Society 134:423–452.

de Muizon, C., D. P. Domning, and D. R. Ketten. 2002. *Odobenocetops peruvianus*, the walrus-convergent delphinoid (Mammalia: Cetacea) from the Early Pliocene of Peru. Smithsonian Contributions to Paleobiology 93:223–261.

Deméré, T. A., M. R. McGowen, A. Berta, and J. Gatesy. 2008. Morphological and molecular evidence for a stepwise evolutionary transition from teeth to baleen in mysticete whales. Systematic Biology 57:15–37.

Ekdale, E. G., T. A. Deméré, and A. Berta. 2015. Vascularization of the gray whale palate (Cetacea, Mysticeti, *Eschrichtius robustus*): soft tissue evidence for an alveolar source of blood to baleen. The Anatomical Record:691–702.

Ekdale, E. G., and R. A. Racicot. 2015. Anatomical evidence for low frequency sensitivity in an archaeocete whale: comparison of the inner ear of *Zygorhiza kochii* with that of crown Mysticeti. Journal of Anatomy 226:22–39.

Emlong, D. R. 1966. A new archaic cetacean from the Oligocene of northwest Oregon. Bulletin of the Oregon University Museum of Natural History 3:1–51.

Fahlke, J., and O. Hampe. 2015. Cranial symmetry in baleen whales (Cetacea, Mysticeti) and the occurrence of cranial asymmetry throughout cetacean evolution. The Science of Nature 102:1–16.

Fahlke, J. M., P. D. Gingerich, R. C. Welsh, and A. R. Wood. 2011. Cranial asymmetry in Eocene archaeocete whales and the evolution of directional hearing in water. Proceedings of the National Academy of Sciences USA 108:14545–14548.

Fish, F. E. 1996. Transitions from drag-based to lift-based propulsion in mammalian swimming. American Zoologist 36:628–641.

Fitzgerald, E. M. G. 2006. A bizarre new toothed mysticete (Cetacea) from Australia and the early evolution of baleen whales. Proceedings of the Royal Society B 273:2955–2963.

Fitzgerald, E. M. G. 2010. The morphology and systematics of *Mammalodon colliveri* (Cetacea: Mysticeti), a toothed mysticete from the Oligocene of Australia. Zoological Journal of the Linnean Society 158:367–476.

Fleischer, G. 1976. Hearing in extinct cetaceans as determined by cochlear structure. Journal of Paleontology 50:133–152.

Fordyce, R. E. 1980. Whale evolution and Oligocene Southern Ocean environments. Palaeogeography, Palaeoclimatology, Palaeoecology 31:319–336.

Fordyce, R. E. 1983. Rhabdosteid dolphins (Mammalia: Cetacea) from the Middle Miocene, Lake Frome area, South Australia. Alcheringa 7:27–40.

Fordyce, R. E. 1994. *Waipatia maerewhenua*, new genus and new species (Waipatiidae, new family), an archaic Late Oligocene dolphin (Cetacea: Odontoceti: Platanistoidea) from New Zealand. Proceedings of the San Diego Society of Natural History 29:147–176.

Fordyce, R. E. 2002. *Simocetus rayi* (Odontoceti: Simocetidae, new family): a bizarre new archaic Oligocene dolphin from the eastern North Pacific. Smithsonian Contributions to Paleobiology 93:185–222.

Fordyce, R. E. 2003. Early crown-group Cetacea in the southern ocean: the toothed archaic mysticete *Llanocetus*. Journal of Vertebrate Paleontology 23 (Suppl. to 3):50A.

Fordyce, R. E., and F. G. Marx. 2013. The pygmy right whale *Caperea marginata*: the last of the cetotheres. Proceedings of the Royal Society B 280:20122645.

Fraser, F. C., and P. E. Purves. 1960. Hearing in cetaceans. Bulletin of the British Museum (Natural History) 7:1–140.

Fudge, D. F., L. J. Szewciw, and A. N. Schwalb. 2009. Morphology and development of blue whale baleen: an annotated translation of Tycho Tullberg's classic 1883 paper. Aquatic Mammals 35:226–252.

Geisler, J. H., M. W. Colbert, and J. L. Carew. 2014. A new fossil species supports an early origin for toothed whale echolocation. Nature 508:383–386.

Geisler, J. H., S. J. Godfrey, and O. Lambert. 2012. A new genus and species of late Miocene inioid (Cetacea, Odontoceti) form the Meherrin River, North Carolina, USA. Journal of Vertebrate Paleontology 32:198–211.

Geisler, J. H., and Z.-X. Luo. 1996. The petrosal and inner ear of *Herpetocetus* sp. (Mammalia; Cetacea) and their implications for the phylogeny and hearing of archaic mysticetes. Journal of Paleontology 70:1045–1066.

Geisler, J. H., M. R. McGowen, G. Yang, and J. Gatesy. 2011. A supermatrix analysis of genomic, morphological, and paleontological data from crown Cetacea. BMC Evolutionary Biology 11:1–33.

Geisler, J. H., A. E. Sanders, and Z.-X. Luo. 2005. A new protocetid whale (Cetacea: Archaeoceti) from the Late Middle Eocene of South Carolina. American Museum Novitates 3480:1–65.

Geisler, J. H., and M. D. Uhen. 2005. Phylogenetic relationships of extinct Cetartiodactyls: results of simultaneous analyses of molecular, morphological, and stratigraphic data. Journal of Mammalian Evolution 12:145–160.

Gingerich, P. D. 2003. Land-to-sea transition in early whales: evolution of Eocene Archaeoceti (Cetacea) in relation to skeletal proportions and locomotion of living semiaquatic mammals. Paleobiology 29:429–454.

Gingerich, P. D., M. S. M. Antar, and I. Zalmout. 2014. Skeleton of new protocetid (Cetacea, Archaeoceti) from the lower Gehannam Formation of Wadi al Hitan in Egypt: survival of a protocetid into the Priabonian Late Eocene. Journal of Vertebrate Paleontology: Program and Abstracts, 138.

Gingerich, P. D., M. Arif, and W. C. Clyde. 1995. New archaeocetes (Mammalia, Cetacea) from the middle Eocene Domanda Formation of the Sulaiman Range, Punjab (Pakistan). Contributions from the Museum of Paleontology, University of Michigan 29:291–330.

Gingerich, P. D., M. u. Haq, I. S. Zalmout, I. H. Khan, and M. S. Malakani. 2001. Origin of whales from early artiodactyls: hands and feet of Eocene Protocetidae from Pakistan. Science 293:2239–2242.

Gingerich, P. D., S. M. Raza, M. Arif, M. Anwar, and X. Zhou. 1994. New whale from the Eocene of Pakistan and the origin of cetacean swimming. Nature 368:844–847.

Gingerich, P. D., and D. E. Russell. 1981. *Pakicetus inachus*, a new archaeocete (Mammalia, Cetacea) from the Early-Middle Eocene Kuldana Formation of Kohat (Pakistan). Contributions from the Museum of Paleontology, University of Michigan 25:235–246.

Gingerich, P. D., B. H. Smith, and E. L. Simons. 1990. Hind limbs of Eocene *Basilosaurus*: evidence of feet in whales. Science 249:154–157.

Gingerich, P. D., M. Ul-Haq, W. von Koenigswald, W. J. Sanders, B. H. Smith, and I. S. Zalmout. 2009. New protocetid whale from the Middle Eocene of Pakistan: birth on land, precocial development, and sexual dimorphism. PLoS One 4:1–20.

Gingerich, P. D., N. A. Wells, D. E. Russell, and S. M. I. Shah. 1983. Origin of whales in epicontinental remnant seas: new evidence from the early Eocene of Pakistan. Science 220:403–406.

Godfrey, S. J., J. H. Geisler, and E. M. G. Fitzgerald. 2013. On the olfactory anatomy in an archaic whale (Protocetidae, Cetacea) and the minke whale *Balaenoptera acutorostrata* (Balaenopteridae, Cetacea). The Anatomical Record 296:257–272.

Gol'din, P., and E. Zvonok. 2013. *Basilotritus uheni*, a new cetacean (Cetacea, Basilosauridae) from the late Middle Eocene of Eastern Europe. Journal of Paleontology 87:254–268.

Gray, N.-M., K. Kainec, S. Madar, L. Tomko, and S. Wolfe. 2007. Sink or swim? Bone density as a mechanism for buoyancy control in early cetaceans. The Anatomical Record 290:638–653.

Gutstein, C. S., M. A. Cozzuol, and N. D. Pyenson. 2014a. The antiquity of riverine adaptations in Iniidae (Cetacea, Odontoceti) documented by a humerus from the Late

Miocene of the Ituzaingó Formation, Argentina. The Anatomical Record 297:1096–1102.

Gutstein, C. S., C. P. Figueroa-Bravo, N. D. Pyenson, R. E. Yury-Yañez, M. A. Cozzuol, and M. Canals. 2014b. High frequency echolocation, ear morphology, and the marine–freshwater transition: a comparative study of extant and extinct toothed whales. Palaeogeography, Palaeoclimatology, Palaeoecology 400:62–74.

Hamilton, H., S. Caballero, A. G. Collins, and R. L. Brownell, Jr. 2001. Evolution of river dolphins. Proceedings of the Royal Society B 268:549–556.

Harlan, R. 1834. Notice of fossil bones found in the Tertiary formation of the state of Louisiana. Transactions of the American Philosophical Society Philadelphia 4:397–403.

Hemilä, S., S. Nummela, and T. Reuter. 2010. Anatomy and physics of the exceptional sensitivity of dolphin hearing (Odontoceti: Cetacea). Journal of Comparative Physiology A 196:165–179.

Heyning, J. E. 1989. Comparative facial anatomy of beaked whales (Ziphiidae) and a systematic revision among the families of extant Odontoceti. Contributions in Science, Natural History Museum of Los Angeles County 405:1–64.

Heyning, J. E. 1997. Sperm whale phylogeny revisited: analysis of the morphological evidence. Marine Mammal Science 13:596–613.

Heyning, J. E., and J. G. Mead. 1990. Evolution of the nasal anatomy of cetaceans; pp. 67–79 in J. Thomas and R. Kastelein (eds.), Sensory Abilities of Cetaceans. Springer, New York.

Houssaye, A., P. Tafforeau, C. de Muizon, and P. D. Gingerich. 2015. Transition of Eocene whales from land to sea: evidence from bone microstructure. PLoS One 10:e0118409.

Hui, C. A. 1981. Seawater consumption and water flux in the common dolphin *Delphinus delphis*. Physiological Zoology 54:430–440.

Hulbert, R. C., Jr. 1998. Postcranial osteology of the North American Middle Eocene protocetid *Georgiacetus*; pp. 235–268 in

J. G. M. Thewissen (ed.), The Emergence of Whales. Plenum Press, New York.

Hulbert, R. C., Jr., R. M. Petkewich, G. A. Bishop, D. Bukry, and D. P. Aleshire. 1998. A new middle Eocene protocetid whale (Mammalia: Cetacea: Archaeoceti) and associated biota from Georgia. Journal of Paleontology 72:907–927.

Ichishima, H., H. Sawamura, H. Ito, S. Otani, and H. Ishikawa. 2008: Do the so-called nutrient foramina on the palate tell us the presence of baleen plates in toothed mysticetes? Paper presented at the Fifth Conference on Secondary Adaptations of Tetrapods to Life in Water, Tokyo, 2008.

Jourdan, C. 1861. Description de restes fossiles de deux grands mammifères constituant deux genres: l'un le genre *Rhizoprion* de l'ordre des Cétacés et du groupe des *Delphinoides*; l'autre le genre *Dynocyon* de l'ordre des Carnassiers et de la famille des Canidés. Annales des Sciences Naturelles, Zoologie 16:369–372.

Kellogg, R. 1936. A Review of the Archaeoceti. Carnegie Institution of Washington Publication 482. Carnegie Institution, Washington, DC.

Kishida, T., and J. G. M. Thewissen. 2012. Evolutionary changes of the importance of olfaction in cetaceans based on the olfactory marker protein gene. Gene 492:349–353.

Klima, M. 1980. Morphology of the pectoral girdle in the Amazon dolphin *Inia geoffrensis* with special reference to the shoulder joint and the movements of the flippers. Zeitschrift für Säugetierkunde 45:288–309.

Köhler, R., and R. E. Fordyce. 1997. An archaeocete whale (Cetacea: Archaeoceti) from the Eocene Waihao Greensand, New Zealand. Journal of Vertebrate Paleontology 17:574–583.

Lambert, O., G. Bianucci, K. Post, C. de Muizon, R. Salas-Gismondi, M. Urbina, and J. Reumer. 2010. The giant bite of a new raptorial sperm whale from the Miocene epoch of Peru. Nature 466:105–108.

Levenson, D. H., and A. Dizon. 2003. Genetic evidence for the ancestral loss of short-wavelength-sensitive cone pigments in mysticete and odontocete cetaceans. Proceedings of the Royal Society B 270:673–679.

Lindberg, D. R., and N. D. Pyenson. 2007. Things that go bump in the night: evolutionary interactions between cephalopods and cetaceans in the tertiary. Lethaia 40:335–343.

Liu, Y., S. J. Rossiter, X. Han, J. A. Cotton, and S. Zhang. 2010. Cetaceans on a molecular fast track to ultrasonic hearing. Current Biology 20:1834–1839.

Lucas, F. A. 1901. The pelvic girdle of zeuglodon *Basilosaurus cetoides* (Owen), with notes on other portions of the skeleton. Proceedings of the United States National Museum 23:327–331.

Luo, Z.-X., and E. R. Eastman. 1995. Petrosal and inner ear of a squalodontoid whale: implications for evolution of hearing in odontocetes. Journal of Vertebrate Paleontology 15:431–442.

Luo, Z.-X., and K. Marsh. 1996. Petrosal (periotic) and inner ear of a Pliocene kogiine whale (Kogiinae, Odontoceti): implications on relationships and hearing evolution of toothed whales. Journal of Vertebrate Paleontology 16:328–348.

MacLeod, C. D., J. S. Reidenberg, M. Weller, M. B. Santos, J. Herman, J. Goold, and G. J. Pierce. 2007. Breaking symmetry: the marine environment, prey size, and the evolution of asymmetry in cetacean skulls. The Anatomical Record 290:539–545.

Madar, S. 1998. Structural adaptations of early archaeocete long bones; pp. 353–378 in J. G. M. Thewissen (ed.), The Emergence of Whales. Plenum Press, New York.

Madar, S. I. 2007. The postcranial skeleton of Early Eocene pakicetid cetaceans. Journal of Paleontology 81:176–200.

Madar, S. I., J. G. M. Thewissen, and S. T. Hussain. 2002. Additional holotype remains of *Ambulocetus natans* (Cetacea, Ambulocetidae), and their implications for locomotion in early whales. Journal of Vertebrate Paleontology 22:405–422.

Marino, L., D. W. McShea, and M. D. Uhen. 2004. Origin and evolution of large brains in toothed whales. The Anatomical Record 281A:1247–1255.

Marples, B. J. 1956. Cetotheres (Cetacea) from the Oligocene of New Zealand. Proceedings of the Zoological Society of London 126:565–580.

Martínez Cáceres, M., and C. de Muizon. 2011. A new basilosaurid (Cetacea, Pelagiceti) from the Late Eocene to Early Oligocene Otuma Formation of Peru. Comptes Rendus Palevol 10:517–526.

Marx, F. G. 2011. The more the merrier? A large cladistic analysis of mysticetes, and comments on the transition from teeth to baleen. Journal of Mammalian Evolution 18:77–100.

McGowen, M. R., C. Clark, and J. Gatesy. 2008. The vestigial olfactory receptor subgenome of odontocete whales: phylogenetic congruence between gene-tree reconciliation and supermatrix methods. Systematic Biology 57:574–590.

McGowen, M. R., M. Spaulding, and J. Gatesy. 2009. Divergence date estimation and a comprehensive molecular tree of extant cetaceans. Molecular Phylogenetics and Evolution 53:891–906.

McKenna, M. F., T. W. Cranford, A. Berta, and N. D. Pyenson. 2012. Morphology of the odontocete melon and its implications for acoustic function. Marine Mammal Science 28:690–713.

Mead, J. G., and R. E. Fordyce. 2009. The therian skull: a lexicon with emphasis on the odontocetes. Smithsonian Contributions to Zoology 627:1–248.

Meredith, R. W., J. Gatesy, C. A. Emerling, V. M. York, and M. S. Springer. 2013. Rod monochromacy and the coevolution of cetacean retinal opsins. PLoS Genetics 9:e1003432.

Milinkovitch, M. C. 1995. Molecular phylogeny of cetaceans prompts revision of morphological transformations. Trends in Ecology & Evolution 10:328–334.

Mitchell, E. D. 1989. A new cetacean from the late Eocene La Meseta Formation, Seymour Island, Antarctic Peninsula. Canadian Journal of Fisheries and Aquatic Science 46:2219–2235.

Montgomery, S. H., J. H. Geisler, M. R. McGowen, C. Fox, L. Marino, and J. Gatesy. 2013. The evolutionary history of cetacean brain and body size. Evolution 67:3339–3353.

Nicol, S., A. Bowie, S. Jarman, D. Lannuzel, K. M. Meiners, and P. Van Der Merwe. 2010. Southern Ocean iron fertilization by baleen whales and Antarctic krill. Fish and Fisheries 11:203–209.

Norris, K. S. 1968. The evolution of acoustic mechanisms in odontocete cetaceans; pp. 297–324 in E. T. Drake (ed.), Evolution and Environment. Yale University Press, New Haven, CT.

Nummela, S., J. E. Kosove, T. E. Lancaster, and J. G. M. Thewissen. 2004a. Lateral mandibular wall thickness in *Tursiops truncatus*: variation due to sex and age. Marine Mammal Science 20:491–497.

Nummela, S., J. G. M. Thewissen, S. Bajpai, S. T. Hussain, and K. Kumar. 2004b. Eocene evolution of whale hearing. Nature 430:776–778.

Nummela, S., J. G. M. Thewissen, S. Bajpai, T. Hussain, and K. Kumar. 2007. Sound transmission in archaic and modern whales: anatomical adaptations for underwater hearing. The Anatomical Record 290:716–733.

Okazaki, Y. 2012. A new mysticete from the upper Oligocene Ashiya Group, Kyushu, Japan and its significance to mysticete evolution. Bulletin of the Kitakyushu Museum of Natural History and Human History, Series A (Natural History) 10:129–152.

Owen, R. 1839. Observations on the *Basilosaurus* of Dr. Harlan (*Zeuglodon cetoides*, Owen). Transactions of the Geological Society of London 6:69–79.

Pihlström, H. 2008. Comparative anatomy and physiology of chemical senses in aquatic mammals; pp. 95–111 in J. G. M. Thewissen, and S. Nummela (eds.), Sensory Evolution on the Threshold. University of California Press, Berkeley.

Pinto, S. J. D., and R. E. Shadwick. 2013. Material and structural properties of fin whale (*Balaenoptera physalus*) Zwischensubstanz. Journal of Morphology 274:947–955.

Pivorunas, A. 1979. The feeding mechanisms of baleen whales. American Scientist 67:432–440.

Pyenson, N. D., J. Vélez-Juarbe, C. S. Gutstein, H. Little, D. Vigil, and A. O'Dea. 2015. *Isthminia panamensis*, a new fossil inioid (Mammalia, Cetacea) from the Chagres Formation of Panama and the evolution of "river dolphins" in the Americas. PeerJ 3:e1227.

Reeves, R. R., and A. R. Martin. 2009. River dolphins; pp. 976–979 in W. F. Perrin, B. Würsig, and J. G. M. Thewissen (eds.), Encyclopedia of Marine Mammals. Academic Press, Burlington, MA.

Reidenberg, J. S., and J. T. Laitman. 2008. Sisters of the sinuses: cetacean air sacs. The Anatomical Record 291:1389–1396.

Rice, D. W. 2009. Baleen; pp. 78–80 in W. F. Perrin, B. Würsig, and J. G. M. Thewissen (eds.), Encyclopedia of Marine Mammals. Academic Press, Burlington, MA.

Roe, L. J., J. G. M. Thewissen, J. Quade, J. R. O'Neil, S. Bajpai, A. Sahni, and S. T. Hussain. 1998. Isotopic approaches to understanding the terrestrial-to-marine transition of the earliest cetaceans; pp. 399–422 in J. G. M. Thewissen (ed.), The Emergence of Whales. Plenum Press, New York.

Roman, J., and J. J. McCarthy. 2010. The whale pump: marine mammals enhance primary productivity in a coastal basin. PLoS One 5:e13255.

Rose, K. D. 1985. Comparative osteology of North American dichobunid artiodactyls. Journal of Paleontology 59:1203–1226.

Sanders, A. E., and L. G. Barnes. 2002. Paleontology of the Late Oligocene Ashley and Chandler Bridge Formations of South Carolina, 3: Eomysticetidae, a new family of primitive mysticetes (Mammalia: Cetacea). Smithsonian Contributions to Paleobiology 93:313–356.

Sawamura, H., H. Ichishima, H. Ito, and H. Ishikawa. 2006. Features implying the beginning of baleen growth in aetiocetids. Journal of Vertebrate Paleontology 26 (Supplement to 3):120A.

Slijper, E. J. 1962. Whales. Basic Books, New York.

Spoor, F. 2003. The semicircular canal system and locomotor behaviour, with special reference to hominin evolution. Courier Forschungsinstitut Senckenberg 243:93–104.

Spoor, F., S. Bajpai, S. T. Hussain, K. Kumar, and J. G. M. Thewissen. 2002. Vestibular evidence for the evolution of aquatic behaviour in early cetaceans. Nature 417:163–166.

Spoor, F., T. Garland, G. Krovitz, T. M. Ryan, M. T. Silcox, and A. Walker. 2007. The primate semicircular canal system and locomotion. Proceedings of the National Academy of Sciences 104:10808–10812.

Spoor, F., and J. G. M. Thewissen. 2008. Comparative and functional anatomy of balance in aquatic mammals; pp. 257–284 in J. G. M. Thewissen, and S. Nummela (eds.), Sensory Evolution on the Threshold. University of California Press, Berkeley.

Steeman, M. E. 2007. Cladistic analysis and a revised classification of fossil and recent mysticetes. Zoological Journal of the Linnean Society 150:875–894.

Steeman, M. E., M. B. Hebsgaard, R. E. Fordyce, S. Y. W. Ho, D. L. Rabosky, R. Nielsen, C. Rhabek, H. Glenner, M. V. Sørensen, and E. Willerslev. 2009. Radiation of extant cetaceans driven by restructuring of the oceans. Systematic Biology 58:573–585.

Tanaka, Y., and R. E. Fordyce. 2014. Fossil dolphin *Otekaikea marplesi* (Latest Oligocene, New Zealand) expands the morphological and taxonomic diversity of Oligocene cetaceans. PLoS One 9:e107972.

Thewissen, J. G. M., and S. Bajpai. 2009. New skeletal material of *Andrewsiphius* and *Kutchicetus*, two Eocene cetaceans from India. Journal of Paleontology 83:635–663.

Thewissen, J. G. M., L. N. Cooper, M. T. Clementz, S. Bajpai, and B. N. Tiwari. 2007. Whales originated form aquatic artiodactyls in the Eocene epoch of India. Nature 450:1190–1195.

Thewissen, J. G. M., and F. E. Fish. 1997. Locomotor evolution in the earliest cetaceans: functional model, modern analogues, and paleontological evidence. Paleobiology 23:482–490.

Thewissen, J. G. M., J. George, C. Rosa, and T. Kishida. 2011a. Olfaction and brain size in the bowhead whale (*Balaena mysticetus*). Marine Mammal Science 27:282–294.

Thewissen, J. G. M., and S. T. Hussain. 1993. Origin of underwater hearing in whales. Nature 361:444–445.

Thewissen, J. G. M., S. T. Hussain, and M. Arif. 1994. Fossil evidence for the origin of aquatic locomotion in archaeocete whales. Science 263:210–212.

Thewissen, J. G. M., D. McBurney, J. C. George, and R. Suydam. 2014. Tooth and baleen development in the bowhead whale; in Seventh Triennial Conference on Secondary Adaptation of Tetrapods to Life in Water: Abstracts, San Diego.

Thewissen, J. G. M., and W. A. McLellan. 2009. *Maiacetus*: displaced fetus or last meal? Comment on: Gingerich, P. D., ul-Haq, M., W. von Koenigswald, W. J. Sanders, B. H. Smith, and I. S. Zalmout. New protocetid whale from the middle Eocene of Pakistan: birth on land, precocial development, and sexual dimorphism. PLoS One 5:e4366.

Thewissen, J. G. M., and S. Nummela. 2008. Towards an integrative approach; pp. 333–340 in J. G. M. Thewissen, and S. Nummela (eds.), Sensory Evolution on the Threshold. University of California Press, Berkeley.

Thewissen, J. G. M., L. J. Roe, J. R. O'Neil, S. T. Hussain, A. Sahni, and S. Bajpai. 1996. Evolution of cetacean osmoregulation. Nature 381:379–380.

Thewissen, J. G. M., J. D. Sensor, M. T. Clementz, and S. Bajpai. 2011b. Evolution of dental wear and diet during the origin of whales. Paleobiology 37:655–669.

Thewissen, J. G. M., E. M. Williams, L. J. Roe, and S. T. Hussain. 2001. Skeletons of terrestrial cetaceans and the relationship of whales to artiodactyls. Nature 413: 277–281.

Uhen, M. D. 2004. Form, function, and anatomy of *Dorudon atrox* (Mammalia: Cetacea): An archaeocete from the Middle to Late Eocene of Egypt. University of Michigan Papers on Paleontology 34. Ann Arbor: University of Michigan.

Uhen, M. D. 2008a. New protocetid whales from Alabama and Mississippi, and a new cetacean clade, Pelagiceti. Journal of Vertebrate Paleontology 28:589–593.

Uhen, M. D. 2008b. A new *Xenorophus*-like odontocete cetacean from the Oligocene of North Carolina and a discussion of the basal odontocete radiation. Journal of Systematic Palaeontology 6:433–452.

Uhen, M. D. 2014. New material of *Natchitochia jonesi* and a comparison of the innominata and locomotor capabilities of Protocetidae. Marine Mammal Science 30:1029–1066.

Valerio, A. L., and C. A. Laurito. 2012. Cetáceos fósiles (Mammalia, Odontoceti, Eurhinodelphionoidea, Inioidea, Physeterioidea) de la Formación Curré, Mioceno Superior (Hemphilliano Temprano Tardío). Revista Geológica de América Central 46:151–160.

Van Valen, L. 1968. Monophyly or diphyly in the origin of whales. Evolution 22:37–41.

Wall, W. P. 1983. The Correlation between high limb-bone density and aquatic habits in recent mammals. Journal of Paleontology 57:197–207.

Werth, A. J. 2000. Feeding in marine mammals; pp. 487–526 in K. Schwenk (ed.), Feeding: Form, Function and Evolution in Tetrapods. Academic Press, San Diego.

Werth, A. J. 2001. How do mysticetes remove prey trapped in baleen? Bulletin of the Museum of Comparative Zoology 156:189–203.

Werth, A. J. 2012. Hydrodynamic and sensory factors governing response of copepods to simulated predation by balaenid whales. International Journal of Ecology 2012:208913.

Whitmore, F. C., Jr., and L. G. Barnes. 2008. The Herpetocetinae, a new subfamily of extinct baleen whales (Mammalia, Cetacea, Cetotheriidae). Virginia Museum of Natural History Special Publication 14:141–180.

Wilson, V. J., R. Boyle, K. Fukushima, P. K. Rose, Y. Shinoda, Y. Sugiuchi, and Y. Uchino. 1995. The vestibulocollic reflex. Journal of Vestibular Research 5:147–170.

6 Fossil Evidence of Cetacean Biology

6.1 Feeding strategies

All cetaceans are marine carnivores of one sort or another, yet their diets vary considerably depending on how they acquire and process their prey, and at what level in the food web they feed. Cetaceans utilize an astonishing breadth of food sources, ranging from zooplankton to variably sized fish, cephalopods, and sometimes even other marine mammals (Figure 6.1). This diversity has given rise to a variety of feeding morphologies, most of which are reflected in the skeleton and/or leave evidence in the form of dental wear, stable isotope compositions, and trace fossils (Clementz *et al.*, 2014; Fahlke *et al.*, 2013; Thewissen *et al.*, 2011). Often, the correlation between a particular feeding strategy and its morphological correlates is tight enough that a predictable, sometimes defining set of features will evolve convergently in several unrelated taxa. Particularly striking examples of this pattern are features associated with raptorial snapping (e.g., an elongated rostrum bearing numerous teeth), which characterize all living "river dolphins," or the reduced dentition commonly found in suction feeders (Werth, 2000).

6.1.1 Archaeocetes

To understand the feeding strategies of the earliest cetaceans, it is informative to study their closest living relatives, the artiodactyls. Most of the latter are strictly **herbivorous**, although some, such as pigs and entelodonts, are **omnivores** (Joeckel, 1990). Interestingly, even grazing hippopotami have occasionally been observed to consume meat, although generally only during times of severe nutritional stress (Dudley, 1998). Notwithstanding these omnivorous tendencies, the closest relatives of cetaceans, the extinct raoellids, probably fed on terrestrial vegetation, as shown by both stable isotope data and their retention of **bunodont** (low with rounded cusps) molars bearing well-developed crushing basins (Thewissen *et al.*, 2007, 2011).

Despite the overall conservative feeding morphology of raoellids, at least one of them (*Indohyus*) differs from other bunodont artiodactyls in the size and shape of its tooth wear. In bunodont herbivores, the tips of the individual tooth cusps are initially abraded through contact with the food itself, resulting in the formation of roughly horizontal, smooth **apical wear** facets. As the jaws continue to close, the lingual sides of the upper molars start to slide down along the buccal sides of the lower ones, thus creating pairs of corresponding **shearing** facets (Phase I). A second pair of facets (Phase II) is formed as the lower jaw is abducted again and the teeth move out of occlusion. *Indohyus* shows all three types of wear, but, unlike in other artiodactyls, Phase I shearing is clearly dominant (Thewissen *et al.*, 2011). A similar, even more pronounced shearing-dominated wear pattern is present in archaeocetes,

Cetacean Paleobiology, First Edition. Felix G. Marx, Olivier Lambert, and Mark D. Uhen.
© 2016 John Wiley & Sons, Ltd. Published 2016 by John Wiley & Sons, Ltd.

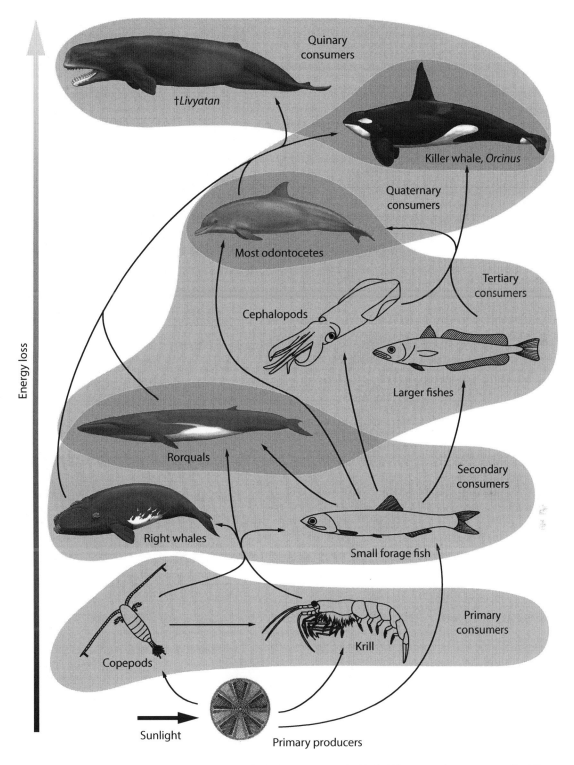

Figure 6.1 Cetacean feeding strategies and food web. Archaeocetes probably fed like most odontocetes and/or killer whales.

even though they markedly differ from *Indohyus* in the gross morphology of their teeth (discussed further in this chapter; and see Figure 6.2). This discrepancy between tooth morphology and wear reveals an interesting decoupling of form, function, and diet at the origin of Cetacea, which inherited a functional (wear) pattern that is suited to their aquatic diet, yet arose before they evolved

a corresponding dental or rostral morphology, or even started to forage in water (Thewissen *et al.,* 2011).

Isotopic and morphological evidence for aquatic foraging implies a major shift in dietary preferences coinciding with the origin of cetaceans (section 5.1.1) (Clementz *et al.,* 2006; O'Leary and Uhen, 1999; Thewissen *et al.,* 2011). Compared to

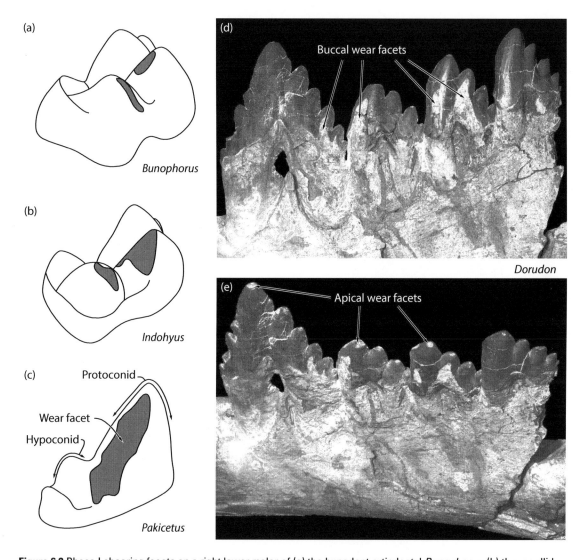

Figure 6.2 Phase I shearing facets on a right lower molar of (a) the bunodont artiodactyl *Bunophorus*, (b) the raoellid *Indohyus*, and (c) the pakicetid *Pakicetus*, all in buccodistal view. (d) Phase I shearing facets (in lateral/buccal view) and (e) apical wear (in dorsolateral/occlusobuccal view) in the basilosaurid *Dorudon atrox*. (a–c) Adapted from Thewissen *et al.* (2011).

Indohyus, the rostrum of archaeocetes is elongate, which forces the incisors to become anteroposteriorly aligned with the cheek teeth (Uhen, 2007). In aquatic tetrapods, a long snout facilitates prey capture via rapid forward or lateral movements of the head or body. Such a **raptorial** feeding strategy is present in several extant odontocetes, and is likely ancestral for cetaceans as a whole (Werth, 2000). Like their artiodactyl forebears, but unlike modern odontocetes, archaeocetes have a **heterodont** dentition and likely chewed their prey (Fahlke *et al.*, 2013; Loch *et al.*, 2015; O'Leary and Uhen, 1999). However, in most archaeocetes, including pakicetids, the lower molars effectively comprise only two relatively high cusps: the mesial protoconid and the distal hypoconid (Figure 6.2). This is in stark contrast to artiodactyls and most other terrestrial mammals, in which the protoconid and the hypoconid form part of much broader, multi-cusped slicing (trigonid) and crushing (talonid) portions of the tooth, respectively.

With the reduction of the trigonid and talonid to one cusp each, the lower molars of pakicetids, ambulocetids, and protocetids are no longer able to perform their ancestral crushing function and become mainly adapted for slicing (Thewissen *et al.*, 2007). Remingtonocetids and basilosaurids go one step further and convergently lose the distinction between the trigonid and the talonid altogether. Instead, their lower molars consist of a series of anteroposteriorly aligned cusps, which creates a jagged cutting surface between opposing teeth that further emphasizes slicing (Figure 6.3) (Thewissen and Bajpai, 2001; Uhen, 1998). The upper molars

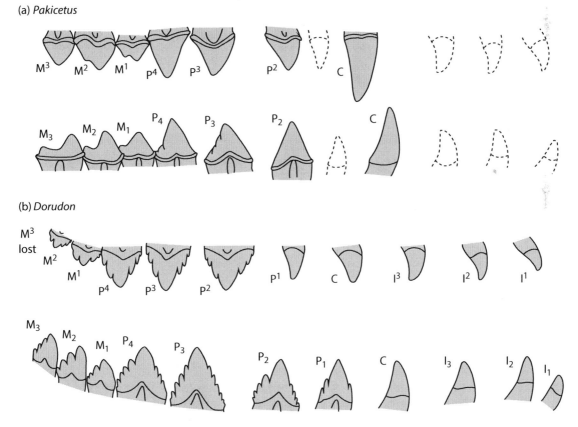

Figure 6.3 Dentition of (a) the pakicetid *Pakicetus* and (b) the basilosaurid *Dorudon*, both in lateral view. The dentition of *Pakicetus* is a composite based on several species. Note the presence of accessory denticles on the cheek teeth in *Dorudon*. Dentition of *Pakicetus* adapted from Cooper *et al.* (2009).

retain three cusps forming a distinct crushing basin—the mesial paracone, the distal metacone and the lingual protocone—in pakicetids and ambulocetids. However, the protocone is reduced over the course of protocetid evolution and convergently lost in both remingtonocetids and basilosaurids (Hulbert *et al.*, 1998; Thewissen and Bajpai, 2001; Thewissen *et al.*, 2011). As in the case of the lower molars, the loss of the protocone has the effect of making the upper molars transversely narrow, and functionally emphasizes slicing over crushing.

Macroscopic tooth wear in all archaeocetes is virtually limited to Phase I shearing, which often extends across the entire length of the tooth and may be indicative of a largely fish-based diet (**piscivory**) (Gingerich *et al.*, 2009; O'Leary and Uhen, 1999; Thewissen *et al.*, 2011). This interpretation is seemingly confirmed by the fossilized stomach contents of two basilosaurids: that of

Basilosaurus cetoides, consisting of various sharks and teleost fishes (Swift and Barnes, 1996); and that of *Dorudon atrox*, consisting entirely of bony fishes (Uhen, 2004). Studies of cetacean microwear show, however, that the diet of early cetaceans was likely much more varied. In general, the microwear pattern of most archaeocetes resembles that of extant pinnipeds, which, besides fish, feed on a variety of relatively hard objects, including shelled gastropods, bivalves, and crustaceans (Figure 6.4) (Fahlke *et al.*, 2013). Other evidence for variable feeding strategies, especially in protocetids and the mysterious Oligocene kekenodontids (see section 6.1.2), comes from stable carbon isotopes and variations in rostral shape (Clementz *et al.*, 2014; Gingerich *et al.*, 2001; Thewissen *et al.*, 2011).

Pakicetids likely were bottom waders and foraged in freshwater systems (Thewissen *et al.*, 2007, 2011). The presence of numerous foramina near the

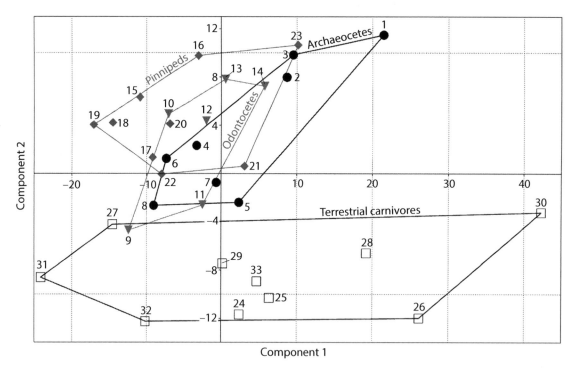

Figure 6.4 Principal components analysis of tooth microwear in extinct archaeocetes, extant odontocetes, pinnipeds, and terrestrial carnivores. Component 1 correlates with the texture and size of ingested food, with negative values indicating soft (in terrestrial mammals) or small hard food items, whereas positive values point to larger, hard items in the diet. Component 2 mostly separates terrestrial from aquatic taxa. Reproduced from Fahlke *et al.* (2013), with permission of Elsevier.

tip of the rostrum suggests that they may have detected their prey via well-developed whiskers or some other kind of mechanoreceptor (section 5.1.1) (Thewissen and Nummela, 2008). *Ambulocetus* and remingtonocetids have both been interpreted as ambush predators foraging in freshwater and coastal environments, respectively (Bajpai *et al.*, 2011; Thewissen *et al.*, 1996). Alternatively, remingtonocetids may have been raptorial **snap feeders** similar to extant "river dolphins" (section 5.3), as indicated by their extremely elongate snouts and tall incisors (Cooper *et al.*, 2014). Judging from the size of their orbits, at least some remingtonocetids had comparatively small eyes, and hence likely poor vision. This again resembles the situation in extant "river dolphins" and may be explained by their habitat preferences, which included the turbid waters of coastal marshes and swamps (Bajpai *et al.*, 2011; Thewissen and Bajpai, 2009). Unlike river dolphins, however, remingtonocetids could not resort to echolocation to detect potential prey. Instead, like pakicetids, they may have relied on mechanoreception, such as in the form of well-developed whiskers (Thewissen and Bajpai, 2009).

Some protocetids and basilosaurids show both micro- and macroscopic dental evidence (e.g., pronounced apical tooth wear and irregular cracking of enamel) of ingesting large, hard objects, such as other marine mammals and birds (Figure 6.4) (Fahlke *et al.*, 2013; Thewissen *et al.*, 2011). This idea is supported by the presence of tell-tale bite marks on juvenile specimens of *D. atrox*, which match the teeth of adult *Basilosaurus isis* in size and shape (Fahlke, 2012; Uhen, 2004). It currently remains unclear whether these bite marks are the result of scavenging or predation, but 3D modeling suggests a bite force for *B. isis* that is well above that needed to crush bone, and indeed the strongest so far recorded in any mammal (Figure 6.5) (Snively *et al.*, 2015). It is also striking that all of the affected *D. atrox* specimens were recovered from shallow marine deposits that have yielded both adult and juvenile *D. atrox*, but only adult *B. isis*. A possible explanation for this distribution may be that *B. isis* invaded the breeding grounds of *Dorudon* in order to prey on its calves (Fahlke, 2012; Uhen, 2004). A similar suggestion, although this time involving cannibalism, has been put forward to explain the preservation of an

extremely juvenile skeleton inside an adult individual of the protocetid *Maiacetus* (Thewissen and McLellan, 2009). This view remains debated, however, with the specimen usually being interpreted as a mother bearing a near-term fetus (section 6.3) (Gingerich *et al.*, 2009).

6.1.2 Mysticeti

Mysticetes are characterized by their loss of teeth and the possession of baleen, which they use to feed on vast amounts of zooplankton and small fish. This prey preference places them near the bottom of the marine food web, and is reflected not only in their filter-feeding strategy, but likely also their often gigantic size (Werth, 2000). Shorter food chains mean that less energy is lost through **trophic fractionation**, which makes filter feeding a rather efficient process capable of supporting sizeable populations of large-bodied predators. Less trophic fraction is in turn reflected in the relatively low stable carbon isotope values of extant mysticetes. Similarly low values, especially when compared with archaeocetes or contemporaneous odontocetes, have been recorded in fossil chaeomysticetes from the Late Oligocene of New Zealand (Figure 5.3), thus demonstrating the early origin of filter feeding during mysticete evolution (Clementz *et al.*, 2014).

Extant mysticetes capture their prey via three distinct strategies (Figure 6.6): (1) **skim feeding** (also known as continuous ram feeding), which is found in balaenids, *Caperea* and the sei whale *Balaenoptera borealis*, and involves long bouts of swimming through prey aggregations with a partially opened mouth (Lambertsen *et al.*, 2005; Werth, 2004b) (Plate 5b); (2) lateralized benthic **suction feeding**, which is employed by eschrichtiids and involves the sifting of invertebrates from mouthfuls of bottom sediment (Ray and Schevill, 1974); and (3) the **lunge** or **gulp feeding** typical of balaenopterids (Lambertsen *et al.*, 1995) (Plate 6a). The latter is the most complex of the three feeding modes and starts with the whale lunging itself into a school of fish or krill. During the lunge, a vast amount of prey-laden water is engulfed as part of a single, enormous gulp, and then briefly stored within an expandable throat pouch that may cover more than half of the ventral side of the body. As soon as the mouth is almost closed again,

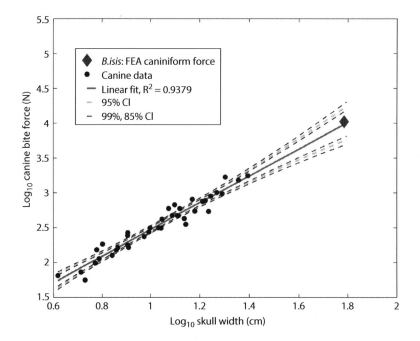

Figure 6.5 Regression of $\log_{10}$-transformed bite force at the canine against skull width in carnivorous mammals. Because the canine of *Basilosaurus* is located further posteriorly than in other mammals, bite force at the caniniform second incisor was estimated instead. *B. isis* has a slightly larger bite force than expected based on its skull width, and the strongest bite of any mammal measured so far. However, the estimate for *B. isis* is also well within the 95% confidence interval of the entire sample, and thus not unusual for a mammal of its size. Reproduced from Snively *et al.* (2015) under a Creative Commons Attribution license.

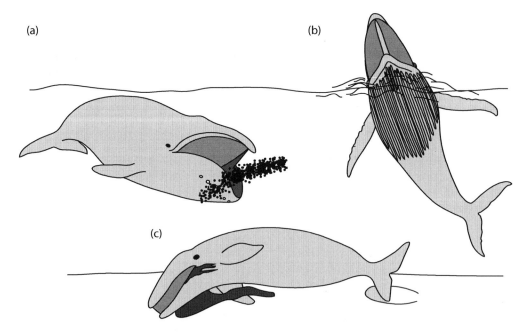

Figure 6.6 Modern mysticete feeding methods: (a) balaenid skim feeding, (b) rorqual lunge feeding, and (c) gray whale benthic suction feeding. (a) Based on drawing by C. Buell; (b,c) adapted from Werth (2000).

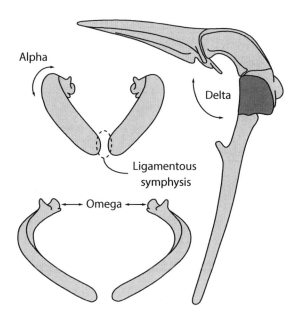

Figure 6.7 Possible movements of the mandible in extant mysticetes, including abduction and adduction (delta rotation), axial rotation (alpha rotation), and lateral displacement (omega rotation). The latter is likely only possible in balaenopterids and, maybe, eschrichtiids. Adapted from Lambertsen *et al.* (1995).

the outer wall of the pouch contracts, which causes the engulfed water to become expelled through the baleen racks. Any prey contained within the water is trapped on the inside of the baleen, and subsequently licked off and swallowed (Pivorunas, 1979; Werth, 2000).

The mandible of mysticetes is highly distinctive, and the largest single bone of any vertebrate (Pyenson *et al.*, 2013). In chaeomysticetes, the mandibular symphysis is always non-sutured (i.e., ligamentous), and the body of the mandible is laterally bowed. The non-sutured symphysis allows each of the lower jaws to rotate in at least two separate planes (Figure 6.7): ventrally (**delta rotation**) and around its own longitudinal axis (**alpha rotation**). In addition, in balaenopterids the usually **synovial** temporomandibular joint, which includes a distinct articular cavity and capsule, has become replaced with a non-synovial, fibrocartilaginous cushion. The latter is highly flexible and allows the joint to **subluxate** (become

dislocated), thus displacing the mandible laterally (**omega rotation**) (Lambertsen *et al.*, 1995). The temporomandibular joint of eschrichtiids has some, but not all, of the features of a synovial joint, and thus appears to be intermediate between that of balaenopterids and other mysticetes (El Adli and Deméré, 2015; Johnston *et al.*, 2010). There is currently no evidence as to whether eschrichtiids are capable of omega rotation.

Because the mandibles of extant mysticetes are laterally bowed, the effects of alpha rotation are comparable to opening and closing an old-fashioned doctor's bag: when the mandibles rotate outward, the size of the mouth increases and vice versa. At the most basic level, alpha rotation ensures that the baleen plates do not become trapped between the upper and lower jaws as the mouth closes (Lillie, 1915). In addition, it likely also helps to regulate the size of the mouth during feeding, and to control the movements of the enormous lips of gray, right, and pygmy right whales (Lambertsen *et al.*, 1995; Ray and Schevill, 1974; Werth, 2004b). In balaenopterids, alpha and omega rotation increase the size of the mouth during the initial gulp, and thus maximize the amount of water that can flow into the ventral throat pouch (Lambertsen *et al.*, 1995). Conversely, moving the mandibles back to their original position as the jaws close may prevent water from bouncing back out of the pouch, although this function has not yet been substantiated by observations on live animals (Arnold *et al.*, 2005). In balaenopterids and possibly eschrichtiids, but not balaenids or *Caperea*, the degree of jaw rotation and (in balaenopterids) ventral pouch expansion is controlled via feedback from a novel sensory organ lodged within the mandibular symphysis (Figure 6.8) (Pyenson *et al.*, 2012).

Bulk feeding characterizes all extant and most fossil mysticetes, but its first appearance is likely not coincident with the origin of the group as a whole (Deméré *et al.*, 2008; Fitzgerald, 2010). Unlike their modern cousins, all of the most basal mysticete families—the Llanocetidae, Mammalodontidae, Aetiocetidae, and some as-yet-undescribed taxa from South Carolina—have well-developed teeth. With few exceptions, the dentition of these archaic mysticetes is heterodont and generally comprises 10–11 teeth in each

Figure 6.8 Symphyseal sensory organ detecting the degree of mandible rotation in balaenopterids. Artwork © C. Buell.

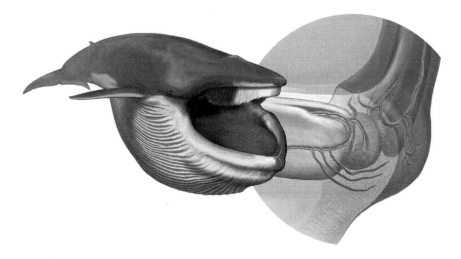

quadrant, with an upper dental formula of 3.1.4.2–3. In some species, the mandible contains one additional posterior tooth, which presumably represents a molar (Barnes *et al.*, 1995; Barnes and Sanders, 1996; Fitzgerald, 2010). The cheek teeth are denticulate and covered by highly ornamented enamel. Unlike in archaeocetes, the (upper) molars are usually separated from each other by variably developed diastemata. Nevertheless, they occlude with their respective counterparts, resulting in the development of shearing facets (Deméré and Berta, 2008; Fitzgerald, 2010; Marx *et al.*, 2015).

Nothing is known of the symphyseal morphology of either *Llanocetus* or *Mammalodon*, but in *Janjucetus* the anterior ends of the mandibles are fused and thus incapable of alpha and omega rotation (Fitzgerald, 2012). In addition, neither llanocetids nor mammalodontids preserve any unequivocal osteological correlates of baleen, although fine grooves around the alveoli of *Llanocetus* may be indicative of well-developed gingival tissues (Fitzgerald, 2010; Fordyce, 2003). Nevertheless, *Llanocetus* in particular has been interpreted as a filter feeder that used its broad, palmate teeth as a sieve to trap small prey items, similar to the extant crabeater seal *Lobodon* (Mitchell, 1989). Similar suggestions have been made regarding other toothed mysticetes (Barnes *et al.*, 1995; Fordyce and Barnes, 1994). More recent studies have, however, questioned whether

the teeth of early mysticetes could indeed have functioned in filter feeding, pointing out they are not only too widely spaced, but also not morphologically intricate enough to form an interlocking lattice capable of capturing small prey (Fitzgerald, 2006; Ichishima, 2005).

In line with the morphological evidence against filter feeding, at least some Oligocene toothed mysticetes have stable carbon isotope values that resemble those of odontocetes, but are higher than those of both contemporaneous toothless mysticetes and, intriguingly, kekenodontids (Figure 6.9) (Clementz *et al.*, 2014). Higher carbon isotope values suggest that early mysticetes were foraging near the top of a comparatively long food chain, and hence likely targeted fewer, much larger prey items than their filter-feeding cousins. In doing so, toothed mysticetes employed a surprising variety of feeding modes: from archaeocete-like raptorial feeding (e.g., in the taxa from South Carolina) to suction feeding in mammalodontids, and incipient filter or suction-assisted raptorial feeding in aetiocetids (Fitzgerald, 2010; Marx *et al.*, 2015). Besides morphology (Figure 6.10), this diversity is again supported by isotopic data, which distinguish between the archaic mysticetes from South Carolina (high δ¹³C values, comparable to those of odontocetes) on the one hand, and mammalodontids (δ¹³C values intermediate between those of odontocetes and mysticetes) on the other (Figure 6.9). Unfortunately, such data

Figure 6.9 Boxplot of bioapatite δ¹³C values for Late Oligocene fossil cetaceans from New Zealand and South Carolina. Statistically significant differences in δ¹³C values are noted by letters above each boxplot; values for odontocetes (A) are significantly higher than those for edentulous mysticetes (B), but neither are significantly distinct from those of toothed mysticetes (AB) or anterior dentition sampled from kekenodontids (AB) (α = 0.05). Reproduced from Clementz *et al.* (2014), with permission of Elsevier.

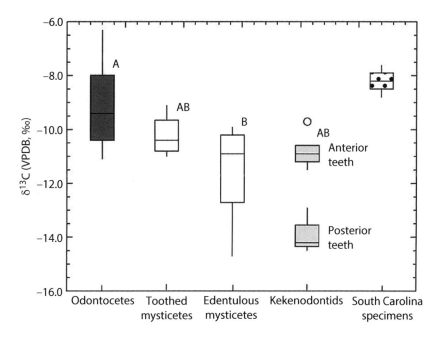

are currently lacking for both llanocetids and the morphologically intermediate, potentially tooth- *and* baleen-bearing aetiocetids (section 5.2.1).

The presence of suction feeding in mammalodontids has been inferred from their markedly foreshortened, blunt jaws, which in extant odontocetes correlate with a relatively small mouth and hence an increased capacity for suction (Werth, 2006). Besides its short rostrum, *Janjucetus* bears many of the hallmarks of a **macrophagous** predator specializing on large prey (Plate 4a), including a robust mandible and strong, sharp, and deeply rooted anterior teeth (Fitzgerald, 2006). By contrast, *Mammalodon* may have relied on benthic suction feeding, as judged from its vestigial upper incisors and large infraorbital and mental foramina (Figure 6.10). The latter may be indicative of highly tactile and/or mobile lips, which would have given the animal a large degree of control over the shape of its mouth (Plate 4b). In addition, the robust manubrium of the sternum

suggests a well-developed sternohyoid muscle, capable of generating considerable negative intraoral pressure. Finally, benthic suction feeding is also supported by the presence of longitudinal grooves on the inside of the teeth, which were likely caused by the ingestion of abrasive sediments (Fitzgerald, 2010). It is unclear whether the striking planar tooth wear of *Mammalodon* is also in some way indicative of benthic foraging. Nevertheless, it does at least exclude tooth-assisted filter feeding as a possible strategy, and suggests a direct crown-to-crown pattern of tooth occlusion that differs from the usually interlocking teeth of raptorial feeders.

Aetiocetids stand out among archaic mysticetes for combining traits otherwise typically associated with either toothed mysticetes or baleen-bearing chaeomysticetes. The most striking of these are the co-occurrence of occluding teeth, palatal nutrient foramina and sulci, and a non-sutured symphysis (Figure 6.10) (section 5.2.1). In

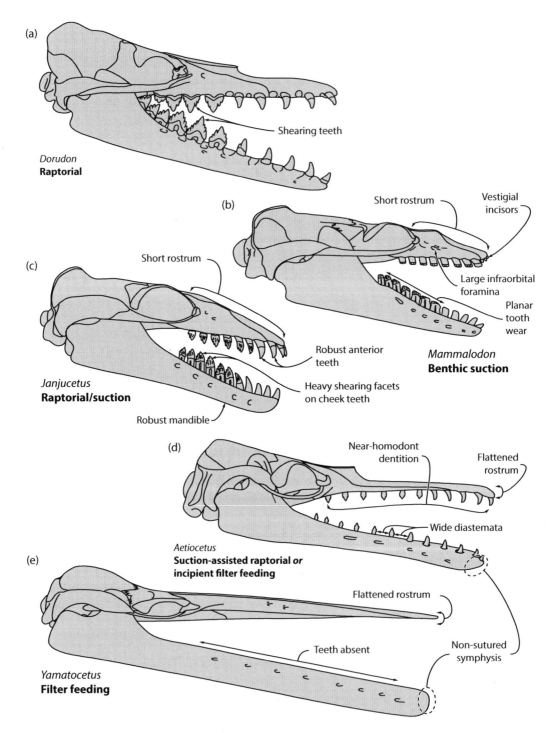

Figure 6.10 Feeding adaptations of (a) the basilosaurid archaeocete *Dorudon*, which resembles the archaic toothed mysticetes from South Carolina, United States; (b) *Mammalodon*; (c) *Janjucetus*; (d) *Aetiocetus*; and (e) the archaic, functionally toothless chaeomysticete *Yamatocetus*. Not to scale.

Figure 6.11 Orientation and relative size of the orbit in chaeomysticetes (circles), toothed mysticetes (squares), pinnipeds (triangles), and the basilosaurid archaeocete *Zygorhiza kochii* (cross). Adapted from Marx (2011), with permission of Springer.

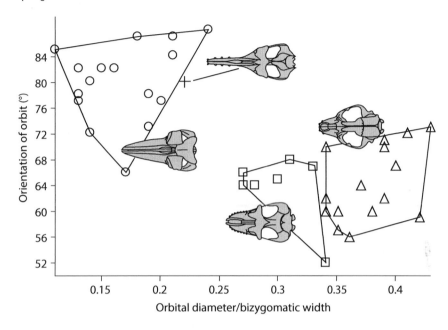

addition, the orbits of both aetiocetids and mammalodontids are remarkably large relative to their body size, which makes them somewhat similar to pinnipeds (Figure 6.11) (Fitzgerald, 2006; Marx, 2011). Within aetiocetids, there are evolutionary tendencies to flatten and broaden the rostrum (**platyrostry**), increase the number of teeth, and simplify the shape of the cheek teeth from broad and denticulate (e.g., *Morawanocetus*) to nearly homodont in some species of *Aetiocetus* (section 8.3.2). All of these trends could plausibly relate to both raptorial and bulk filter feeding: platyrostry, either by decreasing resistance during rapid sideways movements of the head or by increasing the volume of the oral cavity; and the loss of dental complexity either to facilitate prey capture or to reflect a reduced need for a functional dentition (Barnes *et al.*, 1995; Deméré *et al.*, 2008; Fitzgerald, 2010; Werth, 2000). Overall, aetiocetids could thus either have employed a (suction-assisted) raptorial strategy, or else have used a combination of active grasping and incipient filter feeding (Deméré *et al.*, 2008; Marx *et al.*, 2015) (Plate 5a).

Eomysticetids are the earliest whales (ca 30 Ma) that lacked functional teeth and almost certainly possessed baleen and, thus, employed a modern style of mysticete bulk filter feeding (Figure 6.10) (Boessenecker and Fordyce, 2015b; Sanders and Barnes, 2002). Nevertheless, their cranial and mandibular morphology markedly differs from that of later taxa, for example, in the presence of an enlarged temporal fossa and correspondingly narrow intertemporal region, as well as a broad coronoid process and hollow angular process (Boessenecker and Fordyce, 2014a; Okazaki, 2012). Most strikingly, eomysticetids possess a narrow, elongate rostrum that seems to become considerably more pronounced during ontogeny. Exactly what feeding strategy these traits correlate with is not yet entirely clear; however, it is possible that eomysticetids employed a form of skim feeding loosely resembling that of balaenids, with the elongation of the rostrum helping to increase the area across which prey can be filtered (Boessenecker and Fordyce, 2015a). Virtually all other described fossil mysticetes seem to form part of the crown group, and accordingly likely foraged in a manner similar to that of the extant species (Kimura, 2002). Specifically, taxa closely resembling living balaenids and balaenopterids have existed since at least the Early and the Middle Miocene, respectively (Cabrera, 1926; Kohno *et al.*, 2007). "Cetotheres" *sensu lato* may also have employed some incipient form of gulp feeding, although it is unclear to what degree their feeding mode really resembled that of modern balaenopterids (Tsai and Fordyce, 2015).

Cetotheriids are the only major group of crown mysticetes to show a peculiar combination of feeding-related characters not found in any living taxon (*Caperea* being a special case; see section 4.3.2). Virtually all cetotheres have a relatively long, narrow, and flattened rostrum, far posteriorly projecting rostral bones, a comparatively small supraoccipital, and an enlarged angular process of the mandible. The basioccipital crest and sternomastoid fossa are also often well-developed, hinting at strong neck muscles and potentially increased neck flexibility

(Bouetel, 2005; El Adli *et al.*, 2014; Gol'din *et al.*, 2014). In addition to these features, herpetocetines in particular stand out for their twisted postglenoid process, greatly reduced coronoid process, and posteriorly elongated angular process, with the latter bearing attachment surfaces for the seemingly well-developed medial pterygoid and superficial masseter muscles (Figure 6.12). The twisted postglenoid process prevents omega rotation of the mandible and furthermore limits abduction, or delta rotation. By contrast, alpha rotation—controlled by the medial pterygoid and superficial masseter muscles—may have been more pronounced than in extant mysticetes (El Adli *et al.*, 2014).

Considering their narrow rostrum, lack of omega rotation and restricted gape, herpetocetines could not achieve the large oral volumes seen in lunge-feeding balaenopterids. Likewise, herpetocetines lack the arched rostrum typical of skim feeders, such as balaenids and *Caperea*.

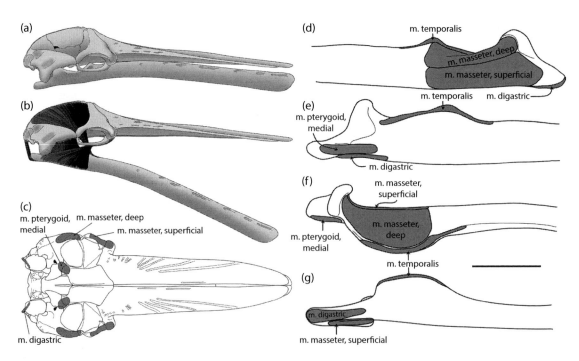

Figure 6.12 Jaw musculature of the cetotheriid mysticete *Herpetocetus morrowi*. (a) Skull and (b) skull with reconstructed musculature in lateral view; (c) muscle attachment sites on ventral side of cranium, in ventral view; and muscle attachment sites on the posterior portion of the mandible in (d) lateral, (e) medial, (f) dorsal, and (g) ventral views. Reproduced from El Adli *et al.* (2014), with permission of the Linnean Society of London.

A small gape and enhanced alpha rotation are, however, consistent with benthic suction feeding, as seen in the gray whale *Eschrichtius robustus* (El Adli *et al.*, 2014). A similar suggestion has been put forward for the non-herpetocetine cetotheriid *Cetotherium riabinini*, based on morphological similarities with suction-feeding ducks and a slight degree of asymmetry in the mandibles and palatal nutrient foramina (Gol'din *et al.*, 2014). The latter might imply that cetotheriids, like the gray whale, fed from the sea bottom while being rolled on their sides. Interestingly, a similar strategy has also been proposed for *Diorocetus hiatus*, a non-cetotheriid fossil chaeomysticete not readily assignable to any of the established families (Beatty and Dooley, 2009). Some form of suction feeding may therefore have been widespread among fossil mysticetes, and possibly represents the ancestral condition for the clade as a whole (Arnold *et al.*, 2005). Alternatively, it may be that cetotheriids in particular followed a feeding strategy no longer employed by any of the living species (El Adli *et al.*, 2014).

6.1.3 Odontoceti

Unlike baleen whales, most odontocetes retain a functional dentition. Archaic odontocetes are generally heterodont and resemble basilosaurids in having conical anterior teeth for capturing prey and denticulate cheek teeth adapted for shearing. Nevertheless, individual cheek teeth tend to be smaller and somewhat more numerous than in basilosaurids, and may be separated from each other by distinct diastemata. By contrast, most crown odontocetes have a roughly homodont dentition, reflecting a drastic decrease in the importance of mastication. Accordingly, most of the living species swallow relatively small, individual prey items in a single piece, with limited or no oral processing (Loch and Simões-Lopes, 2013).

Extant odontocetes prey on fish and cephalopods via raptorial feeding (e.g., "river dolphins"), suction feeding (sperm whales, beaked whales, monodontids, porpoises, and some delphinids), or a combination of the two (some delphinids) (Johnston and Berta, 2011). Specifically, raptorial feeding may involve **ram feeding** (i.e., engulfment of prey via rapid forward movements of the whole body) and **snapping** (i.e., rapid movements of the

body, head, and usually elongate jaws) (Werth, 2000). Initial prey capture via ram feeding may be followed by suction to move the ingested item to the back of the mouth (intraoral transport). Alternatively, suction alone may be used to capture and swallow prey (**capture suction feeding**) (Figure 6.13). In addition to these relatively common strategies, the killer whale (*Orcinus*) and, to a lesser degree, the false and pygmy killer whales (*Pseudorca* and *Feresa*) also prey on larger animals, including other marine mammals, via macrophagous **biting and tearing**.

Adaptations related to any of these feeding styles primarily affect the anterior, tooth-bearing portion of the mandible, as well as, in suction feeders, the hyoid apparatus (Bloodworth and Marshall, 2007; Werth, 2000, 2006). By contrast, the morphology of the mandibular ramus is more closely tied to sound reception (Barroso *et al.*, 2012). As a result of homodonty, modern odontocete teeth lack the morphological variety that in other mammals often reflects feeding styles and preferred types of prey. Nevertheless, the number of teeth and dental wear still to some degree correlate with feeding ecology. Thus, killer whales (*Orcinus orca*) tend to show little dental wear when specializing on mysticetes, heavy dental wear when feeding primarily on bony fish (e.g., herring and mackerel) and seals, and extreme wear when consuming Pacific sleeper sharks, owing to their abrasive skin (Foote *et al.*, 2009; Ford *et al.*, 2011). Such clear associations between wear and prey type are promising in terms of interpreting tooth wear in fossils. However, the great variety of dental wear seen here within a single species also suggests caution, although killer whales are perhaps extreme in their ecomorphological plasticity. In general, odontocete wear patterns remain poorly studied, and further work in this area (e.g., on how wear increases and/or changes with age) is necessary to establish whether there are any more broadly applicable patterns (Loch and Simões-Lopes, 2013).

In terms of tooth number and jaw morphology, raptorial taxa, and especially snappers, rely on elongate pincer jaws bearing numerous, conical teeth to grasp and pierce prey (Figure 6.14). The opposite is true for suction feeders, which instead work more like vacuum cleaners by using negative pressure to

Figure 6.13 Suction feeding process and related hyoid morphology. (a) Combination (top) and capture (bottom) suction feeding, as employed by extant odontocetes; (b) hyoid apparatus of the harbor porpoise *Phocoena phocoena*, shown in its "flexed" (top) and "extended" (bottom) state. Extension of the hyoid apparatus via contraction of the sternohyoideus results in the depression of the tongue, and thus the creation of negative intraoral pressure. Lowercase letters signify joints and uppercase letters associated movements during extension: a, basihyal and ceratohyal; b, ceratohyal and stylohyal; and c, stylohyal and tympanohyal. (a) Reproduced from Werth (2000), with permission of Elsevier; (b) adapted from Werth (2007).

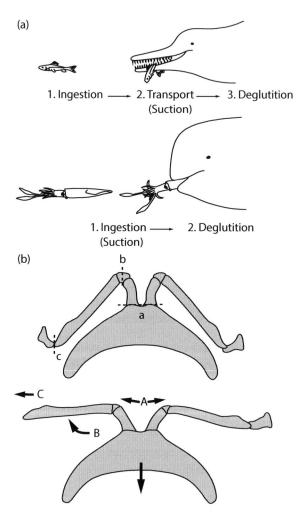

draw prey items—mostly cephalopods, but also fish—directly into the mouth. As a result, they have little use for, and often lose, most or all of their teeth. At the same time, their jaws tend to be relatively short and blunt (**amblygnathous**), which reduces the size of the mouth and thereby improves suction-feeding performance (Figure 6.14) (Werth, 2006). While amblygnathy and a reduced dentition are common among suction feeders, it is important to note that the presence of these features does not always imply suction-feeding capabilities, and vice versa. Specifically, there are at least three important exceptions that muddy the water.

First, killer whales are markedly amblygnathous and, at least as calves, seem to be capable of suction feeding. However, in line with their macrophagous

Figure 6.14 Head shape of extant odontocetes in relation to feeding style, shown for (a) raptorial snap feeders, (b) raptorial ram feeders, (c) capture suction feeders, and (d) macrophagous bite and tear feeders. Note the markedly blunter head shape of capture suction and macrophagous feeders. Life reconstructions © C. Buell.

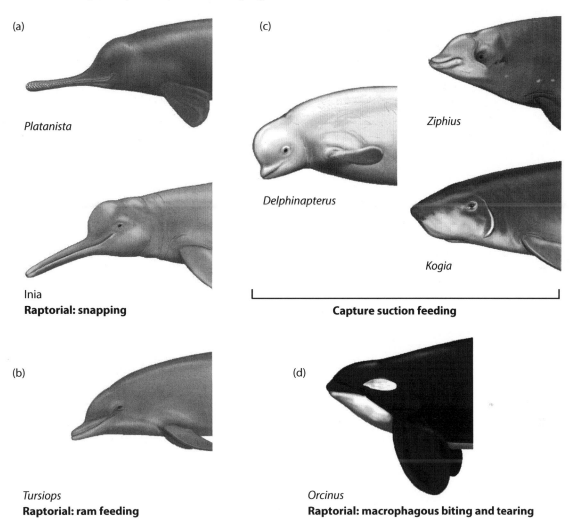

(a)

Platanista

Inia
Raptorial: snapping

(b)

Tursiops
Raptorial: ram feeding

(c)

Delphinapterus

Ziphius

Kogia

Capture suction feeding

(d)

Orcinus
Raptorial: macrophagous biting and tearing

feeding strategy they also retain numerous, robust teeth, and it is currently unclear to what degree adults retain the ability to generate negative intraoral pressure (Werth, 2006). Second, many ziphiids retain elongate jaws despite being suction feeders. Nevertheless, they manage to create an efficient suction opening by partially closing off the sides of their mouth with a distinct **precoronoid crest** and/or adjacent soft tissues (Lambert *et al.*, 2009; Werth, 2006). Finally, the giant sperm whale

is a prime example of a suction feeder, yet retains an elongate lower jaw bearing numerous teeth. Sperm whales have no lips or cheeks, and hence effectively lack an oral cavity whose opening could be adapted for suction. Instead, food is drawn directly into the **oropharyngeal isthmus**, which separates the (nonexistent) oral cavity from the pharynx (Werth, 2004a).

Suction feeding is achieved by momentarily expanding the volume of the oral cavity through rapid and repeated, piston-like retractions of the tongue. This movement creates a negative pressure inside the mouth, which is almost immediately equalized by drawing in water and prey. In many cases, tongue-powered suction is aided by the expansion of 2–10 longitudinal throat grooves (especially in beaked whales), which likely helps to accommodate the ingested water and may increase suction pressure. As the jaws close, prey items are swallowed, whereas most or all of the ingested water is forced back out (Werth, 2000, 2007). The retraction of the tongue is powered by well-developed gular muscles inserting on to the hyoid, which upon contraction is pulled poster-oventrally toward the sternum (Figure 6.13) (Reidenberg and Laitman, 1994; Werth, 2007). The hyoid apparatus of suction feeding taxa is therefore often sturdier than in purely raptorial species, with broader muscle attachment areas and fusion of the basihyal and thyrohyals (Bloodworth and Marshall, 2007; Heyning and Mead, 1996; Johnston and Berta, 2011).

Archaic odontocetes are morphologically diverse and include some short-snouted forms, such as *Prosqualodon* and *Simocetus* (Figure 6.15). With its down-turned jaws and delicate cheek teeth, the latter may have been a bottom feeder preying upon soft-bodied invertebrates (Fordyce, 2002). The rostra of most Oligocene and Miocene odontocetes are elongated, and thus may have been employed in raptorial (snap) feeding. Among the long-snouted Miocene dolphins, eurhinodelphinids stand out for having an anterior portion of the rostrum that projects far beyond the mandibles and may have been used to probe soft sediment for hidden prey (Myrick, 1979). Exactly the opposite condition occurs in the Pliocene phocoenid *Semirostrum*, which has a prolonged, virtually edentulous symphyseal region of the mandible projecting far beyond the rostrum. Well-developed mental canals suggest that the soft tissue covering this portion of the mandible was highly innervated, and thus plausibly tactile (Figure 6.16). A further clue to the benthic feeding habits of this taxon comes from the teeth, which show pronounced abrasion probably caused by ingested sediment (Racicot *et al.*, 2014).

Some form of incipient suction feeding was likely employed by the most archaic ziphiids,

such as *Ninoziphius*. The latter retains a complete set of functional teeth, which it likely used to catch prey (Plate 8). Nevertheless, it also has a well-developed precoronoid crest closing off part of the lateral gape, which suggests suction capabilities (Lambert *et al.*, 2013). Although *Ninoziphius* is about 10 Ma younger than the oldest reliably dated ziphiids, its basal position within the family nonetheless implies that beaked whales have used suction feeding since at least the early Middle Miocene. Heavy apical tooth wear in *Ninoziphius* may point to the ingestion of abrasive sediment, and thus a benthic feeding strategy similar to that of some extant ziphiids (Auster and Watling, 2010; MacLeod *et al.*, 2003). Other likely suction feeders include the walrus-like delphinoid *Odobenocetops*, as judged from its edentulous, vaulted palate, and seemingly well-developed upper lip musculature (de Muizon, 1993; de Muizon *et al.*, 2002) (Plate 11); and the delphinid *Australodelphis*, which resembles ziphiids in its lack of teeth and the presence of a distinct precoronoid crest (section 7.7) (Fordyce *et al.*, 2002). Indirect evidence is also provided by the discovery of fossil ambergris, including several permineralized squid beaks and organic matter indicative of degraded cellular lipids, from the Pleistocene of Italy (Baldanza *et al.*, 2013). Ambergris is produced in the intestines of sperm whales, and its presence suggests that the latter had become adapted to a largely squid-based diet by circa 1.7 Ma (Plate 6b).

Macrophagous biting and tearing also has a long and distinguished history among odontocetes, but, curiously, was originally not the preserve of large delphinids. Instead, Miocene seas were haunted by a series of "killer" sperm whales, such as *Acrophyseter*, *Brygmophyseter*, *Livyatan*, and *Zygophyseter* (Bianucci and Landini, 2006; Lambert *et al.*, 2010b). Unlike their extant, suction-feeding relatives *Physeter* and *Kogia*, these animals had large upper and lower teeth, robust jaws, and vast temporal fossae for well-developed jaw adductor muscles. Their enormous bite force is reflected in unusual bony outgrowths next to the posterior upper tooth alveoli in a specimen of *Acrophyseter*, which likely buttressed the upper teeth against extreme occlusal forces (Figure 6.17) (Lambert *et al.*, 2014). The largest of

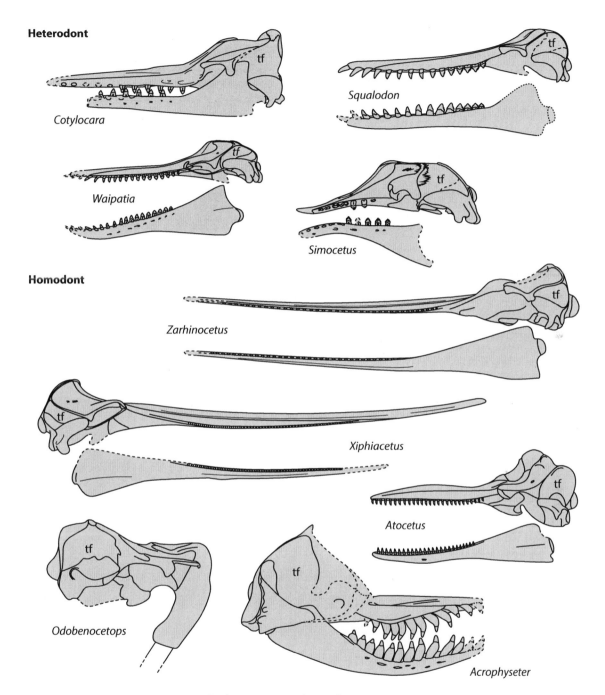

Figure 6.15 Skull of various heterodont (top) and homodont (bottom) extinct odontocetes illustrating the diversity of feeding adaptations within this group. Tf, temporal fossa. Skulls of *Atocetus*, *Odobenocetops*, *Simocetus*, and *Waipatia* adapted from de Muizon (1988, 1993) and Fordyce (2002, 1994), respectively.

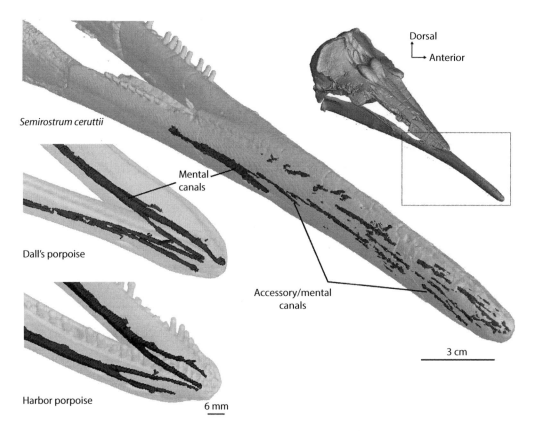

Figure 6.16 Mental and accessory mandibular canals in the Pliocene porpoise *Semirostrum ceruttii*, compared with the extant species *Phocoenoides dalli* (Dall's porpoise) and *Phocoena phocoena* (harbor porpoise), all in right dorsolateral view. Reproduced from Racicot (2014) under a Creative Commons Attribution license.

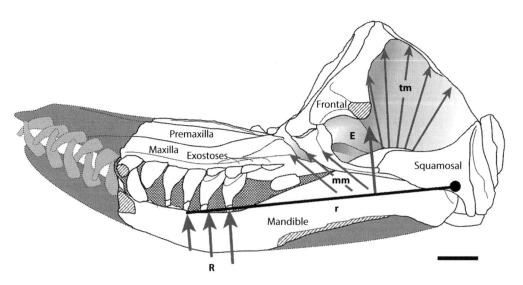

Figure 6.17 Lateral view of the partially reconstructed skull of the stem physeteroid *Acrophyseter* (parallel hatching indicates breakage, cross hatching sediment). Large arrows illustrate (E) the effort force resulting from the adduction of the jaw by the masseter (mm) and temporalis (tm) muscles, and (R) the resistance force acting on the upper teeth during a bite. Note the bony outgrowths (exostoses) buttressing the posterior upper teeth. r, resistance lever arm from lower jaw joint to a given upper tooth. Scale bar equals 100 mm. Reproduced from Lambert *et al.* (2014), with permission of Springer.

Figure 6.18 Comparison of the skulls of (a) *Livyatan*, (b) the extant giant sperm whale *Physeter*, and (c) the extant killer whale *Orcinus*, all in lateral view and to the same scale. Note the large temporal fossa (in dark gray) and teeth of *Livyatan*. Reproduced from Lambert *et al.* (2010), with permission of Macmillan Publishers.

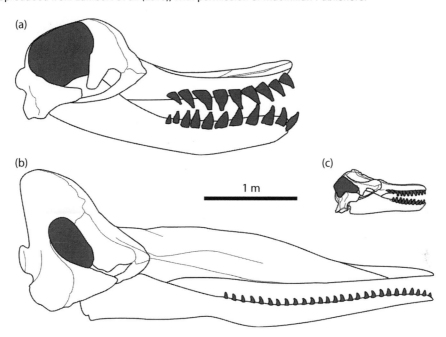

these macropredatorial stem physeteroids, *Livyatan*, lived during the late Middle–Late Miocene and may have reached a total body length of 14–17 m, similar to the extant *Physeter* (Figure 6.18 and Plate 7). This makes *Livyatan* one of the largest raptorial predators ever found, and probably enabled it to feed on prey as large as medium-sized mysticetes (Lambert *et al.*, 2010b).

6.2 Cetaceans as a source of food

6.2.1 Active predation

The only living cetacean that regularly preys on other marine mammals is the killer whale, *Orcinus orca*, with more occasional attacks being carried out by its smaller cousins, the false (*Pseudorca*) and pygmy (*Feresa*) killer whales (Jefferson *et al.*, 1991; Werth, 2000). A similar strategy was presumably followed by several Miocene macropredatorial physeteroids, such as *Livyatan* (section 6.1.3) (Lambert *et al.*, 2010b). However, to date no direct evidence for other cetaceans falling

prey to these killer sperm whales has been found. By contrast, there are several specimens of the basilosaurid *Dorudon atrox*, generally representing young individuals that died in the midst of dental eruption, that display bite marks likely attributable to other, larger archaeocetes, such as *Basilosaurus isis* (Figure 6.19) (Fahlke, 2012; Uhen, 2004).

Besides their own kind, the most obvious potential predators of cetaceans are sharks. Most modern shark consumption of cetaceans is either by larger sharks preying on smaller odontocetes (Heithaus, 2001; Heithaus and Dill, 2002) or by sharks scavenging on a variety of species (Dudley *et al.*, 2000; Fallows *et al.*, 2013; Leclerc *et al.*, 2011). The cetacean fossil record is rich in shark-produced bite marks, shark teeth found alongside whale fossils, and even shark teeth embedded in the bones themselves (Deméré and Cerutti, 1982; Ehret *et al.*, 2009; Govender and Chinsamy, 2013; Uhen, 2004). It is likely that most of these finds reflect scavenging, rather than active predation, although the difference between the two is often difficult to tell (Cicimurri and Knight, 2009; Cigala Fulgosi, 1990).

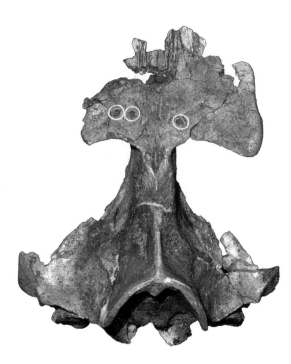

Figure 6.19 Skull of a juvenile *Dorudon atrox*, (University of Michigan Museum of Paleontology, USA, specimen 94814) White circles enclose a series of bite marks on the frontal shield. The bite marks are very shallow, and the two on the left side of the skull are very closely spaced and far from the third mark.

Potential clues as to the nature of a particular shark–whale interaction are provided by the size of the bite marks, as well as their location and orientation. Scavengers tend to be relatively small or young, and thus usually leave much fainter traces than their larger, predatory counterparts, such as the giant shark *Carcharocles megalodon* (Aguilera *et al.*, 2008). In addition, bite marks arising from scavenging are often produced by several animals over prolonged periods of time, and hence oriented randomly. By contrast, active predation results in fewer bite marks whose orientation is determined by the nature of the attack (Bianucci *et al.*, 2010). Finally, bite marks located deep inside the body more likely reflect scavenging than predation, since the bite that produced them could only have been delivered after the victim had already been partially dismembered (Lambert and Gigase, 2007).

Some of the best evidence for a likely shark attack is found in an extremely well-preserved delphinid specimen from the Pliocene of Italy (Figure 6.20). Here, multiple tooth marks roughly align across a series of successive ribs, suggesting a strong and likely fatal bite into the flank of the animal. The size and spacing of the tooth marks are most compatible with those of *Cosmopolitodus hastalis*, an extinct relative of the great white shark (Bianucci *et al.*, 2010). Similar bite marks are also found on two ribs of a Middle Eocene protocetid from Egypt, but the smaller number and more restricted distribution of the traces make their origin more uncertain (Bianucci and Gingerich, 2011). Finally, maybe the most reliable evidence for active (and failed) predation consists of partially healed bones showing evidence of trauma, such as a fragmentary cetacean rib from the Pliocene of the eastern United States (Kallal *et al.*, 2012). This specimen preserves a string of three evenly spaced bony lesions, which likely formed in response to a strong bite damaging the bone and surrounding periosteum. The fact that the bone had time to start the healing process proves that the animal survived the attack, and therefore that the bite was delivered as part of a failed predation attempt. There is no evidence as to the identity of the attacker, which could have been either a shark or a large physeteroid. Another possible case is recorded by a large, partially healed whale vertebra that was subject to a compression fracture (Godfrey and Altmann, 2005). The latter likely resulted from the sudden flexion of the vertebral column beyond its natural limits, as may occur during the attack of a large predator such as *Carcharocles megalodon*; however, other explanations, such as convulsions or a collision of the whale with a floating log, cannot be excluded in this case.

6.2.2 Whale falls

Whales suffering a natural death tend to be in poor nutritional condition and thus, with some exceptions, negatively buoyant. As a result, they rapidly sink to the ocean floor upon death, creating a **whale fall**. If the site where the carcass comes to rest is sufficiently deep (≥1000 m), the overlying pressure will prevent it from becoming refloated by the buildup of decompositional gas (Allison *et al.*, 1991). At shallower depths, microbial decomposition can cause the carcass to rise again to the

Figure 6.20 Fossil evidence for a Pliocene shark attack on a dolphin. (a) Lateral view of a skeleton of *Astadelphis gastaldii* (Museo Regionale di Storia Naturale di Torino specimen PU13884) showing major shark bites. The stippled line represents the main first bite of the shark. Scale bar represents 100 mm. (b) Hypothesized attack sequence as judged from the bite marks. The shark approaches the dolphin from the rear and delivers a major, fatal bite to its flank. The injured dolphin rolls over, turning its back to the shark, which delivers a second, smaller bite to the area near the dorsal fin. Reproduced from Bianucci *et al.* (2010), with permission of the Palaeontological Association.

(a)

(b)

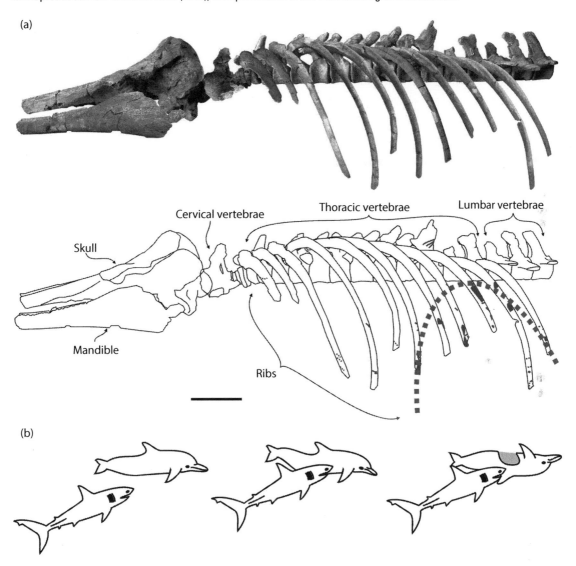

surface, where it may start to disarticulate (Schäfer, 1972). However, intense scavenging at the original site, leading to the opening of the abdominal cavity and the release of decompositional gas, may prevent this from happening, and relatively intact carcasses have been found at depths as shallow as 150 m (Smith, 2006; Smith and Baco, 2003).

Whale falls supply large quantities of organic material to the otherwise relatively barren deep sea: a single 40 t whale provides as much organic carbon as an entire hectare of abyssal sea floor receives via normal background sinking over 100–200 years (Smith and Demopoulos, 2003). These brief, yet intense, bursts in food availability lead to the creation of highly localized "island" ecosystems on which a considerably portion of the local biota ultimately depends. In the deep sea, a whale carcass goes through at least three stages of decay, including: (1) the removal of the soft tissue by mobile scavengers; (2) the colonization of the skeleton and surrounding, organically enriched sediments by polychaetes and crustaceans; and (3) the breakdown of the bone lipids into sulfides, which directly sustains a diverse, **chemoautotrophic** assemblage of invertebrates (Smith and Baco, 2003). This last stage can last for several decades, and it is exploited by both a range of whale-fall specialists and several species commonly found in other sulfide-producing deep-sea settings, such as hydrothermal vents and cold seeps (Bennett *et al.*, 1994; Smith, 2006). Whale falls may thus serve as dispersal and evolutionary "stepping stones" for a variety of sulfide-dependent, chemosynthetic deep-sea organisms, whose offspring need to move from one ephemeral source of sulfide to the next (Smith and Baco, 2003; Smith *et al.*, 1989).

Shelled members of whale fall communities, such as gastropods and bivalves, can be preserved in intimate association with partially decayed fossil whale bones. Numerous such associations have been reported, with the oldest dating from the Late Eocene (Amano and Little, 2005; Kiel and Goedert, 2006). The importance of chemosynthesis seems to increase with depth, with shallow whale falls containing no or few chemoautotrophic taxa (Danise and Dominici, 2014; Dominici *et al.*, 2009). Among deep-water falls, the earliest examples from the Eocene and Oligocene also lack taxa heavily dependent on sulfide, thus indicating a lower supply of skeletal oil (Kiel and Goedert, 2006). This pattern was initially attributed to the generally smaller body size of archaic cetaceans compared to that of modern whales. However, the occurrence of modern-looking whale fall communities on small-sized Miocene mysticetes indicates that a relative increase in skeletal oil content—possibly during the Early Miocene—may have been a more important

factor (Kiel and Goedert, 2006; Pyenson and Haasl, 2007).

From a paleocetological point of view, prolonged colonization of a skeleton by whale fall communities often leads to considerable bone damage. One of the major culprits in this regard is the annelid *Osedax*, which bores into the bone via a set of enlarged "roots" (Higgs *et al.*, 2014; Rouse *et al.*, 2004). Trace fossils likely attributable to *Osedax* have been identified in several fossil whale falls, some of them as old as the Early Oligocene (Boessenecker and Fordyce, 2014b; Higgs *et al.*, 2012; Kiel *et al.*, 2010). Indeed, the diversification of *Osedax* may even be linked to that of cetaceans themselves (Vrijenhoek *et al.*, 2009). Given the high degree of destruction *Osedax* can cause, it is likely that it had a large impact on the quantity and quality of whale fossils available for study (Kiel *et al.*, 2010).

6.3 Reproduction

Reproduction is one of the most fundamental behaviors an animal has to engage in, yet often little evidence of it survives in the fossil record. Unlike seals and sea lions, cetaceans mate in the water. Mating strategies vary, ranging from direct competition between males or groups of males (as hypothesized for beaked whales), to the production of copious amounts of sperm to displace that of any previous mates (e.g., in right whales), to frequent and peaceful copulation that may serve as much a social purpose as it does reproduction (e.g., in spinner dolphins) (Mesnick and Ralls, 2009). Aggressive male–male interactions may at least partially explain the presence of sexual dimorphism in many species, especially in cases where males are larger, develop distinctive weaponry (e.g., tusks), and show signs of intraspecific fights (section 6.4).

Because cetaceans give birth in the water, their calves have to be able to breathe, suckle, swim, and maintain their body temperature from the moment they are born. This need for **precocial** (mature and mobile) offspring explains why females of all extant species give birth to just a single calf, following a relatively long gestation period of about a year (Chivers, 2009). Unlike large

Figure 6.21 Different position of the fetus in (a) artiodactyls (cephalic presentation) and (b) extant cetaceans (caudal presentation). Adapted from Gingerich *et al.* (2009).

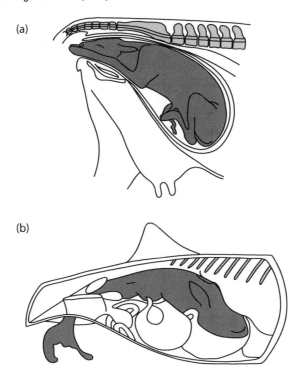

terrestrial ungulates, which are born head first (**cephalic presentation**), cetacean calves emerge from their mother's body in the opposite direction (i.e., tail first, or **caudal presentation**) (Figure 6.21). This change in the position of the fetus is thought to reduce the risk of drowning, especially when labor is lengthy or difficult (Slijper, 1956). Unlike cetaceans, modern pinnipeds (seals, sea lions, and walruses) still return to shore or pack ice to breed and give birth to their pups. Pinnipeds have often been used as a behavioral model for the transition from land to sea in cetaceans, which, if correct, would suggest that reproduction was one of the last behaviors to change in the process.

A potential insight into this transition is provided by the protocetid *Maiacetus inuus*. Judging from its large hind limbs and robust pelvis, *Maiacetus* was clearly capable of moving on land, yet, like all protocetids, was recovered from fully marine rocks and likely spent much of its time in the sea. This taxon earned its evocative name, the "mother whale," because its holotype—identified

as a female—appears to preserve a near-term fetus inside the abdomen (Plate 15a). The fetal skeleton includes a partially mineralized M^1, which is the first adult tooth to form in archaeocetes (Uhen, 2000) and indicates that the fetus was both near birth and comparatively precocial compared to other mammals (Gingerich *et al.*, 2009). Crucially, the position of the fetus in the mother indicates that it was positioned head first, as in artiodactyls. This suggests that *Maiacetus* probably still gave birth on land, and, consequently, that parturition at sea was indeed one of the last aquatic behaviors to evolve in cetaceans. It should be noted, however, that the status of this presumed fetus is still debated (Thewissen and McLellan, 2009).

Besides the insights provided by *Maiacetus*, the mating and reproductive behaviors of fossil cetaceans remain largely unknown. Relevant anatomical observations are scarce and restricted to largely speculative hypotheses, such as the idea that the tiny external hind limbs of the snake-like

Basilosaurus may have acted as guides during copulation (Antar *et al.*, 2010; Gingerich *et al.*, 1990). Further information is, however, provided by stable isotope data. Thus, differences in the stable carbon and oxygen isotope composition of tooth enamel that developed (1) before birth, (2) after birth but before weaning, and (3) post weaning have been demonstrated in *Pakicetus*, *Zygorhiza*, and *Dorudon*, and provide the only direct evidence of milk consumption and suckling (Clementz *et al.*, 2006; Clementz and Uhen, 2012; Uhen and Clementz, 2010). Two additional aspects related to reproduction—migration to breeding grounds and sexual dimorphism—are detailed further in sections 6.4 and 6.5.

6.4 Migration

Migration describes a large-scale, largely linear movement between different sites of an animal's home range. Among cetaceans, true migration occurs in mysticetes, most of which travel to high-latitude feeding areas at the beginning of summer and return to low-latitude breeding areas during the winter. High-latitude productivity provides an obvious explanation for the summer distribution, but it is less clear why whales migrate as far as they do to reach their breeding grounds. Reproduction in relatively warm waters may help to minimize attacks by killer whales or to avoid thermal stress, although the latter seems unlikely given that mysticete calves are relatively large and well insulated (Corkeron and Connor, 1999). Alternatively, the retention of low-latitude breeding grounds may simply be an evolutionary holdover. After the end of the last glacial period, the Northern and Southern Hemisphere polar fronts, and thus most of the prime mysticete feeding areas, started to shift poleward. As a result, baleen whales were forced to migrate across the ever larger distances that now separated their ancestral, low-latitude breeding grounds from the seasonally productive polar seas (Marx and Fordyce, 2015; Stern, 2009).

The process of migration generally does not leave any lasting traces, and hence is nigh impossible to observe in the fossil record. Accumulations of juvenile specimens (e.g., *Dorudon*) may indicate the existence of distinct breeding grounds

(Uhen, 2004), but they provide no evidence as to whether adult animals had to migrate in order to get there. Similarly, large body size likely correlates with the ability of an animal to migrate over large distances, but does not necessitate that migration actually happens (Corkeron and Connor, 1999; Millar and Hickling, 1990). By contrast, intra individual variations in stable oxygen isotope values (e.g., across different teeth) could potentially record movements between water masses of different isotopic composition, and thus provide direct evidence of migration (Clementz and Uhen, 2012; Roe *et al.*, 1998). However, this aspect of stable isotope research has so far barely been explored in the context of cetacean evolution.

Luckily, however, modern mysticete migration has produced at least one potential proxy. The skin of extant baleen whales is often host to whale barnacles, which, though firmly attached, are continuously shed—for example, during breaching (Félix *et al.*, 2006; Seilacher, 2005). Shedding of barnacles commonly occurs as the whales move within breeding areas or along their migratory routes, thus leaving behind a trail of breadcrumbs that can potentially become fossilized (Bianucci *et al.*, 2006). A good example of this is provided by fossil occurrences of the whale barnacle *Coronula*, whose global distribution largely matches that of the breeding grounds and migration routes of the extant humpback whale, *Megaptera novaeangliae* (Figure 6.22). The high degree of geographical overlap between the fossil barnacles and the living whales suggests that some form of large-scale mysticete migration may have occurred since at least the Pliocene. In addition, an abundance of fossil *Coronula* shells from the Mediterranean indicates that, unlike today, this sea may once have served as a mysticete breeding area (Bianucci *et al.*, 2006).

6.5 Sexual dimorphism

Sexual dimorphism, the development of morphological differences in males and females of the same species, is a common occurrence among extant cetaceans. Phenotypic sex differences may be functional, the outcome of sexual selection, or both. Among extant mysticetes, females are typically

Figure 6.22 Fossil occurrences of the whale barnacle *Coronula* (circles) compared to the global distribution of the humpback whale *Megaptera novaeangliae*. Arrows indicate migratory routes between tropical breeding and polar feeding grounds (in dark gray). Stippled line shows a possible fossil migration route of humpback or related whales between the North Atlantic (feeding area) and the Mediterranean Sea (breeding area). Reproduced from Bianucci *et al.* (2006), with permission of The Royal Society of New Zealand.

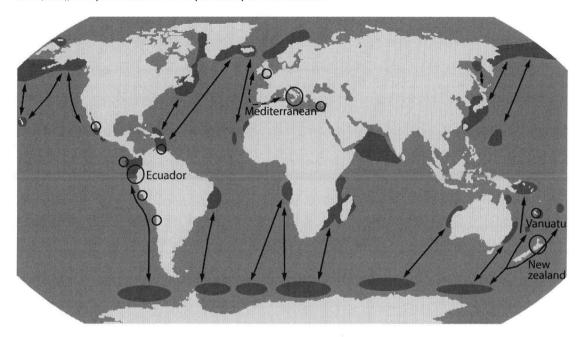

around 5% longer than males, likely out of the need to cope with pregnancy and lactation during the time away from feeding grounds. The pattern is more complicated among odontocetes: some species (e.g., *Phocoena*, *Platanista*, *Pontoporia*, and some ziphiids) share the mysticete pattern of larger females, whereas in others the sexes are equally sized or males are larger—sometimes considerably so. Thus, for example, male sperm whales (*Physeter macrocephalus*) can grow up to 5 m longer than females (Fordyce and de Muizon, 2001; Ralls and Mesnick, 2009). Size and shape differences between the sexes often also specifically affect particular parts of the body, such as the pelvic bones (Dines *et al.*, 2014) or, in *Physeter*, the supracranial basin and the spermaceti organ. The latter may be used for acoustic display, as a battering ram, or both (Carrier *et al.*, 2002; Cranford, 1999).

Besides body size, sexual dimorphism is most clearly expressed in the morphology of the teeth and the rostrum. Usually only male narwhals

(*Monodon monoceros*) feature the long, spiraled tusk to which the species owes its name. The tusk is a greatly modified, left maxillary tooth likely used in visual display or as a sensory tool (Nweeia *et al.*, 2009). Similarly, the mandibular tusks characteristic of many beaked whales usually only erupt in males (MacLeod and Herman, 2004). This feature is especially well developed in the strap-toothed whale, *Mesoplodon layardii*, in which a pair of lower teeth fully encircles the rostrum. Some ziphiids engage in intraspecific fights, during which the tusks are used to rake the opponent's body (Heyning, 1984). Extant beaked whales also often display sexual dimorphism in the degree of compactness and thickening of some of the rostral bones. Thus, male bottlenose whales (*Hyperoodon ampullatus*) bear high, voluminous, and spongy rostral maxillary crests that are at least occasionally used for head-butting, while a thickened, dense vomer fills the mesorostral groove in males of several species of *Mesoplodon* (de Buffrénil et al., 2000; Heyning, 1984). The function of these bizarre rostral structures is still a matter of debate (Lambert et al., 2011). Although hidden underneath soft tissues in life, they might nonetheless be detectable by biosonar, and thus could potentially be used for sexual display (Gol'din, 2014).

Despite the prevalence of sexually dimorphic features among the living species, they are often difficult to recognize in the fossil record. Soft tissue and behavioral characters, such as differences in the shape and size of the flukes or male vocalizations, cannot be assessed at all in extinct taxa. In addition, minor differences in the proportions of the skull (e.g., Frandsen and Galatius, 2013) often go undetected because of small sample sizes and difficulties in distinguishing sexual from intra- and even interspecific variation. Nevertheless, there are some notable exceptions. The preservation of a fetus allowed the tentative identification of male and female individuals of the protocetid Maiacetus inuus, with the male being somewhat larger and having slightly larger canines (Gingerich et al.,

2009). Similarly, a bimodal distribution in the size of several femora attributable to Basilosaurus may represent a sexual difference, and provides support for a function of the hind limb during copulation (Antar et al., 2010; Gingerich et al., 1990).

Among fossil neocetes, the expanded rostral tip of some squalodontids has been linked with the development of procumbent incisors and proposed to vary between sexes (Dooley, 2005). If correct, this interpretation could potentially extend to several other fossil odontocetes characterized by anteriorly projected apical teeth, such as Neosqualodon, several "kentriodontids," and some platanistoids. Males of the Miocene stem ziphiid Messapicetus gregarius seem to possess mandibles characterized by a robust apex and relatively large tusks (Lambert et al., 2010a). Several other extinct ziphiids collected from the sea floor (e.g., Globicetus and Tusciziphius) display enlarged crests and prominences that are somewhat reminiscent of those of male Hyperoodon, but made of much more compact bone (Figure 6.23). Similarly,

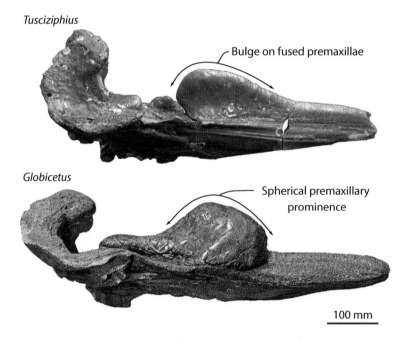

Tusciziphius

Bulge on fused premaxillae

Globicetus

Spherical premaxillary prominence

100 mm

Figure 6.23 Partial skulls of the fossil ziphiids *Tusciziphius* (Museo da Natureza, Ferrol, Spain, specimen MA0632) and *Globicetus* (Museu da Lourinhã, Portugal, specimen 1361), from Neogene deep-sea deposits off Portugal and Spain, in right lateral view. The bulge and spherical prominence on the rostrum are tentatively interpreted as sexually dimorphic features. Photos courtesy of I. Miján.

variations in the extent of the rostral premaxillary crest in *Tusciziphius atlanticus* may also represent sexual dimorphism (Bianucci *et al.*, 2013).

The fossil record of tusk-bearing monodontids only goes back to the Pleistocene (Newton, 1891), but they are rivaled by the aberrant, Miocene fossil delphinoid *Odobenocetops*. Members of this genus are unique among cetaceans in having a pair of posteroventrally oriented, premaxillary tusks. In some specimens—presumably males—the tusks are developed asymmetrically, with the right tusk being considerably larger (Plate 11). The tusks may have had a social function, possibly including visual display. Nevertheless, similarities with the walrus suggest that they originally evolved to aid in benthic feeding, before becoming secondarily involved in social interactions (de Muizon *et al.*, 1999, 2002).

6.6 Diving

Unlike their terrestrial ancestors, ocean-going cetaceans are free to move in three dimensions. Many of the extant species are capable of diving dozens or even hundreds of metres when escaping predators or in search of food (Schreer and Kovacs, 1997). Both mysticetes and odontocetes readily exploit the sunlit waters of the photic zone, which extends to about 200 m in the open ocean. The depths beyond this limit are largely the preserve of echolocating odontocetes, as darkness makes the detection of prey much more difficult for animals relying purely on sight. The most extreme divers are beaked whales, sperm whales, and a few delphinoids, all of which regularly dive below 500 m (record at 2992 m for *Ziphius*), sometimes for up to 2 hours (Schorr *et al.*, 2014; Tyack *et al.*, 2006). Most of these taxa are suction feeders, whose deep-diving abilities gave them access to rich and comparatively little exploited mesopelagic feeding grounds (Lindberg and Pyenson, 2007).

Diving presents mammals with a unique set of physiological challenges, the most important of which include (1) the maintenance of an adequate supply of oxygen during prolonged periods of **apnea** (breath-holding); and (2) dealing with the direct and indirect effects of increasing pressure on air spaces inside the body. Information on how marine mammals have adapted to these challenges is still patchy and often limited to relatively few taxa, which suggests caution when making generalizations across species (Hooker *et al.*, 2012). As far as is known, cetaceans are able to exchange most of their lung volume in a single breath and seem to dive on inspiration (Ponganis, 2011; Ridgway *et al.*, 1969). Nevertheless, the lungs at best serve as a minor source of oxygen while underwater. Instead, oxygen is primarily stored in the blood and muscles, where it binds to **hemoglobin** and **myoglobin**, respectively. As a result, cetaceans tend to have considerably higher concentrations of these proteins (especially myoglobin) than their terrestrial cousins (Noren and Williams, 2000; Ponganis, 2011).

Lungs become ineffective as an oxygen store at depth because the increase in ambient pressure causes them to compress and, eventually, collapse, thus preventing any gas exchange (Kooyman, 2009). Cetaceans have adapted to this process not by attempting to resist pressure, but by facilitating lung collapse via a flexible rib cage and reinforced upper airways capable of holding the compressed air (Scholander, 1940). At what depth collapse actually occurs varies between species, but its effects are likely to be universal below 200 m (Fahlman *et al.*, 2014). Allowing gas exchange to cease has the major advantage of limiting the buildup of nitrogen in the blood and tissues, and thus helps to avoid two of the major negative side effects of breathing pressurized air: **decompression sickness** (discussed further in this section) and **nitrogen narcosis**—a state of mind that is often likened to being drunk (Ponganis, 2011). Other air spaces typically present in the mammalian body, and thus also subject to compression during a dive, include the facial sinuses and the middle ear cavity. Cetaceans circumvent this problem by having lost the facial sinuses altogether, and by supplying the middle ear with additional air from the accessory sinus system. The latter allows the maintenance of an air pocket around the auditory ossicles, which enables them, and thus also odontocete echolocation, to function under increased hydrostatic pressure (Cranford *et al.*, 2008; Fraser and Purves, 1960). To prevent compression damage, the air sinuses themselves are lined by a network of blood vessels—the **fibrous venous plexus**—which expand and fill the sinus cavities as the air they contain is squeezed out during a dive.

Like other behaviors, diving does not fossilize, but in rare cases it may leave recognizable traces. Thus, for example, a Late Miocene specimen of the archaic ziphiid *Messapicetus gregarius* was found in association with the remains of several newly ingested, shallow-water fish in its throat and stomach area. Because these fish live relatively close to the surface, they prove that *Messapicetus*, unlike modern ziphiids, was not a specialized deep diver (Lambert *et al.*, 2015). Such predator–prey associations are highly informative, but unfortunately also extremely rare. In the vast majority of cases, diving behavior can only be inferred indirectly, and thus uncertainly, from a variety of morphological proxies.

In general, larger body mass correlates with a slower metabolic rate, and thus also a slower rate of oxygen depletion and, consequently, greater diving capacity (Brischoux *et al.*, 2008; Schreer and Kovacs, 1997). Unfortunately, this correlation does not completely apply to modern cetaceans, since mysticetes generally dive less deep and shorter than would be expected based on their size. This is especially true for large rorquals, whose diving capacity is limited by the high energetic cost of lunge feeding (Goldbogen *et al.*, 2012). The large spermaceti organ of sperm whales and the pachyosteosclerotic rostrum of beaked whales have also been correlated with diving ability, and are sometimes thought to function as ballast; however, this suggestion remains controversial (de Buffrénil *et al.*, 2000; Gol'din, 2014).

More directly related to diving may be the enlargement of the pterygoid sinus (as an air reservoir supplying the middle ear cavity) and the concurrent reduction of the surrounding bony laminae, as seen in both ziphiids and physeteroids (Cranford *et al.*, 2008; Fraser and Purves, 1960). A large pterygoid sinus fossa is present in stem ziphiids like *Ninoziphius*, suggesting that deep-diving abilities evolved early in the history of this family. This idea may be further supported by the surface structure of the hamular process. In extant *Ziphius*, this region is covered by soft tissue crests and trabeculae forming part of the fibrous venous plexus. Remarkably similar, albeit bony, structures are found on the hamular process of *Ninoziphius*, where they may have served a similar, diving-related function (Figure 6.24) (Lambert *et al.*, 2013). Also linked to diving may be the presence of retia mirabilia in the braincase and along the

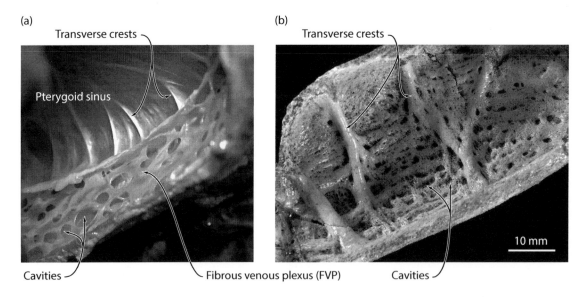

Figure 6.24 Comparison of the soft tissue hamular region of the extant *Ziphius* (a) with the bony hamular process of the extinct *Ninoziphius* (b), revealing similar crests and cavities. Reproduced from Lambert *et al.* (2013).

vertebral column (section 3.4.4). The exact function of the retia in this context is still debated, but they may help to protect the brain from fluctuations in blood pressure (Nagel *et al.*, 1968; Vogl and Fisher, 1982).

Another way to trace deep divers in the fossil record is to search for clues of decompression sickness. Although they have adapted to deal with the effects of diving, extant cetaceans can still be affected by a buildup of nitrogen in their blood and tissues as the air in their lungs is compressed during a dive (Fahlman *et al.*, 2014; Hooker *et al.*, 2012). Under certain circumstances, such as an uncontrolled ascent, the nitrogen is released too fast and may form bubbles blocking the blood supply to parts of the body. Among other, immediately life-threatening effects, this process may lead to the permanent destruction of bone tissues (**avascular osteonecrosis**), which in turn can be detected in fossils. Avascular osteonecrosis has been shown to occur in archaic odontocetes (xenorophids), the chaeomysticete *Aglaocetus*, and an undetermined Pliocene balaenopterid. None of these taxa are related to any of the extant extreme divers, and may therefore represent early steps on several independent, ultimately unsuccessful, paths toward deep-diving behavior (Beatty and Rothschild, 2008).

Besides morphology and pathology, a third method to trace the history of diving capacity focuses on genetic correlates, with case studies so far having focused on oxygen storage capacity and vision at depth. The skeletal muscle concentration of myoglobin, the major oxygen-binding protein in cetaceans, cannot be assessed directly for extinct species, but can still be estimated via ancestral sequence reconstruction (section 1.4). This technique has revealed a 3.5-fold increase in myoglobin concentration between the earliest and the most crownward archaeocetes, followed by further increases in different modern lineages (Mirceta *et al.*, 2013). Combined with reconstructed body masses, these data allow the estimation of maximal active dive times, from 1.6 minutes in *Pakicetus* to 17.4 minutes in *Basilosaurus*. This increase is impressive, but still far from the performance of extant extreme divers, such as *Ziphius*.

In adapting their sense of vision to diving, modern cetaceans have invariably lost one or both of the cone-based retinal pigments that provide most mammals with dichromatic color vision (section 5.1.2). The loss of the cone pigment sensitive to short wavelengths occurred first, and is characteristic of all living whales and dolphins. Its long-wavelength counterpart was, however, only lost in a select few lineages, including most mysticetes, *Physeter*, *Kogia*, and at least one species of *Mesoplodon*. In several mysticetes, this loss was accompanied by convergent red shifts of the single remaining (rod-based) pigment, whereas the opposite (i.e., a blue shift) occurred in the last common ancestors of ziphiids and physeteroids.

The switch to an entirely rod-based mode of vision likely represents an adaptation to extremely low light levels, under which cone-based pigments can interfere with the function of the more effective rods (Meredith *et al.*, 2013). In *Physeter*, *Kogia*, and *Mesoplodon*, the loss of cone-based vision and the blue shift affecting the rod opsin likely arose as a result of foraging at depths where sunlight is virtually absent and the only source of light consists of bluish **bioluminescence**. The fact that the blue shift occurred along the stem lineages of both ziphiids and physeteroids implies a long history of deep diving for these clades. Mysticetes forage much closer to the surface, and thus may have lost their cones to facilitate the location of bioluminescent prey at night. This idea is supported by the red shift in the rod pigment, which is characteristic of shallow-water taxa (Bischoff *et al.*, 2012; Meredith *et al.*, 2013).

6.7 Ontogenetic age

One of the most commonly employed methods to gauge the ontogenetic age of an individual involves the assessment of sutural closure (Mead and Potter, 1990). At birth, the sutures between the individual skull bones, as well as those between vertebral and long bone epiphyses and their respective main elements, are open. Depending on the distance between the individual elements, this may create large openings (**fontanelles**) in the skull roof, which are also apparent in human infants (Walsh and Berta, 2011). As the animal matures, the individual sutures gradually close and, in old individuals, may even fuse. In the vertebral column, this process generally starts in the neck and tail regions, and from there gradually moves toward the center of the body (Moran *et al.*, 2015).

Assessing the degree of sutural closure is a subjective exercise, and as such it can at best provide a rough idea of individual age. In addition, matters are complicated further by the fact that the onset of fusion may be delayed until long after sexual maturity (Moran *et al.*, 2015). Nevertheless, the method is valuable because of its relative simplicity and noninvasive nature, and because it can be equally applied to both living and fossil species. More precise ways to estimate age involve the study of tissues or substances that are deposited in a periodical fashion, and thus form growth layers. Examples of such structures include the waxy earplugs and the continuously growing baleen plates of mysticetes (Hohn, 2009). Unfortunately, neither of them tends to be preserved in the fossil record.

A more durable record is provided by teeth. In odontocetes, deposition of dentin inside the pulp cavity continues throughout life, although in old individuals the cavity may eventually become occluded. Regular variations in the rate of growth result in the formation of differently sized bands similar to tree rings, which combine into repeated, predictable **growth layer groups** (GLGs). The age of an individual can be established by simply counting the GLGs in a sectioned tooth. The starting point for this count is the **neonatal line**, a somewhat less mineralized, easily recognizable layer deposited at birth. It is important to bear in mind that individual GLGs do not necessarily reflect annular cycles, even though they are often defined as such (Hohn, 2009; Myrick, 1980). This is particularly important when studying fossils, for which little independent evidence is available that could be used to calibrate GLGs against time.

Unlike bone, teeth are not remodeled and thus usually preserve a growth layer record spanning the entire life of an individual. However, problems may arise in taxa with continuously growing teeth, such as sperm whales, in which tooth wear may gradually remove the oldest, outermost layers of dentin. In such cases, counting the number of GLGs can only provide a minimum age estimate. Somewhat surprisingly, the assessment of GLGs has so far hardly been applied to fossil

Cement-dentine boundary
Sediment in pulp cavity

20 mm

Figure 6.25 Sectioned tooth of a fossil sperm whale (Institut royal des sciences naturelles de Belgique specimen M.512). Note the prominent growth layer groups.

cetaceans, with the exception of the basilosaurid *Dorudon* and a very limited number of fossil sperm whale teeth (Figure 6.25) (Gilbert *et al.*, 2010; Uhen, 2004).

6.8 Suggested readings

de Muizon, C., and D. P. Domning. 2002. The anatomy of *Odobenocetops* (Delphinoidea, Mammalia), the walrus-like dolphin from the Pliocene of Peru and its palaeobiological implications. Zoological Journal of the Linnean Society 134:423–452.

Gatesy, J., J. H. Geisler, J. Chang, C. Buell, A. Berta, R. W. Meredith, M. S. Springer, and M. R. McGowen. 2013. A phylogenetic blueprint for a modern whale. Molecular Phylogenetics and Evolution 66:479–506.

Gingerich, P. D., W. von Koenigswald, W. J. Sanders, B. H. Smith, and I. S. Zalmout. 2009. New protocetid whale from the middle Eocene of Pakistan: birth on land, precocial development, and sexual dimorphism. PLoS ONE 4:e4366.

References

Aguilera, O., L. García, and M. Cozzuol. 2008. Giant-toothed white sharks and cetacean trophic interaction from the Pliocene Caribbean Paraguaná Formation. Paläontologische Zeitschrift 82:204–208.

Allison, P. A., C. R. Smith, H. Kukert, J. W. Deming, and B. A. Bennett. 1991. Deep-water taphonomy of vertebrate carcasses: a whale skeleton in the bathyal Santa Catalina Basin. Paleobiology 17:78–89.

Amano, K., and C. T. S. Little. 2005. Miocene whale-fall community from Hokkaido, northern Japan. Palaeogeography, Palaeoclimatology, Palaeoecology 215:345–356.

Antar, M., I. Zalmout, and P. D. Gingerich. 2010. Sexual dimorphism in hind limbs of late Eocene *Basilosaurus isis* (Mammalia, Cetacea), Wadi Al Hitan World Heritage Site, Egypt. Journal of Vertebrate Paleontology: Program and Abstracts, 54A.

Arnold, P. W., R. A. Birtles, S. Sobtzick, M. Matthews, and A. Dunstan. 2005. Gulping behaviour in rorqual whales: underwater observations and functional interpretation. Memoirs of the Queensland Museum 51:309–332.

Auster, P. J., and L. Watling. 2010. Beaked whale foraging areas inferred by gouges in the seafloor. Marine Mammal Science 26:226–233.

Bajpai, S., J. G. M. Thewissen, and R. W. Conley. 2011. Cranial anatomy of Middle Eocene *Remingtonocetus* (Cetacea, Mammalia) from Kutch, India. Journal of Paleontology 85:703–718.

Baldanza, A., R. Bizzarri, F. Famiani, P. Monaco, R. Pellegrino, and P. Sassi. 2013. Enigmatic, biogenically induced structures in Pleistocene marine deposits: a first record of fossil ambergris. Geology 41:1075–1078.

Barnes, L. G., M. Kimura, H. Furusawa, and H. Sawamura. 1995. Classification and distribution of Oligocene Aetiocetidae (Mammalia; Cetacea; Mysticeti) from western North America and Japan. The Island Arc 3:392–431.

Barnes, L. G., and A. E. Sanders. 1996. The transition from archaeocetes to mysticetes: Late Oligocene toothed mysticetes from near Charleston, South Carolina. Paleontological Society Special Publication 8:24.

Barroso, C., T. W. Cranford, and A. Berta. 2012. Shape analysis of odontocete mandibles: functional and evolutionary implications. Journal of Morphology 273:1021–1030.

Beatty, B. L., and A. C. Dooley, Jr. 2009. Injuries in a mysticete skeleton from the Miocene of Virginia, with a discussion of buoyancy and the primitive feeding mode in the Chaeomysticeti. Jeffersoniana 20:1–28.

Beatty, B. L., and B. M. Rothschild. 2008. Decompression syndrome and the evolution of deep diving physiology in the Cetacea. Naturwissenschaften 95:793–801.

Bennett, B. A., C. R. Smith, B. Glaser, and H. L. Maybaum. 1994. Faunal community structure of a chemoautotrophic assemblage on whale bones in the deep northeast Pacific Ocean. Marine Ecology Progress Series 108:205–223.

Bianucci, G., and P. D. Gingerich. 2011. *Aegyptocetus tarfa*, n. gen. et sp. (Mammalia, Cetacea), from the middle Eocene of Egypt: clinorhynchy, olfaction, and hearing in a protocetid whale. Journal of Vertebrate Paleontology 31:1173–1188.

Bianucci, G., and W. Landini. 2006. Killer sperm whale: a new basal physeteroid (Mammalia, Cetacea) from the Late Miocene of Italy. Zoological Journal of the Linnean Society 148:103–131.

Bianucci, G., W. Landini, and J. Buckeridge. 2006. Whale barnacles and Neogene cetacean migration routes. New Zealand Journal of Geology and Geophysics 49:115–120.

Bianucci, G., I. Miján, O. Lambert, K. Post, and O. Mateus. 2013. Bizarre fossil beaked whales (Odontoceti, Ziphiidae) fished from the Atlantic Ocean floor off the Iberian Peninsula. Geodiversitas 35:105–153.

Bianucci, G., B. Sorce, T. Storai, and W. Landini. 2010. Killing in the Pliocene: shark attack on a dolphin from Italy. Palaeontology 53:457–470.

Bischoff, N., B. Nickle, T. W. Cronin, S. Velasquez, and J. I. Fasick. 2012. Deep-sea and pelagic rod visual pigments identified in the mysticete whales. Visual Neuroscience 29:95–103.

Bloodworth, B. E., and C. D. Marshall. 2007. A functional comparison of the hyolingual complex in pygmy and dwarf sperm whales (*Kogia breviceps* and *K. sima*), and bottlenose dolphins (*Tursiops truncatus*). Journal of Anatomy 211:78–91.

Boessenecker, R. W., and R. E. Fordyce. 2014a. A new eomysticetid (Mammalia: Cetacea) from the Late Oligocene of New Zealand and a re-evaluation of "*Mauicetus*" *waitakiensis*. Papers in Palaeontology 1:107–140.

Boessenecker, R. W., and R. E. Fordyce. 2014b. Trace fossil evidence of predation upon bone-eating worms on a baleen whale skeleton from the Oligocene of New Zealand. Lethaia 48:326–341.

Boessenecker, R. W., and R. E. Fordyce. 2015a. Anatomy, feeding ecology, and ontogeny of a transitional baleen whale: a new genus and species of Eomysticetidae (Mammalia: Cetacea) from the Oligocene of New Zealand. PeerJ 3:e1129.

Boessenecker, R. W., and R. E. Fordyce. 2015b. A new genus and species of eomysticetid (Cetacea: Mysticeti) and a reinterpretation of "*Mauicetus*" *lophocephalus* Marples, 1956: transitional baleen whales from the upper Oligocene of New Zealand. Zoological Journal of the Linnean Society 175:607–660.

Bouetel, V. 2005. Phylogenetic implications of skull structure and feeding behavior in balaenopterids (Cetacea, Mysticeti). Journal of Mammalogy 86:139–146.

Brischoux, F., X. Bonnet, T. R. Cook, and R. Shine. 2008. Allometry of diving capacities: ectothermy vs. endothermy. Journal of Evolutionary Biology 21:324–329.

Cabrera, Á. 1926. Cetáceos fósiles del Museo de La Plata. Revista del Museo de La Plata 29:363–411.

Carrier, D. R., S. M. Deban, and J. Otterstrom. 2002. The face that sank the Essex: potential function of the spermaceti organ in aggression. Journal of Experimental Biology 205:1755–1763.

Chivers, S. J. 2009. Cetacean life history; pp. 215–220 in W. F. Perrin, B. Würsig, and J. G. M. Thewissen (eds.), Encyclopedia of Marine Mammals. Academic Press, Burlington, MA.

Cicimurri, D. J., and J. L. Knight. 2009. Two shark-bitten whale skeletons from coastal plain deposits of South Carolina. Southeastern Naturalist 8:71–82.

Cigala Fulgosi, F. 1990. Predation (or possible scavenging) by a great white shark on an extinct species of bottlenosed dolphin in the Italian Pliocene. Tertiary Research 12:17–36.

Clementz, M. T., R. E. Fordyce, S. Peek, L., and D. L. Fox. 2014. Ancient marine isoscapes and isotopic evidence of bulk-feeding by Oligocene cetaceans. Palaeogeography, Palaeoclimatology, Palaeoecology 400:28–40.

Clementz, M. T., A. Goswami, P. D. Gingerich, and P. L. Koch. 2006. Isotopic records from early whales and sea cows: contrasting patterns of ecological transition. Journal of Vertebrate Paleontology 26:355–370.

Clementz, M. T., and M. D. Uhen. 2012. Ontogenetic variationin dental stable isotope values of two species of basilosaurids (*Zygorhiza kochii* and *Dorudon atrox*). Journal of Vertebrate Paleontology: Program and Abstracts, 80.

Cooper, L. N., T. L. Hieronymus, C. J. Vinyard, S. Bajpai, and J. G. M. Thewissen. 2014. New applications for constrained ordination: reconstructing feeding behaviors in fossil Remingtonocetinae (Cetacea: Mammalia); pp. 89–107 in D. I. Hembree, B. F. Platt, and J. J. Smith (eds.), Experimental Approaches to Understanding Fossil Organisms. Springer, Dordrecht.

Cooper, L. N., J. G. M. Thewissen, and S. T. Hussain. 2009. New middle Eocene archaeocetes (Cetacea: Mammalia) from the Kuldana Formation of northern Pakistan. Journal of Vertebrate Paleontology 29:1289–1299.

Corkeron, P. J., and R. C. Connor. 1999. Why do baleen whales migrate? Marine Mammal Science 15:1228–1245.

Cranford, T. W. 1999. The sperm whale's nose: sexual selection on a grand scale? Marine Mammal Science 15:1133–1157.

Cranford, T. W., M. F. McKenna, M. S. Soldevilla, S. M. Wiggins, J. A. Goldbogen, R. E. Shadwick, P. Krysl, J. A. St. Leger, and J. A. Hildebrand. 2008. Anatomic geometry of sound transmission and reception in Cuvier's beaked whale (*Ziphius cavirostris*). The Anatomical Record 291:353–378.

Danise, S., and S. Dominici. 2014. A record of fossil shallow-water whale falls from Italy. Lethaia 47:229–243.

de Buffrénil, V., L. Zylberberg, W. Traub, and A. Casinos. 2000. Structural and mechanical characteristics of the hyperdense bone of the rostrum of *Mesoplodon densirostris* (Cetacea, Ziphiidae): summary of recent observations. Historical Biology 14:57–65.

de Muizon, C. 1988. Les vertébrés fossiles de la Formation Pisco (Pérou). Troisième partie: Les Odontocètes (Cetacea, Mammalia) du Miocène. Travaux de l'Institut Français d'Etudes Andines 42:1–244.

de Muizon, C. 1993. Walrus-like feeding adaptation in a new cetacean from the Pliocene of Peru. Nature 365:745–748.

de Muizon, C., D. P. Domning, and D. R. Ketten. 2002. *Odobenocetops peruvianus*, the walrus-convergent delphinoid (Mammalia: Cetacea) from the Early Pliocene of Peru. Smithsonian Contributions to Paleobiology 93:223–261.

de Muizon, C., D. P. Domning, and M. Parrish. 1999. Dimorphic tusks and adaptive strategies in a new species of walrus-like dolphin (Odobenocetopsidae) from the Pliocene of Peru. Comptes Rendus de l'Académie des Sciences, Serie II. Sciences de la Terre et des planètes 329:449–455.

Deméré, T. A., and A. Berta. 2008. Skull anatomy of the Oligocene toothed mysticete *Aetiocetus weltoni* (Mammalia; Cetacea): implications for mysticete evolution and functional anatomy. Zoological Journal of the Linnean Society 154:308–352.

Deméré, T. A., and R. A. Cerutti. 1982. A Pliocene shark attack on a cethotheriid whale. Journal of Paleontology 56:1480–1482.

Deméré, T. A., M. R. McGowen, A. Berta, and J. Gatesy. 2008. Morphological and molecular evidence for a stepwise evolutionary transition from teeth to baleen in mysticete whales. Systematic Biology 57:15–37.

Dines, J. P., E. Otárola-Castillo, P. Ralph, J. Alas, T. Daley, A. D. Smith, and M. D. Dean. 2014. Sexual selection targets cetacean pelvic bones. Evolution 68:3296–3306.

Dominici, S., E. Cioppi, S. Danise, U. Betocchi, G. Gallai, F. Tangocci, G. Valleri, and S. Monechi. 2009. Mediterranean fossil whale falls and the adaptation of mollusks to extreme habitats. Geology 37:815–818.

Dooley, A. C., Jr. 2005. A new species of *Squalodon* (Mammalia, Cetacea) from the Middle Miocene of Virginia. Virginia Museum of Natural History Memoirs 8:1–14.

Dudley, J. P. 1998. Reports of carnivory by the common hippo *Hippopotamus amphibius*. South African Journal of Wildlife Research 28:58–59.

Dudley, S. F. J., M. D. Anderson-Reade, G. S. Thompson, and P. B. McMullen. 2000. Concurrent scavenging off a whale carcass by great white sharks, *Carcharodon carcharias*, and tiger sharks, *Galeocerdo cuvier*. Fishery Bulletin 98:646–649.

Ehret, D. J., B. J. MacFadden, and R. Salas-Gismondi. 2009. Caught in the act: trophic interactions between a 4-million-year-old white shark (*Carcharodon*) and mysticete whale from Peru: PALAIOS 24:329–333.

El Adli, J. J., and T. A. Deméré. 2015. On the anatomy of the temporomandibular joint and the muscles that act upon it: observations on the gray whale, *Eschrichtius robustus*. The Anatomical Record 298:680–690.

El Adli, J. J., T. A. Deméré, and R. W. Boessenecker. 2014. *Herpetocetus morrowi* (Cetacea: Mysticeti), a new species of diminutive baleen whale from the Upper Pliocene (Piacenzian) of California, USA, with observations on the evolution and relationships of the Cetotheriidae. Zoological Journal of the Linnean Society 170:400–466.

Fahlke, J. M. 2012. Bite marks revisited—evidence for middle-to-late Eocene *Basilosaurus isis* predation on *Dorudon atrox* (both Cetacea, Basilosauridae). Palaeontologia Electronica 15:32A.

Fahlke, J. M., K. A. Bastl, G. M. Semprebon, and P. D. Gingerich. 2013. Paleoecology of archaeocete whales throughout the Eocene: dietary adaptations revealed by microwear analysis. Palaeogeography, Palaeoclimatology, Palaeoecology 386:690–701.

Fahlman, A., P. L. Tyack, P. J. O. Miller, and P. H. Kvadsheim. 2014. How man-made interference might cause gas bubble emboli in deep diving whales. Frontiers in Physiology 5:13.

Fallows, C., A. J. Gallagher, and N. Hammerschlag. 2013. White sharks (*Carcharodon carcharias*) scavenging on whales and its potential role in further shaping the ecology of an apex predator. PLoS One 8:e60797.

Félix, F., B. Bearson, and J. Falconí. 2006. Epizoic barnacles removed from the skin of a humpback whale after a period of intense surface activity. Marine Mammal Science 22:979–984.

Fitzgerald, E. M. G. 2006. A bizarre new toothed mysticete (Cetacea) from Australia and the early evolution of baleen whales. Proceedings of the Royal Society B 273:2955–2963.

Fitzgerald, E. M. G. 2010. The morphology and systematics of *Mammalodon colliveri* (Cetacea: Mysticeti), a toothed mysticete from the Oligocene of Australia. Zoological Journal of the Linnean Society 158:367–476.

Fitzgerald, E. M. G. 2012. Archaeocete-like jaws in a baleen whale. Biology Letters 8:94–96.

Foote, A. D., J. Newton, S. B. Piertney, E. Willerslev, and M. T. P. Gilbert. 2009. Ecological, morphological and genetic divergence of sympatric North Atlantic killer whale populations. Molecular Ecology 18:5207–5217.

Ford, J. K. B., G. M. Ellis, C. O. Matkin, M. H. Wetklo, L. G. Barrett-Lennard, and R. E. Withler. 2011. Shark predation and tooth wear in a population of northeastern Pacific killer whales. Aquatic Biology 11:213–224.

Fordyce, R. E. 1994. *Waipatia maerewhenua*, new genus and new species (Waipatiidae, new family), an archaic Late Oligocene dolphin (Cetacea: Odontoceti: Platanistoidea) from New Zealand. Proceedings of the San Diego Society of Natural History 29:147–176.

Fordyce, R. E. 2002. *Simocetus rayi* (Odontoceti: Simocetidae, new family): a bizarre new archaic Oligocene dolphin from the eastern North Pacific. Smithsonian Contributions to Paleobiology 93:185–222.

Fordyce, R. E. 2003. Early crown-group Cetacea in the southern ocean: the toothed archaic mysticete *Llanocetus*. Journal of Vertebrate Paleontology 23 (Suppl. to 3):50A.

Fordyce, R. E., and L. G. Barnes. 1994. The evolutionary history of whales and dolphins. Annual Review of Earth and Planetary Sciences 22:419–455.

Fordyce, R. E., and C. de Muizon. 2001. Evolutionary history of cetaceans: a review; pp. 169–223 in J.-M. Mazin and V. de Buffrénil (eds.), Secondary Adaptation of Tetrapods to Life in Water. Verlag Dr. Friedrich Pfeil, München.

Fordyce, R. E., P. G. Quilty, and J. Daniels. 2002. *Australodelphis mirus*, a bizarre new toothless ziphiid-like fossil dolphin (Cetacea: Delphinidae) from the Pliocene of Vestfold Hills, East Antarctica. Antarctic Science 14:37–54.

Frandsen, M. S., and A. Galatius. 2013. Sexual dimorphism of Dall's porpoise and harbor porpoise skulls. Mammalian Biology—Zeitschrift für Säugetierkunde 78:153–156.

Fraser, F. C., and P. E. Purves. 1960. Hearing in cetaceans. Bulletin of the British Museum (Natural History) 7:1–140.

Gilbert, K., L. Ivany, and M. D. Uhen. 2010. Killer sperm whales: exploring the life history and ecology of Neogene physeterids from the Atlantic Coastal Plain. Geological Society of America Abstracts with Programs 42:321.

Gingerich, P. D., B. H. Smith, and E. L. Simons. 1990. Hind limbs of Eocene *Basilosaurus*: evidence of feet in whales. Science 249:154–157.

Gingerich, P. D., M. Ul-Haq, I. H. Khan, and I. S. Zalmout. 2001. Eocene stratigraphy and archaeocete whales (Mammalia, Cetacea) of Drug Lahar in the eastern Sulaiman Range,

Balochistan (Pakistan). Contributions from the Museum of Paleontology, University of Michigan 30:269–319.

Gingerich, P. D., M. Ul-Haq, W. von Koenigswald, W. J. Sanders, B. H. Smith, and I. S. Zalmout. 2009. New protocetid whale from the Middle Eocene of Pakistan: birth on land, precocial development, and sexual dimorphism. PLoS One 4:1–20.

Godfrey, S. J., and J. Altmann. 2005. A Miocene cetacean vertebra showing a partially healed compression fracture, the result of convulsions or failed predation by the giant white shark, *Carcharodon megalodon*. Jeffersoniana 16:1–12.

Gol'din, P. 2014. "Antlers inside": are the skull structures of beaked whales (Cetacea: Ziphiidae) used for echoic imaging and visual display? Biological Journal of the Linnean Society 113:510–515.

Gol'din, P., D. Startsev, and T. Krakhmalnaya. 2014. The anatomy of the Late Miocene baleen whale *Cetotherium riabinini* from Ukraine. Acta Palaeontologica Polonica 59:795–814.

Goldbogen, J. A., J. Calambokidis, D. A. Croll, M. F. McKenna, E. Oleson, J. Potvin, N. D. Pyenson, G. Schorr, R. E. Shadwick, and B. R. Tershy. 2012. Scaling of lunge-feeding performance in rorqual whales: mass-specific energy expenditure increases with body size and progressively limits diving capacity. Functional Ecology 26:216–226.

Govender, R., and A. Chinsamy. 2013. Early Pliocene (5 Ma) shark-cetacean trophic interaction from Langebaanweg, western coast of South Africa. PALAIOS 28:270–277.

Heithaus, M. R. 2001. Predator–prey and competitive interactions between sharks (order Selachii) and dolphins (suborder Odontoceti): a review. Journal of Zoology 253:53–68.

Heithaus, M. R., and L. M. Dill. 2002. Food availability and tiger shark predation risk influence bottlenose dolphin habitat use. Ecology 83:480–491.

Heyning, J. E. 1984. Functional morphology involved in intraspecific fighting of the beaked whale, *Mesoplodon carlhubbsi*. Canadian Journal of Zoology/Revue Canadienne de Zoologie 62:1645–1654.

Heyning, J. E., and J. G. Mead. 1996. Suction feeding in beaked whales: morphological and observational evidence. Contributions in Science, Natural History Museum of Los Angeles County 464:1–12.

Higgs, N. D., A. G. Glover, T. G. Dahlgren, C. R. Smith, Y. Fujiwara, F. Pradillon, S. B. Johnson, R. C. Vrijenhoek, and C. T. S. Little. 2014. The morphological diversity of *Osedax* worm borings (Annelida: Siboglinidae). Journal of the Marine Biological Association of the United Kingdom 94:1429–1493.

Higgs, N. D., C. T. S. Little, A. G. Glover, T. G. Dahlgren, C. R. Smith, and S. Dominici. 2012. Evidence of *Osedax* worm borings in Pliocene (~3 Ma) whale bone from the Mediterranean. Historical Biology 24:269–277.

Hohn, A. A. 2009. Age estimation; pp. 11–17 in W. F. Perrin, B. Würsig, and J. G. M. Thewissen (eds.), Encyclopedia of Marine Mammals. Academic Press, Burlington, MA.

Hooker, S. K., A. Fahlman, M. J. Moore, N. Aguilar de Soto, Y. Bernaldo de Quirós, A. O. Brubakk, D. P. Costa, A. M. Costidis, S. Dennison, K. J. Falke, A. Fernandez, M. Ferrigno, J. R. Fitz-Clarke, M. M. Garner, D. S. Houser, P. D. Jepson, D. R. Ketten, P. H. Kvadsheim, P. T. Madsen, N. W. Pollock, D. S. Rotstein, T. K. Rowles, S. E. Simmons, W. Van Bonn, P. K. Weathersby, M. J. Weise, T. M. Williams, and P. L. Tyack. 2012. Deadly diving? Physiological and behavioural management of decompression stress in diving mammals. Proceedings of the Royal Society B 279:1041–1050.

Hulbert, R. C., Jr., R. M. Petkewich, G. A. Bishop, D. Bukry, and D. P. Aleshire. 1998. A new middle Eocene protocetid whale (Mammalia: Cetacea: Archaeoceti) and associated biota from Georgia. Journal of Paleontology 72:907–927.

Ichishima, H. 2005. Notes on the phyletic relationships of the Aetiocetidae and the feeding ecology of toothed mysticetes. Bulletin of the Ashoro Museum of Paleontology 3:111–117.

Jefferson, T. A., P. J. Stacey, and R. W. Baird. 1991. A review of killer whale interactions with other marine mammals: predation to co-existence. Mammal Review 21:151–180.

Joeckel, R. M. 1990. A functional interpretation of the masticatory system and paleoecology of entelodonts. Paleobiology 16:459–482.

Johnston, C., and A. Berta. 2011. Comparative anatomy and evolutionary history of suction feeding in cetaceans. Marine Mammal Science 27:493–513.

Johnston, C., T. A. Deméré, A. Berta, J. Yonas, and J. S. Leger. 2010. Observations on the musculoskeletal anatomy of the head of a neonate gray whale (*Eschrichtius robustus*). Marine Mammal Science 26:186–194.

Kallal, R. J., S. J. Godfrey, and D. J. Ortner. 2012. Bone reactions on a Pliocene cetacean rib indicate short-term survival of predation event. International Journal of Osteoarchaeology 22:253–260.

Kiel, S., and J. L. Goedert. 2006. Deep-sea food bonanzas: early Cenozoic whale-fall communities resemble wood-fall rather than seep communities. Proceedings of the Royal Society B 273:2625–2632.

Kiel, S., J. L. Goedert, W.-A. Kahl, and G. W. Rouse. 2010. Fossil traces of the bone-eating worm *Osedax* in early Oligocene whale bones. Proceedings of the National Academy of Sciences 107:8656–8659.

Kimura, T. 2002. Feeding strategy of an Early Miocene cetothere from the Toyama and Akeyo Formations, central Japan. Paleontological Research 6:179–189.

Kohno, N., H. Koike, and K. Narita. 2007. Outline of fossil marine mammals from the Middle Miocene Bessho and Aoki Formations, Nagano Prefecture, Japan. Research Report of the Shinshushinmachi Fossil Museum 10:1–45.

Kooyman, G. L. 2009. Diving physiology; pp. 327–332 in W. F. Perrin, B. Würsig, and J. G. M. Thewissen (eds.), Encyclopedia of Marine Mammals. Academic Press, Burlington, MA.

Lambert, O., G. Bianucci, and B. Beatty. 2014. Bony outgrowths on the jaws of an extinct sperm whale support macroraptorial feeding in several stem physeteroids. Naturwissenschaften 101:517–521.

Lambert, O., G. Bianucci, and K. Post. 2009. A new beaked whale (Odontoceti, Ziphiidae) from the middle Miocene of Peru. Journal of Vertebrate Paleontology 29:910–922.

Lambert, O., G. Bianucci, and K. Post. 2010a. Tusk-bearing beaked whales from the Miocene of Peru: sexual dimorphism in fossil ziphiids? Journal of Mammalogy 91:19–26.

Lambert, O., G. Bianucci, K. Post, C. de Muizon, R. Salas-Gismondi, M. Urbina, and J. Reumer. 2010b. The giant bite of a new raptorial sperm whale from the Miocene epoch of Peru. Nature 466:105–108.

Lambert, O., A. Collareta, W. Landini, K. Post, B. Ramassamy, C. Di Celma, M. Urbina, and G. Bianucci. 2015. No deep diving: evidence of predation on epipelagic fish for a stem beaked whale from the Late Miocene of Peru. Proceedings of the Royal Society B 282.

Lambert, O., V. de Buffrénil, and C. de Muizon. 2011. Rostral densification in beaked whales: diverse processes for a similar pattern. Comptes Rendus Palevol 10:453–468.

Lambert, O., C. de Muizon, and G. Bianucci. 2013. The most basal beaked whale *Ninoziphius platyrostris* Muizon, 1983: clues on the evolutionary history of the family Ziphiidae (Cetacea: Odontoceti). Zoological Journal of the Linnean Society 167:569–598.

Lambert, O., and P. Gigase. 2007. A monodontid cetacean from the Early Pliocene of the North Sea. Bulletin de l'Institut Royal des Sciences Naturelles de Belgique: Sciences de la Terre 77:197–210.

Lambertsen, R., N. Ulrich, and J. Straley. 1995. Frontomandibular stay of Balaenopteridae: a mechanism for momentum recapture during feeding. Journal of Mammalogy 76:877–899.

Lambertsen, R. H., K. J. Rasmussen, W. C. Lancaster, and R. J. Hintz. 2005. Functional morphology of the mouth of the bowhead whale and its implications for conservation. Journal of Mammalogy 86:342–352.

Leclerc, L.-M., C. Lydersen, T. Haug, K. A. Glover, A. T. Fisk, and K. M. Kovacs. 2011. Greenland sharks (*Somniosus microcephalus*) scavenge offal from minke (*Balaenoptera acutorostrata*) whaling operations in Svalbard (Norway). Polar Research 2011 30:7342.

Lillie, D. G. 1915. Cetacea. British Antarctic ("Terra Nova") Expedition 1910. Natural History Report, Zoology 1:85–152.

Lindberg, D. R., and N. D. Pyenson. 2007. Things that go bump in the night: evolutionary interactions between cephalopods and cetaceans in the tertiary. Lethaia 40:335–343.

Loch, C., J. A. Kieser, and R. E. Fordyce. 2015. Enamel ultrastructure in fossil cetaceans (Cetacea: Archaeoceti and Odontoceti). PLoS One 10:e0116557.

Loch, C., and P. C. Simões-Lopes. 2013. Dental wear in dolphins (Cetacea: Delphinidae) from southern Brazil. Archives of Oral Biology 58:134–141.

MacLeod, C. D., and J. S. Herman. 2004. Development of tusks and associated structures in *Mesoplodon bidens* (Cetaceae, Mammalia). Mammalia 68:175–184.

MacLeod, C. D., M. B. Santos, and G. J. Pierce. 2003. Review of data on diets of beaked whales: evidence of niche separation and geographic segregation. Journal of the Marine Biological Association of the United Kingdom 83:651–665.

Marx, F. G. 2011. The more the merrier? A large cladistic analysis of mysticetes, and comments on the transition from teeth to baleen. Journal of Mammalian Evolution 18:77–100.

Marx, F. G., and R. E. Fordyce. 2015. Baleen boom and bust: a synthesis of mysticete phylogeny, diversity and disparity. Royal Society Open Science 2:140434.

Marx, F. G., Tsai, C.-H., and Fordyce R. E. 2015. A new Early Oligocene toothed "baleen" whale (Mysticeti: Aetiocetidae) from western North America—one of the oldest and the smallest. Royal Society Open Science 2:150476.

Mead, J. G., and C. W. Potter. 1990. Natural history of bottlenose dolphins along the central Atlantic coast of the United Sates; pp. 165–195 in S. Leatherwood, and R. Reeves, R. (eds.), The Bottlenose Dolphin. Academic Press, San Diego.

Meredith, R. W., J. Gatesy, C. A. Emerling, V. M. York, and M. S. Springer. 2013. Rod monochromacy and the coevolution of cetacean retinal opsins. PLoS Genetics 9:e1003432.

Mesnick, S. L., and K. Ralls. 2009. Mating systems; pp. 712–719 in W. F. Perrin, B. Würsig, and J. G. M. Thewissen (eds.), Encyclopedia of Marine Mammals. Academic Press, Burlington, MA.

Millar, J. S., and G. J. Hickling. 1990. Fasting endurance and the evolution of mammalian body size. Functional Ecology 4:5–12.

Mirceta, S., A. V. Signore, J. M. Burns, A. R. Cossins, K. L. Campbell, and M. Berenbrink. 2013. Evolution of mammalian diving capacity traced by myoglobin net surface charge. Science 340:1234192.

Mitchell, E. D. 1989. A new cetacean from the late Eocene La Meseta Formation, Seymour Island, Antarctic Peninsula. Canadian Journal of Fisheries and Aquatic Science 46:2219–2235.

Moran, M., S. Bajpai, J. C. George, R. Suydam, S. Usip, and J. G. M. Thewissen. 2015. Intervertebral and epiphyseal fusion in the postnatal ontogeny of cetaceans and terrestrial mammals. Journal of Mammalian Evolution 22:93–109.

Myrick, A. C., Jr. 1979. Variation, Taphonomy, and Adaptation of the Rhabdosteidae (=Eurhinodelphidae) (Odontoceti, Mammalia) from the Calvert Formation of Maryland and Virginia. PhD Thesis, University of California, Los Angeles.

Myrick, A. C., Jr. 1980. Some approaches to calibration of age in odontocetes using layered hard tissues. Reports of the International Whaling Commission (Special Issue) 3:95–97.

Nagel, E. L., P. J. Morgane, W. L. McFarland, and R. E. Galllano. 1968. Rete mirabile of dolphin: its pressure-damping effect on cerebral circulation. Science 161:898–900.

Newton, E. T. 1891. The Vertebrata of the Pliocene deposits of Britain. Memoirs of the Geological Survey of the United Kingdom 1891:1–137.

Noren, S. R., and T. M. Williams. 2000. Body size and skeletal muscle myoglobin of cetaceans: adaptations for maximizing dive duration. Comparative Biochemistry and Physiology Part A: Molecular & Integrative Physiology 126:181–191.

Nweeia, M. T., C. Nutarak, F. C. Eichmiller, N. Eidelman, A. A. Giuseppetti, J. Quinn, J. G. Mead, K. K'issuk, P. V. Hauschka, E. M. Tyler,

C. Potter, J. R. Orr, R. Avike, P. Nielsen, and
D. Angnatsiak. 2009. Considerations of
anatomy, morphology, evolution, and function
for narwhal dentition; pp. 223–240 in I.
Krupnik, M. A. Lang, and S. E. Miller (eds.),
Smithsonian at the Poles: Contributions to
International Polar Year Science. Smithsonian
Institution Scholarly Press, Washington, DC.

O'Leary, M. A., and M. D. Uhen. 1999. The time
of origin of whales and the role of behavioral
changes in the terrestrial-aquatic transition.
Paleobiology 25:534–556.

Okazaki, Y. 2012. A new mysticete from the
upper Oligocene Ashiya Group, Kyushu, Japan
and its significance to mysticete evolution.
Bulletin of the Kitakyushu Museum of
Natural History and Human History, Series A
(Natural History) 10:129–152.

Pivorunas, A. 1979. The feeding mechanisms
of baleen whales. American Scientist
67:432–440.

Ponganis, P. J. 2011. Diving mammals.
Comprehensive Physiology 1:447–465.

Pyenson, N. D., J. A. Goldbogen, and R. E.
Shadwick. 2013. Mandible allometry in extant
and fossil Balaenopteridae (Cetacea: Mammalia):
the largest vertebrate skeletal element and its
role in rorqual lunge feeding. Biological Journal
of the Linnean Society 108:586–599.

Pyenson, N. D., J. A. Goldbogen, A. W. Vogl,
G. Szathmary, R. L. Drake, and R. E.
Shadwick. 2012. Discovery of a sensory organ
that coordinates lunge feeding in rorqual
whales. Nature 485:498–501.

Pyenson, N. D., and D. M. Haasl. 2007. Miocene
whale-fall from California demonstrates that
cetacean size did not determine the evolution
of modern whale-fall communities. Biology
Letters 3:709–711.

Racicot, R. A. 2014. Fossil focus: porpoises.
Palaeontology Online 4:10.

Racicot, R. A., T. A. Deméré, B. L. Beatty, and
R. W. Boessenecker. 2014. Unique feeding
morphology in a new prognathous extinct
porpoise from the Pliocene of California.
Current Biology 24:774–779.

Ralls, K., and S. L. Mesnick. 2009. Sexual
dimorphism; pp. 1005–1011 in W. F. Perrin,
B. Würsig, and J. G. M. Thewissen (eds.),
Encyclopedia of Marine Mammals. Academic
Press, Burlington, MA.

Ray, G. C., and W. E. Schevill. 1974. Feeding of a
captive gray whale Eschrichtius robustus.
Marine Fisheries Review 36:31–38.

Reidenberg, J. S., and J. T. Laitman. 1994.
Anatomy of the hyoid apparatus in Odontoceti
(toothed whales): specializations of their
skeleton and musculature compared with
those of terrestrial mammals. The Anatomical
Record 240:598–624.

Ridgway, S. H., B. L. Scronce, and J. Kanwisher.
1969. Respiration and deep diving in the
bottlenose porpoise. Science 166:1651–1654.

Roe, L. J., J. G. M. Thewissen, J. Quade, J. R.
O'Neil, S. Bajpai, A. Sahni, and S. T. Hussain.
1998. Isotopic approaches to understanding the
terrestrial-to-marine transition of the earliest
cetaceans; pp. 399–422 in J. G. M. Thewissen
(ed.), The Emergence of Whales. Plenum Press,
New York.

Rouse, G. W., S. K. Goffredi, and R. C.
Vrijenhoek. 2004. Osedax: bone-eating marine
worms with dwarf males. Science
305:668–671.

Sanders, A. E., and L. G. Barnes. 2002.
Paleontology of the Late Oligocene Ashley
and Chandler Bridge Formations of South
Carolina, 3: Eomysticetidae, a new family of
primitive mysticetes (Mammalia: Cetacea).
Smithsonian Contributions to Paleobiology
93:313–356.

Schäfer, W. 1972. Ecology and Palaeoecology of
Marine Environments. University of Chicago
Press, Chicago.

Scholander, P. F. 1940. Experimental investigations
on the respiratory function in diving mammals
and birds. Hvalrådets Skrifter 22:1–131.

Schorr, G. S., E. A. Falcone, D. J. Moretti, and
R. D. Andrews. 2014. First long-term
behavioral records from Cuvier's beaked
whales (Ziphius cavirostris) reveal record-
breaking dives. PLoS One 9:e92633.

Schreer, J. F., and K. M. Kovacs. 1997. Allometry
of diving capacity in air-breathing vertebrates.
Canadian Journal of Zoology 75:339–358.

Seilacher, A. 2005. Whale barnacles: exaptational
access to a forbidden paradise. Paleobiology
31:27–35.

Slijper, E. J. 1956. Some remarks on gestation and birth in Cetacea and other aquatic mammals. Hvalrådets Skrifter 41:1–62.

Smith, C. 2006. Bigger is better—the role of whales as detritus in marine ecosystems; pp. 286–302 in J. A. Estes, D. P. DeMaster, D. F. Doak, T. M. Williams, and R. L. Brownell (eds.), Whales, Whaling, and Ocean Ecosystems. University of California Press, Berkeley.

Smith, C. R., and A. R. Baco. 2003. Ecology of whale falls at the deep-sea floor. Oceanography and Marine Biology: An Annual Review 41:311–354.

Smith, C. R., and A. W. J. Demopoulos. 2003. The deep Pacific Ocean floor; pp. 181–220 in P. A. Tyler (ed.), Ecosystems of the World Volume 28: Ecosystems of the Deep Ocean. Elsevier, Amsterdam.

Smith, C. R., H. Kukert, R. A. Wheatcroft, P. A. Jumars, and J. W. Deming. 1989. Vent fauna on whale remains. Nature 341:27–28.

Snively, E., J. M. Fahlke, and R. C. Welsh. 2015. Bone-breaking bite force of *Basilosaurus isis* (Mammalia, Cetacea) from the Late Eocene of Egypt estimated by Finite Element Analysis. PLoS One 10:e0118380.

Stern, S. J. 2009. Migration and movement patterns; pp. 726–730 in W. F. Perrin, B. Würsig, and J. G. M. Thewissen (eds.), Encyclopedia of Marine Mammals. Academic Press, Burlington, MA.

Swift, C. C., and L. G. Barnes. 1996. Stomach contents of *Basilosaurus cetoides*: implications for the evolution of cetacean feeding behavior, and evidence for vertebrate fauna of epicontinental Eocene seas. Paleontological Society Special Publication 8:380.

Thewissen, J. G. M., and S. Bajpai. 2001. Dental morphology of Remingtonocetidae (Cetacea, Mammalia). Journal of Paleontology 75:463–465.

Thewissen, J. G. M., and S. Bajpai. 2009. New skeletal material of *Andrewsiphius* and *Kutchicetus*, two Eocene cetaceans from India. Journal of Paleontology 83:635–663.

Thewissen, J. G. M., L. N. Cooper, M. T. Clementz, S. Bajpai, and B. N. Tiwari. 2007. Whales originated form aquatic artiodactyls in the Eocene epoch of India. Nature 450:1190–1195.

Thewissen, J. G. M., S. I. Madar, and S. T. Hussain. 1996. *Ambulocetus natans*, an Eocene cetacean (Mammalia) from Pakistan. Courier Forschungsinstitut Senckenberg 191:1–86.

Thewissen, J. G. M., and W. A. McLellan. 2009. *Maiacetus*: displaced fetus or last meal? Comment on: Gingerich, P. D., ul-Haq, M., W. von Koenigswald, W. J. Sanders, B. H. Smith, and I. S. Zalmout. New protocetid whale from the middle Eocene of Pakistan: birth on land, precocial development, and sexual dimorphism. PLoS One 5:e4366.

Thewissen, J. G. M., and S. Nummela. 2008. Towards an integrative approach; pp. 333–340 in J. G. M. Thewissen, and S. Nummela (eds.), Sensory Evolution on the Threshold. University of California Press, Berkeley.

Thewissen, J. G. M., J. D. Sensor, M. T. Clementz, and S. Bajpai. 2011. Evolution of dental wear and diet during the origin of whales. Paleobiology 37:655–669.

Tsai, C.-H., and R. E. Fordyce. 2015. The earliest gulp-feeding mysticete (Cetacea: Mysticeti) from the Oligocene of New Zealand. Journal of Mammalian Evolution 22:535–560.

Tyack, P. L., M. Johnson, N. A. Soto, A. Sturlese, and P. T. Madsen. 2006. Extreme diving of beaked whales. Journal of Experimental Biology 209:4238–4253.

Uhen, M. D. 1998. Middle to Late Eocene basilosaurines and dorudontines; pp. 29–61 in J. G. M. Thewissen (ed.), The Emergence of Whales. Plenum Press, New York.

Uhen, M. D. 2000. Replacement of deciduous first premolars and dental eruption in archaeocete whales. Journal of Mammalogy 81:123–133.

Uhen, M. D. 2004. Form, function, and anatomy of *Dorudon atrox* (Mammalia: Cetacea): An archaeocete from the Middle to Late Eocene of Egypt. University of Michigan Papers on Paleontology 34. Ann Arbor: University of Michigan Press.

Uhen, M. D. 2007. Evolution of marine mammals: back to the sea after 300 million years. The Anatomical Record 290:514–522.

Uhen, M. D., and M. T. Clementz. 2010. Life history and ecological information inferred from stable isotope analysis of the dentition of *Zygorhiza kochii* (Cetacea: Basilosauridae). Journal of Vertebrate Paleontology:Program and Abstracts, 179A.

Vogl, A. W., and H. D. Fisher. 1982. Arterial retia related to supply of the central nervous system in two small toothed whales– narwhal (*Monodon monoceros*) an beluga (*Delphinapterus leucas*). Journal of Morphology 174:41–56.

Vrijenhoek, R., S. Johnson, and G. Rouse. 2009. A remarkable diversity of bone-eating worms (*Osedax*; Siboglinidae; Annelida). BMC Biology 7:1–13.

Walsh, B. M., and A. Berta. 2011. Occipital ossification of balaenopteroid mysticetes. The Anatomical Record 294:391–398.

Werth, A. J. 2000. Feeding in marine mammals; pp. 487–526 in K. Schwenk (ed.), Feeding: Form, Function and Evolution in Tetrapods. Academic Press, San Diego.

Werth, A. J. 2004a. Functional morphology of the sperm whale (*Physeter macrocephalus*) tongue, with reference to suction feeding. Aquatic Mammals 30:405–418.

Werth, A. J. 2004b. Models of hydrodynamic flow in the bowhead whale filter feeding apparatus. Journal of Experimental Biology 207:3569–3580.

Werth, A. J. 2006. Mandibular and dental variation and the evolution of suction feeding in Odontoceti. Journal of Mammalogy 87:579–588.

Werth, A. J. 2007. Adaptations of the cetacean hyolingual apparatus for aquatic feeding and thermoregulation. The Anatomical Record 290:546–568.

7 Macroevolutionary Patterns

7.1 Patterns in cetacean diversity: radiations and extinctions

Biological diversity is a broad concept describing the number, range, and variability of different taxa, morphologies, and ecologies (section 1.4). Most often, diversity is simply understood to mean the number of different kinds of organisms at any given time (taxonomic diversity). The simplest available metric to measure taxonomic diversity is richness (i.e., a count of the taxa existing in a given community, region, or globally). Unless stated otherwise, it is in this sense that we use the term *diversity* throughout this chapter.

In most studies of modern ecology, richness is measured by counting the number of species. By contrast, paleontologists often count the number of genera instead. This is in part because the biological species concept, which states that individuals are members of the same species if they can interbreed, is impossible to apply to extinct organisms. As a result, paleontological characterization of species has to rely purely on morphological similarity, and hence has varied over historical time, from researcher to researcher, and from one taxonomic group to another. The idea of what constitutes a genus is similarly subjective, and itself biologically largely meaningless—except insofar as the number of genera correlates with the true count of species. However, genera have the advantage of being broader in scope than a species, and should therefore in theory be less affected by taxonomic practices. In addition, genera are more accommodating of intraspecific and geographical variation, which, when ignored, can lead to artificially inflated species counts. In the following discussion, diversity should thus be taken to mean generic diversity, unless stated otherwise.

To assess diversity within a paleontological context, taxa first need to be summarized in discrete temporal units, which are often made to coincide with geological stages. This may be problematic for two reasons: first, the temporal resolution offered by geological stages may be too crude. Although stages are the smallest subdivision of the geological time scale, they still span several million years. In the worst case, this may lead to species being lumped together that in reality never actually coexisted. Second, geological stages vary in duration, which may bias diversity counts by having relatively more taxa fall into the longer stages (Marx and Uhen, 2010; Uhen and Pyenson, 2007). Unfortunately, such time averaging is largely unavoidable, because many fossil occurrences cannot be more precisely dated than to within a range of 2–3 million years. New phylogenetic techniques involving morphological clocks may help to ameliorate this issue to some extent, but

Cetacean Paleobiology, First Edition. Felix G. Marx, Olivier Lambert, and Mark D. Uhen.
© 2016 John Wiley & Sons, Ltd. Published 2016 by John Wiley & Sons, Ltd.

their potential applications and problems have yet to be fully explored (Beck and Lee, 2014; Marx and Fordyce, 2015; Pyron, 2011).

7.1.1 Paleogene

Cetaceans originated around 53 Ma in the eastern Tethys (Bajpai and Gingerich, 1998), covering what is now Indo-Pakistan. During the earliest phase of their evolution, the Ypresian, cetacean diversity was low and restricted to just a single family of archaeocetes: the pakicetids (Figure 7.1). This seemingly slow start is unsurprising, since, as a monophyletic group, cetaceans must have arisen from a single species, belonging to a single genus.

Archaeocetes go on to constitute virtually all of cetacean diversity until the end of the Eocene, with the exception of some latest Eocene occurrences of archaic mysticetes (Acosta Hospitaleche and Reguero, 2010; Kiel and Goedert, 2006; Martínez Cáceres et al., 2011; Mitchell, 1989). During the Lutetian, archaeocetes reached their heyday and gave rise to three new families of semiaquatic forms (the ambulocetids, remingtonocetids, and protocetids) which, at least for a while, lived alongside both each other and their pakicetid forebears (Figure 7.2 and Plate 16a). By the end of the Lutetian, however, pakicetids, ambulocetids, and remingtonocetids had disappeared, and were replaced by basilosaurids—the first truly oceanic whales. Protocetids persisted throughout the Bartonian, but eventually went into decline and disappeared almost entirely by the end of that stage. The youngest reported protocetid is an as-yet-undescribed Priabonian individual that lived alongside basilosaurids in what is now Egypt (Gingerich et al., 2014). Thereafter, no semiaquatic whales are known from the fossil record (section 7.2.1).

The Eocene–Oligocene boundary (ca 34 Ma) marks a major transition in the diversification of cetaceans, with archaeocetes (except for the enigmatic kekenodontids) giving way to the earliest neocetes (Figure 7.2). This period also roughly coincides with a rapid drop in global temperatures and the initiation of **Antarctic glaciation** (Figure 7.3), as well as the development of an incipient **Antarctic Circumpolar Current** (ACC) (Katz et al., 2011). The ACC, which owes its existence to the final separation of South America from Antarctica via the opening of the Drake Passage, is the strongest and most important current of the modern ocean. Because of its great depth, it is capable of mixing deep, nutrient-rich bottom waters into the upper layers of the ocean, from where they go on to support a staggering three-quarters of global marine export production (DeConto and Pollard, 2003; Sarmiento et al., 2004). The onset of the ACC thus likely resulted in heightened ocean productivity and may have been at least partially responsible for the initial diversification of neocetes (Berger, 2007; Fordyce, 1980; Marx and Fordyce, 2015). In particular, neocete evolution may have been linked to that of diatoms, which proliferated during the Late Eocene in response to the onset of the ACC (Egan et al., 2013; Marx and Uhen, 2010). The importance of diatoms in this context rests on their large size compared to other types of phytoplankton, which makes them capable of supporting short food chains dominated by large apex predators. For the same reason, the rise of diatoms may also have facilitated the emergence of cetacean mass-feeding adaptations, such as baleen (Berger, 2007).

Detailed analyses of a time-calibrated, molecular phylogeny of extant cetaceans revealed elevated rates of diversification around the time of the Eocene–Oligocene boundary, followed by a second rate increase during the Neogene (section 7.1.2). Both of these events coincide with major changes in ocean circulation, and thus provide support for a potential link between neocete evolution and the establishment of the ACC (Steeman et al., 2009). However, a reanalysis of the same data using a different methodology did not identify the Eocene–Oligocene rate shift. Instead, the results of this second study point to initially elevated rates of speciation followed by a slow, steady decline (Figure 7.4), which is more consistent with an early adaptive radiation than the effects of ocean restructuring (Rabosky, 2014; Steeman et al., 2009). Additional caveats lie in the still-controversial timing of the development of the ACC itself, as well as the question of whether the young Eocene–Oligocene current was already deep enough to allow the kind of vigorous mixing that drives ACC-related productivity today (Hill

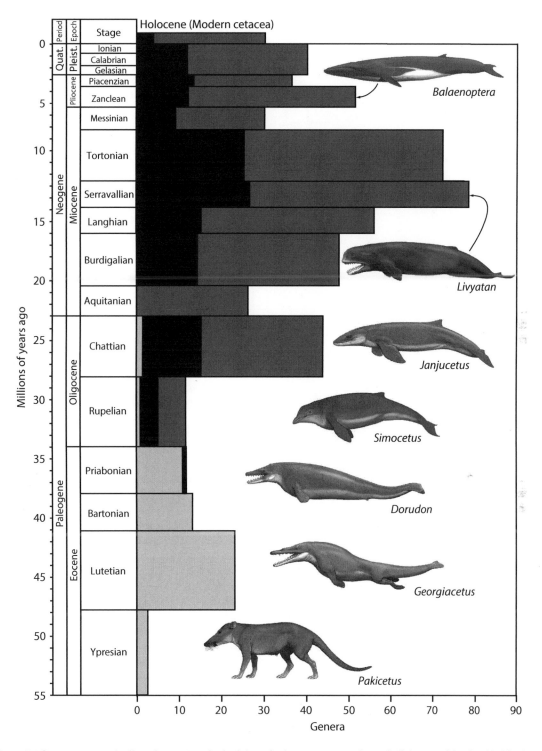

Figure 7.1 Cetacean generic diversity over geological time. Archaeocetes are shown in light gray, Mysticeti in black, and Odontoceti in dark gray. Note the obvious sampling gap during the Aquitanian.

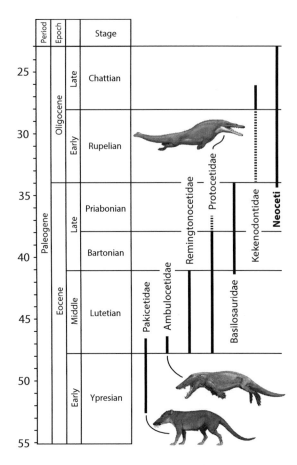

Figure 7.2 Stratigraphic ranges of the major groups of archaeocetes. Stippled lines indicate range extensions based on tentatively identified material.

et al., 2013; Pfuhl and McCave, 2005). The evidence for any immediate effect of the ACC on early neocete evolution thus remains contradictory and in need of further study.

The large-scale glaciation of Antarctica around 34 Ma was likely accompanied by a major fall in sea level, which abruptly reduced the size of the continental shelf. Along with subsequent erosion, this event might help to explain why so few fossil cetaceans have been described from the Early Oligocene. Nevertheless, both the fossil record and molecular/morphological divergence date estimates suggest that mysticetes and odontocetes had not only evolved by this time, but also had started to diversify (Barnes *et al.*, 2001; Marx and Fordyce,

2015; McGowen *et al.*, 2009). By the Chattian (Late Oligocene), several groups of toothed mysticetes, archaic chaeomysticetes (Figure 7.5), and odontocetes (Figure 7.6) had been established, and diversity rose to its highest point to date (Fordyce and de Muizon, 2001). For some of these early taxa, success was short-lived. Nearly all of the toothed mysticetes, several odontocete families (Agorophiidae, Ashleycetidae, Mirocetidae, Patriocetidae, and Xenorophidae) and the last remaining archaeocetes—kekenodontids— apparently did not make it beyond the end of the Chattian, which also marks the end of the Oligocene and the Paleogene.

7.1.2 Neogene

Global cetacean diversity initially dropped during the earliest Miocene, possibly because a new pulse of glaciation—this time centered in the Northern Hemisphere—limited sampling opportunities (Hyeong *et al.*, 2014). Following the Aquitanian, diversity increased across the **Mid-Miocene Climatic Optimum** (Figure 7.3) and finally peaked during the Serravallian and Tortonian (Figure 7.1). Like the origin of neocetes itself, much of this diversification may have been driven by a combination of high, diatom-driven marine productivity and climate change (Figure 7.7) (Marx and Uhen, 2010).

Around the time of the global diversity peak, delphinids experienced a burst in speciation rates (Figure 7.4). This event represents the only consistently identified rate shift in cetacean diversification, and may be linked to either ocean restructuring (e.g., closure of the Tethys seaway) or the emergence of a key innovation shaping early dolphin evolution (Rabosky, 2014; Steeman *et al.*, 2009). Both the rate shift and the molecular estimate for the divergence of delphinids roughly coincide with the first appearance of this clade in the fossil record, and mark the beginning of an initial, moderate rise in species-level diversity (Bianucci, 2013; Murakami *et al.*, 2014). Overall, however, the Late Miocene record of fossil delphinids is sparse, with the vast majority of specimens being restricted to the Pliocene. The discrepancy between the time of origin of the family and the age of most of the available fossil material may imply that delphinids underwent two separate phases of diversification: one shortly

Figure 7.3 Global stable oxygen isotope records over the past 65 million years. Most of the data are derived from deep-sea foraminifera recovered from Deep Sea Drilling Project and Ocean Drilling Program sites. Higher $\delta^{18}O$ values generally correlate with lower global temperatures, but are also driven by global ice volume. The temperature scale on the right assumes an ice-free ocean and therefore applies only to the time preceding the development of large-scale Antarctic ice sheets around 35 Ma. Reproduced from Zachos *et al.* (2008), with permission of Macmillan Publishers.

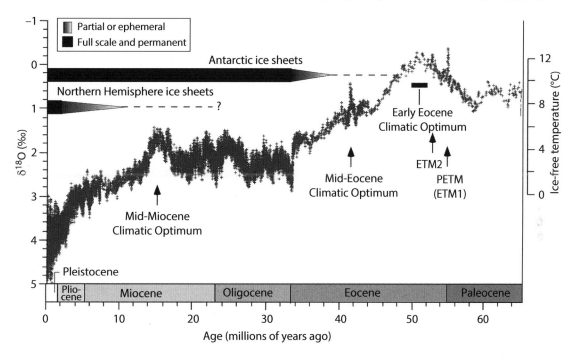

after their origin, which led to the establishment of the major delphinid lineages; and a second during the Pliocene, which was marked by heightened speciation and turnover, and saw the emergence of most of the extant delphinid fauna (Bianucci, 2013).

Following the Middle–Late Miocene global diversity peak, the number of cetacean genera sharply drops toward the Messinian (ca 6 Ma). This may, at least in part, reflect a sampling artefact caused by the repeated drying up of the Mediterranean Sea during the **Messinian Salinity Crisis** (Krijgsman *et al.*, 2010). The sediments surrounding the Mediterranean Basin are both highly fossiliferous and somewhat overrepresented in the cetacean fossil record—the latter primarily because of their location in a densely populated, relatively affluent region of the world. The absence of cetaceans from this area during the Messinian thus likely has an undue effect on

apparent global diversity, causing it to decline much more steeply than was actually the case (Uhen and Pyenson, 2007).

Nevertheless, the Messinian sampling artefact only seems to be superimposed on an equally pronounced, real drop in diversity during the Pliocene, which culminated in the relatively small number of genera still alive today. The reasons for this decline are not entirely clear, but may be related to global cooling and the onset of long-term **Northern Hemisphere glaciation** around 3.5 Ma (De Schepper *et al.*, 2014) (Figure 7.3). The latter not only caused shifts in the strength and location of major zones of productivity, such as the sub-Antarctic Ocean (Martínez-García *et al.*, 2014; Martínez-García and Winckler, 2014), but also repeatedly changed the amount of available shallow-water habitats (Pyenson and Lindberg, 2011). Both of these developments likely were bad news for many small-sized species that were adapted to a

Figure 7.4 Dynamics of cetacean diversification through time, based on the phylogeny of extant cetaceans recovered by Steeman *et al.* (2009). The first two panels show the phylogeny with branch lengths drawn proportional to (a) the rate of speciation and (b) the probability that they contain a rate shift (numbers above branches denote branch-specific shift probabilities). Note the increased rate of speciation and high probability of a rate shift along the lineage leading to Delphinidae. The posterior distribution of the number of distinct processes (c) indicates that speciation was shaped by two processes, including (d) a weak slowdown in speciation rate through time and a burst and subsequent slowdown around 7.5 Ma, corresponding to the early radiation of delphinids. Unlike speciation, extinction rates (e) remain relatively stable and low throughout cetacean evolutionary history. Reproduced from Rabosky (2014) under a Creative Common Attribution license.

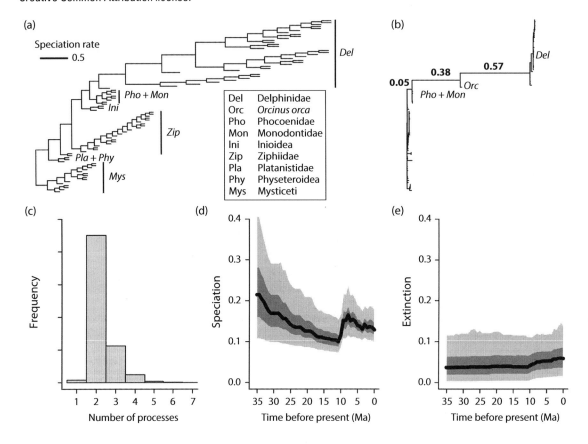

shallow-water existence and unable to undertake the large-scale migrations required to reach alternative, far-away feeding grounds. It was likely under such circumstances that many small-sized cetaceans, such as dwarf balaenids and herpetocetine cetotheres, finally disappeared (Marx and Fordyce, 2015).

The Pliocene and Pleistocene were largely dominated by families that are still alive today, with the exception of albireonids, odobenocetopsids, and herpetocetine cetotheriids (Figures 7.5 and 7.6). Logic dictates that younger rocks should in general be more abundant and better preserved than older ones, and thus exert a large influence on fossil diversity estimates (Smith, 2001). In the case of cetaceans, however, comparatively little is known about the Pleistocene, presumably because many of the marginal marine deposits that formed at the time were flooded when meltwater drove up sea levels at the end of the last glacial period. Overall, circa 70% of modern cetacean genera are known as fossils from Pleistocene rocks.

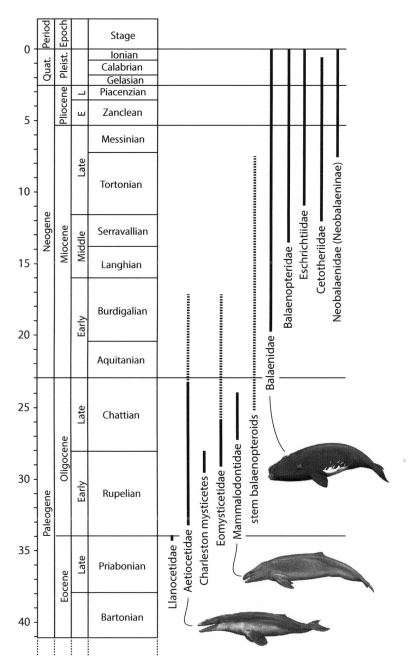

Figure 7.5 Stratigraphic ranges of the major groups of mysticetes. Stippled lines indicate range extension based on uncertainly identified or as-yet-undescribed material. Miocene range extensions of Aetiocetidae and Eomysticetidae are based on Rivin (2010).

Figure 7.6 Stratigraphic ranges of the major groups of odontocetes. Stippled lines indicate range extensions based on tentatively identified or poorly dated material.

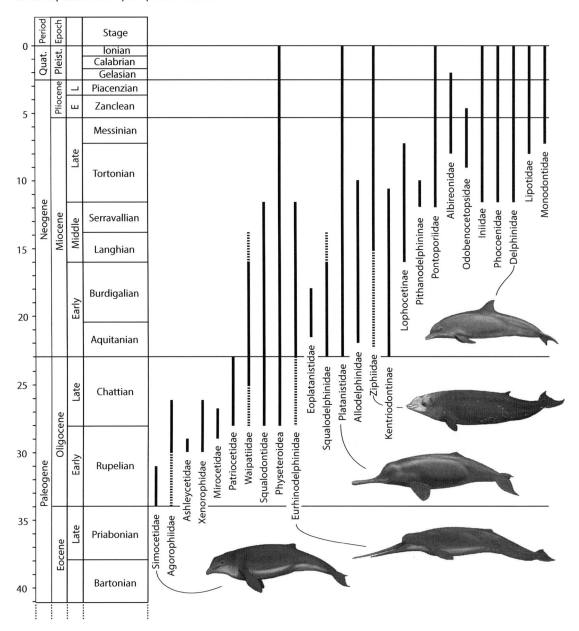

7.2 Major turnover events

Cetacean evolution is marked by a series of pronounced turnover events, during which one major group came to replace another in quick succession. The best defined of these include (1) the transition from archaeocetes to neocetes around the Eocene–Oligocene boundary; (2) the decline of toothed mysticetes and concurrent diversification of the baleen-bearing chaeomysticetes during the Late Oligocene and Early

Figure 7.7 Comparison of (a) neocete, (b) mysticete, and (c) odontocete genus-level taxonomic diversity with (d) diatom diversity and (e) global δ18O values. Gray and black curves denote raw data downloaded from the Paleobiology Database (Uhen, 2015) and a ranged through estimate, respectively. For the latter, a genus was scored as present in ("ranged through") a particular time bin even if it had not actually been sampled, as long as the time in question lay in between confirmed younger and older occurrences of the same taxon. Error for the δ18O represents mean standard error (SE) multiplied by 100. Adapted from Marx and Uhen (2010), with permission of the American Association for the Advancement of Science.

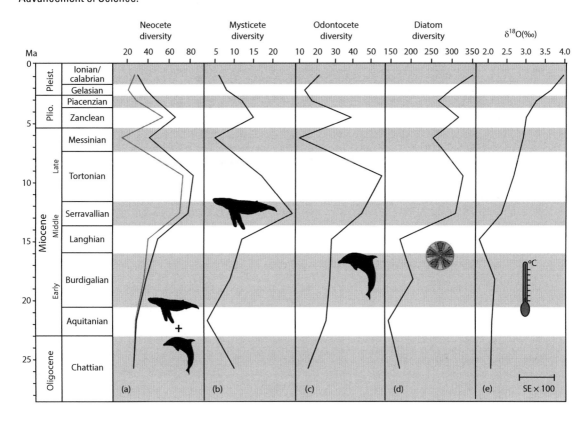

Miocene; (3) the replacement of platanistoids by modern delphinoids; and (4) the establishment of the modern cetacean fauna during the Pleistocene (Figure 7.8). There is generally too little information to tell whether these turnovers happened as a result of direct competition. In most cases, the decline of the older group could equally have been driven by external factors, such as environmental change, which was then followed by the opportunistic radiation of the replacing group into the newly vacant ecospace. In addition, some of the patterns interpreted to be turnovers might really be misinterpretations of the fossil record, arising from sampling bias and/or poor temporal resolution.

7.2.1 Archaeocetes to neocetes

The archaeocete–neocete transition has traditionally been interpreted as rather abrupt, with the disappearance of archaeocetes at the end of the Eocene almost exactly coinciding with the emergence of the earliest neocetes (Mitchell, 1989). This picture is, however, being blurred by the discovery of an increasing number of mysticetes, including a potential chaeomysticete, possibly dating to the latest Eocene (Fordyce, 2003; Goedert et al., 2007; Kiel and Goedert, 2006), as well as the idea that the Late Oligocene kekenodontids may in fact be archaeocetes (Clementz et al., 2014; Fordyce, 2002). If correct, this would indicate a substantial amount of temporal overlap—perhaps as much as 10 Ma.

Figure 7.8 Major faunal turnover events (highlighted in gray) illustrated via genus-level taxonomic diversity plots, including (a) the transition from archaeocetes to Neoceti, (b) the decline of toothed mysticetes, and (c) the transition from platanistoids to modern delphinoids. Although the content of the superfamily Platanistoidea is still debated, the general trend is retained independent of whether squalodontids are included. The delphinoid curve includes counts of Delphinidae, Phocoenidae, and Monodontidae only. All data are from the Paleobiology Database (Uhen, 2015) and represent raw taxon counts—that is, taxa are only shown as present in a particular time bin if they have actually been sampled (as opposed to "ranged through"). No fossils of the single living platanistoid, the South Asian river dolphin *Platanista*, have yet been found, which explains the apparent absence of platanistoids during the Pliocene and Pleistocene.

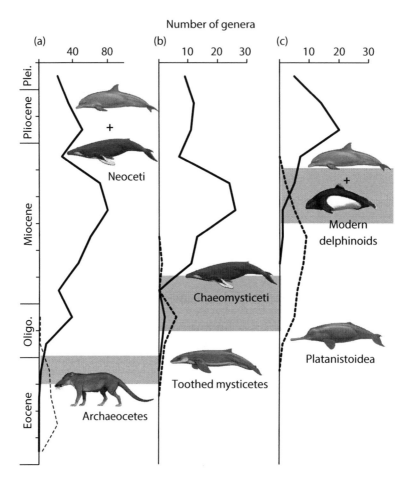

Determining the dynamics of the archaeocete–neocete transition mainly rests on two questions: How much potential archaeocete diversity is obscured by poor Early Oligocene sampling? And do kekenodontids really represent archaeocetes, or are they stem neocetes/mysticetes that may have taken part in the original neocete radiation? There is little published information to bring to bear on either issue, although some rather complete, as-yet-undescribed kekenodontid material from New Zealand will likely help to answer the second question (Clementz *et al.*, 2014). Early Oligocene archaeocete material is virtually unknown, except for an isolated, uncertainly identified tooth from France originally described as *Phococetus vasconum*. *Phococetus* is unfortunately too incomplete to allow any confident taxonomic assignment; nevertheless, its sheer existence is tantalizing, because it might imply a global distribution for Oligocene archaeocetes (Kellogg, 1936).

Overall, there is not enough information at present to determine what happened to archaeocetes at the Eocene–Oligocene boundary. Nevertheless, well-sampled Late Oligocene assemblages from New Zealand and North America show that archaeocetes younger than 30 Ma were at best a minor component of the global cetacean fauna, which was dominated by a variety of highly disparate mysticetes and odontocetes (Fordyce, 2006; Geisler *et al.*, 2014; Geisler and Sanders, 2003). This decline may seem profound, but does not necessarily indicate a major replacement event. In this context, it is important to remember that archaeocetes themselves are merely a paraphyletic group of families succeeding each other both phylogenetically and stratigraphically (Figure 7.2). Four of the five archaeocete families had already gone extinct by the time neocetes first appeared, and it was likely only the basilosaurids that could have been faced with any sort of competitive pressure.

Basilosaurids vary considerably in size, but appear to have employed similar (raptorial) prey capture and feeding strategies. Interestingly, these strategies were conserved in both kekenodontids and some of the earliest mysticetes, neither of which had any immediately obvious competitive advantage (e.g., echolocation) over basilosaurids (Barnes and Sanders, 1996; Fitzgerald, 2006). It thus seems plausible to view basilosaurids as just one link in an ongoing chain of increasingly oceanic raptorial taxa, which, in the form of toothed mysticetes, continued past the archaeocete–neocete transition. The chain only broke with the extinction of archaic mysticetes during the Late Oligocene or Early Miocene, which led to the final partitioning of marine ecospace between filter-feeding baleen whales and echolocating odontocetes (Marx and Fordyce, 2015).

7.2.2 Decline of toothed mysticetes

Compared to the archaeocete–neocete transition, the decline of archaic toothed mysticetes seems more difficult to explain. Judging from their morphology and isotopic data, toothed mysticetes had adapted to a variety of feeding strategies, ranging from raptorial snap feeding to benthic suction

feeding and maybe even incipient filter feeding (section 6.1). Most of these strategies existed alongside and had little or no overlap with fully fledged chaeomysticete filter feeding, which likely arose around the same time. Nevertheless, toothed mysticetes disappeared by the end of the Late Oligocene or, possibly, the Early Miocene, whereas chaeomysticetes flourished and continued to diversify (Figure 7.8).

It is possible that the early diversification of archaic mysticetes simply reflects a short-lived evolutionary experiment, although this still leaves unclear what ultimately caused the demise of such a range of specialized taxa. Alternatively, the decline of toothed mysticetes may have been driven by competition with other marine predators, such as odontocetes and pinnipeds. Odontocetes in particular may increasingly have been at an advantage as their echolocation abilities continued to evolve. If so, the disappearance of tooth mysticetes could be interpreted as the delayed replacement of the archaeocete (i.e., non-filter-feeding, non-echolocating) morphotype by its more efficient, echolocating descendants (Marx and Fordyce, 2015). However, there is currently no evidence supporting direct competition between odontocetes and archaic mysticetes.

7.2.3 Delphinoids and platanistoids—ships passing in the night?

The Middle and early Late Miocene saw the disappearance of several lineages of archaic odontocetes, most of which have at least once been referred to the highly debated superfamily Platanistoidea (Fordyce and de Muizon, 2001). In particular, this extinction event, if indeed synchronous and attributable to a single cause, may have affected the heterodont waipatiids and squalodontids, as well as the homodont allodelphinids, squalodelphinids and eurhinodelphinids. Curiously, the disappearance of these taxa seems to have roughly coincided with the origin of modern delphinoids (delphinids, phocoenids, and monodontids), as well as that of iniids and pontoporiids (Figure 7.8). The overlap of putative platanistoids and modern delphinoids actually recorded in the fossil record is rather short, with the sole exception of the living South Asian river dolphin, *Platanista*, and its immediate relatives (Figure 7.6). Nevertheless, fossil delphinids and phocoenids are known from the earliest Late Miocene, suggesting that the initial diversification of modern delphinoids likely occurred somewhat earlier, in the Middle Miocene (Murakami *et al.*, 2014). This is further supported by molecular clock divergence dating, which places the origin of modern delphinoids around or even before 15 Ma (Chen *et al.*, 2011; Dornburg *et al.*, 2012; McGowen *et al.*, 2009).

It is striking that most of the archaic platanistoids are characterized by an extremely elongated rostrum and a flexible neck similar to that of extant "river dolphins," including the coastal Franciscana (*Pontoporia*). By contrast, delphinids and, to a lesser extent, phocoenids seem to be more adapted to an open ocean habitat. Assuming that these differences indeed reflect habitat choice, it is tempting to speculate that platanistoids fell victim to some form of environmental change that negatively affected shallow-water species, but promoted—or at least did not hinder—the diversification of open-water forms. Interestingly, physeteroids and ziphiids made it through the Middle and Late Miocene without being markedly affected by the contemporary faunal turnover. Considering the open-water habitat and deep diving habit of living beaked and sperm whales, it seems plausible that a similar habitat choice may have shielded their ancestors from whatever drove the demise of platanistoids.

There is an obvious candidate for a mechanism that could have been behind the Late Miocene faunal turnover. Following the Mid-Miocene Climatic Optimum, global temperatures entered a phase of prolonged decline, which culminated in the establishment of Northern Hemisphere ice sheets (Figure 7.3) (Zachos *et al.*, 2008). The beginning of this temperature drop (ca 13–14 Ma) aligns nicely with the disappearance of most platanistoids, and likely would have affected shallow-water habitats in a profound way. However, testing this idea will have to await the publication of relevant stable isotope data, as well as functional studies of extinct platanistoids that could reveal more about their habitat preferences. In addition, there needs to be a reassessment of the phylogenetic affinities and ecology of kentriodontids (kentriodontines, lophocetines, and pithanodelphinines), which are often regarded as morphologically intermediate between modern

delphinoids and archaic odontocetes (de Muizon, 1988; Fordyce and de Muizon, 2001).

7.2.4 Establishment of the modern fauna

Based on both molecular studies and the fossil record, most of the living cetacean genera originated during the Late Miocene and Pliocene, although so far only a few fossils older than the Pleistocene have been assigned confidently to any of the extant species (Marx, 2011; McGowen *et al.*, 2009; Murakami *et al.*, 2014; Steeman *et al.*, 2009). Besides modern taxa, Pliocene rocks also preserve a rich assemblage of now-extinct forms, such as herpetocetine cetotheriids, dwarf balaenids, several species of small-sized balaenopterids, archaic phocoenids, odobenocetopsids, and albireonids (Figures 7.5 and 7.6). Remarkably, all of the latter apparently disappeared before the early Pleistocene, with the exception of a single specimen of *Herpetocetus* from western North America (Boessenecker, 2013a, 2013b).

The decline of nearly all but the extant cetacean lineages around the Pliocene–Pleistocene boundary accompanies similar turnover events in pinnipeds and, somewhat later, sea birds (Boessenecker, 2013a; Valenzuela-Toro *et al.*, 2013). In general, Pliocene cetacean assemblages appear to have been more localized than the modern fauna. This is especially true for the eastern North and South Pacific, where the emergence of modern genera was seemingly delayed relative to the western North Pacific, the North Atlantic, and Australia (Boessenecker, 2013a; Fitzgerald, 2005; Whitmore, 1994). The reasons for this regionalization may lie in a series of geographical barriers that hampered taxon dispersal, such as a closed Bering Strait, the closing of the Panama seaway, and permanent El Niño conditions in the equatorial Pacific during the Early Pliocene. In part, the Pliocene–Pleistocene turnover might thus simply reflect the removal of some of these barriers, which facilitated the spread of modern, cosmopolitan taxa at the expense of geographically restricted species. Regional factors (e.g., tectonically driven local sea-level change), ocean restructuring, and pronounced global cooling likely also played an important role; however, there are currently too few data, especially on the poorly sampled Pleistocene, to draw a clear picture of exactly how they contributed (Boessenecker, 2013a, 2013b; Valenzuela-Toro *et al.*, 2013).

7.3 Disparity and evolutionary rates

Compared to taxonomic richness, the study of cetacean morphological diversity (**disparity**) has received surprisingly little attention. Work so far has focused on quantifying disparity based on (1) body size, scored for extant species only (Slater *et al.*, 2010), and (2) discrete morphological characters, scored for a range of extinct and extant mysticetes (Marx and Fordyce, 2015). A further study quantified mysticete skull shape via two-dimensional landmarks, but used those data mainly to discern phylogenetic relationships (Hampe and Baszio, 2010). Note that body size in this section will be discussed purely as a measure of morphological variation (i.e., one of many possible variables describing shape). A more detailed discussion of cetacean body size evolution is provided in section 7.4.

Based on inferences made from extant cetaceans, body size evolved along separate lines in taxa following different dietary strategies (fish-based, squid-based, and filter feeding). Ancient fish-eating species seem to have been larger than their living counterparts, whereas the opposite is true for squid eaters. Crucially, cetaceans started to occupy a range of size classes (i.e., became highly disparate) early in their history, during the latest Eocene and Oligocene. Judging from the correlation between body size evolution and diet, this early partitioning of size niches may be reflective of equally diverse feeding strategies, and thus potentially an early **adaptive radiation** (Slater *et al.*, 2010). This scenario is supported by initially high, but gradually slowing, rates of speciation (section 7.1). The latter are typical of radiation events but, in the case of cetaceans, have so far only been identified in a single study (Rabosky, 2014). After the Oligocene, body size evolution began to slow, with the exception of a brief burst around 6–11 Ma likely related to the diversification of delphinids (Rabosky, 2014; Slater *et al.*, 2010). This general decline in the rate of morphological evolution may be a result of **ecological niche filling**, during which the gradual crowding of ecospace gives rise to ever more similar niches, which are consequently occupied by species ever more closely resembling their own ancestors (Freckleton and Harvey, 2006).

At least in the case of mysticetes, quantifying disparity based on discrete morphological characters yields similar results (Figure 7.9). When both living and fossil taxa are considered, mysticetes show strongly elevated levels of disparity during the Oligocene, accompanied by accelerated rates of phenotypic and genomic change (Marx and Fordyce, 2015). In qualitative terms, this early peak in morphological diversity is reflected in a variety of morphotypes and feeding strategies, such as benthic suction feeding (*Mammalodon*), macrophagy (*Janjucetus*), and filter feeding (eomysticetids and stem balenopteroids). From about 28 Ma, disparity started to decline and evolutionary rates began to level out, which is consistent with an adaptive radiation followed by ecological niche filling. During the Early Miocene, both rates and disparity became essentially static and have largely remained so until the present. The reasons behind this pattern likely involve a combination of (1) the decline of toothed mysticetes, which prevented mysticetes from accessing ecological niches now mostly occupied by pinnipeds and odontocetes (section 7.2); and (2) the persistence of a narrow set of ecological niches that were actually open to mysticetes, with few or no major new adaptations that could have created ecological opportunity (Marx and Fordyce, 2015).

7.4 Body size

Modern cetaceans span no less than five orders of magnitude in terms of their body mass, from small porpoises weighing about 43 kg to the blue whale *Balaenoptera musculus*—which, at up to 190 000 kg, is the largest animal that ever lived (Evans *et al.*, 2012). In general, mysticetes tend to be considerably bigger than odontocetes, with even the smallest baleen whale, the pygmy right whale *Caperea*, reaching 6.5 m and weighing around 3.5 t (Kemper, 2009). Nevertheless, given that the average size of living mammals is around that of a rat, all cetaceans are comparatively gigantic.

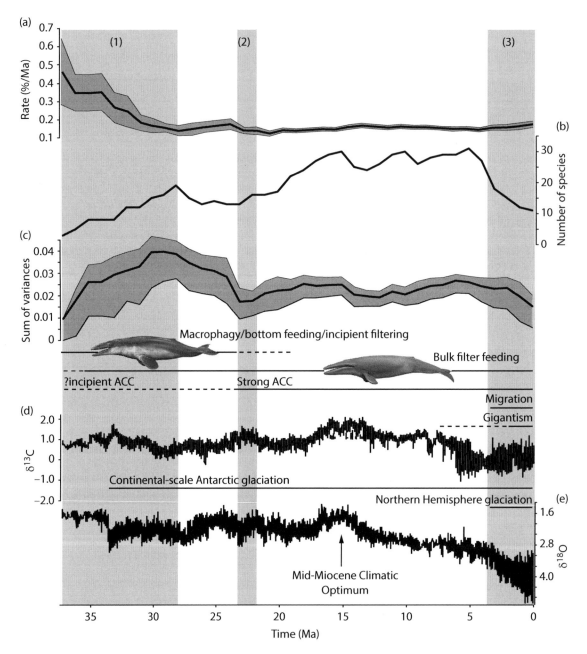

Figure 7.9 Major events in mysticete evolution, as illustrated by (a) rates of phenotypic and genomic evolution, (b) species-level diversity, and (c) disparity, all compared to global carbon (d) and oxygen (e) isotopic records. Evolutionary rates are initially high and coincide with an increase in taxonomic diversity and disparity, thus marking an adaptive radiation (1), around the time of development of the Antarctic Circumpolar Current (ACC). Rates and disparity subsequently decreased, possibly as a result of ecological niche filling, and then became stable from the Early Miocene onward as the ACC developed its full strength and toothed mysticetes largely disappeared (2). Diversity rose until the end of the Early Miocene and then remained stable, before crashing in tandem with the deterioration of the global climate around 3.5 Ma (3). This event led to the disappearance of most small-sized mysticetes and may be associated with the beginnings of large-scale migration. Adapted from Marx and Fordyce (2015) under a Creative Commons Attribution license. Isotope data are from Zachos *et al.*, (2008).

There are many, not mutually exclusive, hypotheses trying to explain why cetaceans have grown so large. Potential contenders include: relaxed constraints on maximum body size because of the increased support provided by water; abundant food derived from a productive ocean; feeding ecology; thermoregulation; predator avoidance; and, at least in some cases, migratory behavior and deep diving (Berger, 2007; Evans *et al.*, 2012; Millar and Hickling, 1990; Montgomery *et al.*, 2013; Schreer and Kovacs, 1997; Slater *et al.*, 2010). Independent of which of these factors actually played a role, it seems that cetaceans are subject to the same macroevolutionary drivers and constraints as their terrestrial cousins. In general, the size of nonaquatic mammals is proposed to be determined by three factors, including (1) a minimum size necessary for effective thermoregulation (2 g in air); (2) a putative general trend toward larger body size (also known as **Cope's rule**), driven by short-term selective advantages related to, for example, resource acquisition or predation; and (3) a heightened risk of extinction of large-sized species caused by longer generation times, lower species abundances, and higher absolute energetic requirements (Clauset and Erwin, 2008). These factors combine to create an **evolutionary trade-off** between the advantages offered by larger size and the higher extinction risk they bring with them.

Because of the minimum size requirement, the trade-off governing body size gives rise to a skewed size distribution with a fixed lower boundary—in other words, a distribution in which most taxa are located closer to the minimum possible size than to the potential maximum (Figure 7.10). The size distribution of cetaceans follows the same pattern, but differs from that of terrestrial mammals in having a considerably larger viable minimum size (ca 7 kg) because of the added difficulty of maintaining a constant body temperature in water (Downhower and Blumer, 1988). This pronounced shift in the viable minimum leads to corresponding increases in the average and maximum expected sizes. As a result, the size of even the largest extant cetacean is not unexpected, and could even be exceeded, under this relatively simple macroevolutionary model (Clauset, 2013). Nevertheless, at least in some cases, maximum body size may be limited by feeding-related energetic demands and the rate at which energy can be delivered (Goldbogen *et al.*, 2012; Potvin *et al.*, 2012).

To extend the study of body size back in time first requires a way to estimate the size of fossil taxa, many of which are known only from incomplete, variably preserved skeletons. In most terrestrial

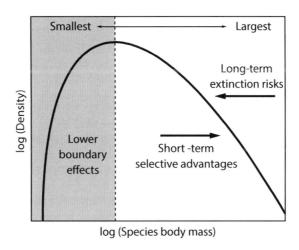

Figure 7.10 The body size distribution characteristic of mammals. The shape of the curve is determined by a minimum viable size constraining a macroevolutionary trade-off between (1) short-term selective advantages creating an upward "pressure" toward larger body size and (2) opposing long-term extinction risks. The size distribution of terrestrial mammals and cetaceans follows the same pattern, except for a larger minimum size in the latter (7 kg vs 2 g). Adapted from Clauset (2013) under a Creative Commons Attribution license.

mammals, the size of the cheek teeth provides a fairly accurate proxy, but this does not work well for cetaceans because of their greatly modified, or absent, dentition (sections 6.1 and 8.3). Instead, the most commonly used measures for whales and dolphins include (1) the width across the occipital condyles and (2) the width across the zygomatic arches (**bizygomatic width**), respectively. Multivariate predictions of body size based on several proxy measurements may yield more accurate results, but are often difficult to apply given the incompleteness of many fossil specimens (Marino *et al.*, 2004; Pyenson and Sponberg, 2011).

The minimum, maximum, and mean body size of cetaceans all increased shortly after the origin of the earliest, dog-sized whales (e.g., *Pakicetus*, weighing maybe as little as 5–10 kg) in the Early Eocene (Figure 7.11) (Uhen *et al.*, in preparation). From the late Middle Eocene onward, minimum and mean body size remained relatively stable, albeit with some marked fluctuations during the Miocene and a more pronounced drop in minimum size during the Pleistocene. Maximum body size also initially plateaued until the end of the Oligocene (with a pronounced but probably sampling-related drop during the Rupelian), but then embarked on a more or less constant increase— from maybe no more than 4 t by the end of the Oligocene, to around 15 t by the end of the Miocene, and up to an extreme of more than 150 t today.

Viewed within a phylogenetic framework (Montgomery *et al.*, 2013), the size evolution of each of the three major cetacean groups appears to be following particular trends: archaeocetes first increase in size (until the emergence of *Basilosaurus*) and then decrease again along the lineage leading to neocetes. By contrast, the picture is more mixed for odontocetes, which show an initial decrease in size along the stem lineage, followed variously by further decreases (e.g., in kogiids) or increases (e.g., giant sperm whales and ziphiids). Mysticete body size has not yet been analyzed in a comprehensive phylogenetic context, but most of the archaic tooth-bearing taxa (aetiocetids and mammalodontids) seem to have been similar in size to the inferred ancestral neocete, that is, the node uniting mysticetes and odontocetes (Montgomery *et al.*, 2013). Indeed, the size difference between archaic toothed mysticetes

and their baleen-bearing relatives is marked (Figure 7.12) and has so far not been sufficiently explained in functional terms—especially given the existence of some extremely small filter-feeding taxa, such as herpetocetines. The only exceptions to this pattern are *Llanocetus* and a *Morawanocetus*-like aetiocetid from the Oligocene of Japan, both of which are distinctly larger than all other archaic mysticetes and hence may imply a more complex pattern of early baleen whale size evolution (Fitzgerald, 2010; Tsai and Ando, 2015). Within Chaeomysticeti, there appears to be a clear trend toward larger body size, although there are likely to have been some reversals (e.g., herpetocetine cetotheriids).

As might be expected from these trends and the size distribution of the extant taxa, fossil chaeomysticetes tend to be considerably larger than both archaeocetes and fossil odontocetes (Figure 7.11), likely because of their adaptation to bulk filter feeding (section 6.1) (Slater *et al.*, 2010). There are, however, some exceptions. On the side of mysticetes, these include several small-sized herpetocetine cetotheriids, as well as certain stem balaenopteroids and balaenids (e.g., *Morenocetus*, *Balaenula*, and *Balaenella*). Similarly, a few large physeteroids (e.g., *Livyatan*) and a trend toward larger body size in ziphiids and a few delphinids demonstrate that odontocetes need not always be small. For most of their history, fossil baleen whales defined the upper limit of the cetacean size range (Figure 7.11). During the Late Miocene, maximum mysticete body size started to increase markedly, and quickly grew to encompass animals comparable to midsized rorquals alive today. This trend continued throughout the Pliocene and Pleistocene, and mysticetes continued to push the boundaries up until the present day. The gigantism of living mysticetes is thus not an entirely recent phenomenon, but kept building up to its present level over at least 10 million years.

Nevertheless, there is a profound change that affected the size distribution of modern cetaceans as little as 3 million years ago: although Late Miocene mysticetes could grow to gigantic proportions, most of them did not. Instead, the majority stayed well within the size range of their mysticete forebears, hardly exceeding 2–3 t in overall weight (Figure 7.11). The size distribution gradually

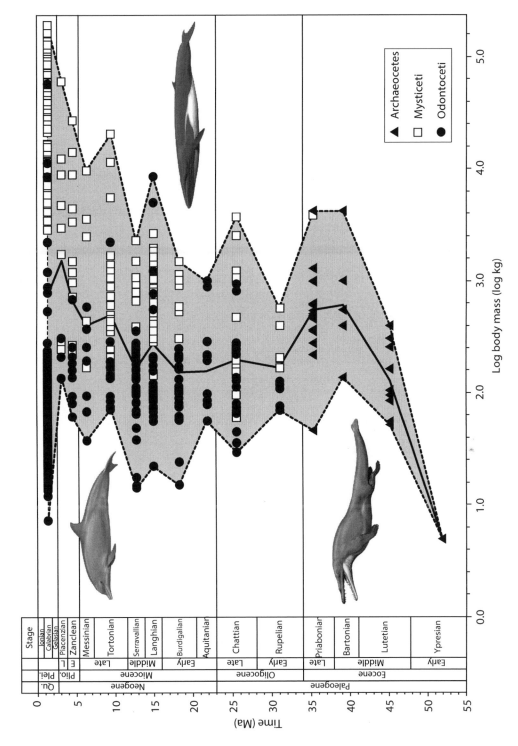

Figure 7.11 Body masses of Cetacea over time. Note that, since the Oligocene, the maximum body mass of cetaceans has increased over time and is almost always occupied by a mysticete. The minimum and the mean have remained relatively steady. Data are from Uhen *et al.* (in preparation).

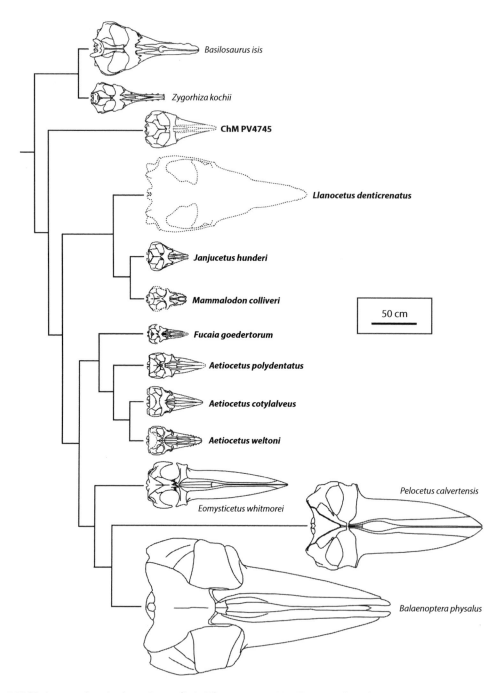

Figure 7.12 Phylogeny of toothed mysticetes (in bold), as proposed by Fitzgerald (2010) with the skulls of various taxa drawn to size. Adapted from Fitzgerald (2010), with permission of the Linnean Society of London.

Within the figure:
- *Basilosaurus isis*
- *Zygorhiza kochii*
- **ChM PV4745**
- **Llanocetus denticrenatus**
- **Janjucetus hunderi**
- **Mammalodon colliveri**
- **Fucaia goedertorum**
- **Aetiocetus polydentatus**
- **Aetiocetus cotylalveus**
- **Aetiocetus weltoni**
- *Eomysticetus whitmorei*
- *Pelocetus calvertensis*
- *Balaenoptera physalus*
- 50 cm

shifted upward during the latest Miocene and Early Pliocene, but small-sized mysticetes still remained a common occurrence. This only changed with the onset of the Late Pliocene and Pleistocene, which saw the extinction of herpetocetines, small balaenids, and some of the smaller balaenopterids, and established gigantism as the dominant mysticete size habit. Perhaps the demise of small baleen whales was driven by the onset of long-term Northern Hemisphere glaciation, which would have resulted in large-scale fluctuations in the availability of shallow-water habitats—with potentially dire consequences for small, shelf-bound species. By contrast, larger mysticetes would have been able to weather these changes more easily, as their size and fat reserves enabled them to undertake large-scale migrations toward productive, high-latitude feeding areas (section 6.4) (Marx and Fordyce, 2015).

7.5 Brain size

Modern cetaceans are remarkable for their enlarged, highly gyrified brains (section 3.4.4) (Hof *et al.*, 2005; Marino, 1998; Worthy and Hickie, 1986). Considering only absolute brain size can, however, be misleading, because the size of the mammalian brain generally correlates with that of the body (Atchley *et al.*, 1984). To factor out the effects of varying body size, the degree of brain development across different taxa is generally expressed in the form of an **encephalization quotient** (EQ). The latter quantifies how much the actual brain size of a given animal departs from that predicted by its body size (Jerison, 1973): a value larger than 1 denotes a brain that is bigger than expected, and a value less than 1 denotes one that is smaller. Several of the living delphinids, including *Delphinus*, *Grampus*, *Pseudorca*, *Sotalia*, *Steno*, *Tursiops*, and several species of *Lagenorhynchus*, have EQs equal to or greater than 4 (Marino, 1998). In other words, these delphinids have brains at least four times as voluminous as one might expect from their size, which makes them the largest-brained of all mammals, except modern humans (EQ = 7.0) (Montgomery *et al.*, 2013). Modern mysticetes, on the other hand, have strikingly low EQs of less than 1.

7.5.1 Trends

The brain is a soft tissue structure, and hence generally not preserved in the fossil record. Luckily, its size and external morphology are reflected in the walls of the endocranial cavity, which can in turn be analyzed via actual or digital endocasts (section 3.4.4). Based on this information, it is possible to gauge the brain size and EQ of well-preserved fossil cetaceans (Figure 7.13), as long as the size of the retia mirabilia is taken into account (Marino *et al.*, 2000). In general, both brain and body size have increased over time in cetaceans, but not in all lineages, and not always to the same degree (Montgomery *et al.*, 2013). Archaeocetes were not highly encephalized, with EQs ranging from about 0.3 to 0.6 (Gingerich, 2015; Marino *et al.*, 2004). With the origin of neocetes, relative brain size rapidly increased to a level equaling, or even exceeding, that of modern chimpanzees (EQs of up to 3) (Figure 7.13). Much of this increase was driven by a concurrent decrease in body mass, but there was also an actual expansion of the cerebral cortex (Manger, 2006; Montgomery *et al.*, 2013).

During much of the Oligocene, the EQs of mysticetes and odontocetes overlapped, but then started to evolve along very different paths. Absolute brain size increased in chaeomysticetes, but was eclipsed by a marked increase in body size that led to an overall decrease in EQ (note, however, that there are currently no data on Miocene fossil mysticetes, because their large size makes computed tomography (CT) scanning impractical). By contrast, brain size remained stable and body size decreased along the odontocete stem lineage, which had exactly the opposite effect (Marino *et al.*, 2004; Montgomery *et al.*, 2013). This trend was accompanied by an **allometric** shift (i.e., an altered scaling factor) in the relationship between brain and body mass, with odontocetes differing from both extant mysticetes and archaeocetes in this regard. Further body size decreases and/or brain size increases within crown Odontoceti—especially within Delphinoidea—led to even higher EQs in some species. Nevertheless, large brains are far from universal among odontocetes, with many of the larger-sized taxa showing low levels of encephalization compared to their smaller relatives (e.g., *Orcinus*) or even mammals as a whole (e.g., *Ziphius* and *Physeter*) (Montgomery *et al.*, 2013).

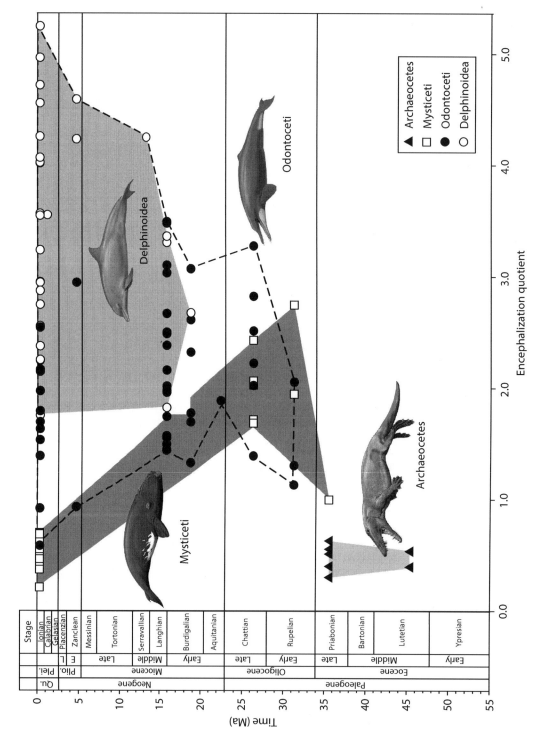

Figure 7.13 Encephalization quotients (EQ) of cetaceans over time. Note the low EQs of archaeocetes and the sudden increase at the origin of Neoceti. Data from Marino *et al.* (2004), plus additional unpublished data.

7.5.2 Potential causes

As with body size, there are several ideas as to what may have driven the relative and absolute size increase of the brain of modern cetaceans. Comparisons with other taxa have centered on the similarly large-brained primates, and suggested that large brains and associated cognitive abilities in both may have arisen from the need to handle complex social interactions and patterns of resource distribution (Connor *et al.*, 1998; Marino *et al.*, 2007). Alternatively, relative brain size in odontocetes may have increased as a result of echolocation and the need to process high-frequency acoustic information (Marino *et al.*, 2004; Worthy and Hickie, 1986). Finally, the large brain of cetaceans has been interpreted as a heat-generating organ enabling them to thermoregulate in water (Manger, 2006).

The idea that the large cetacean brain reflects some form of heightened intelligence is closely tied to the apparent intellectual prowess of many of the living species, as expressed, for example, in the cultural transmission of information, problem-solving skills, cooperative feeding, and, possibly, self-awareness (Marino *et al.*, 2007). The extent and uniqueness of these cognitive abilities are, however, still debated, as are the environmental and/or biotic factors that would have led to their initial emergence (Güntürkün, 2014; Manger, 2013; Marino, 2005). It is also unclear to what degree such abilities are really reflected in relative, rather than absolute, brain size. For example, it appears that the low EQs of modern mysticetes are mainly the result of a disproportionate increase in body mass, rather than a decrease in absolute brain size (Montgomery *et al.*, 2013). Despite their low EQs, mysticetes are known to exhibit cognitively demanding behaviors, and they resemble both odontocetes and primates in several other features usually associated with a complex brain. These include an enlarged, highly convoluted cortex and cerebellum, a similar number of cerebral neurons (within the range of an average chimpanzee), and the presence of a specialized class of cells known as **Von Economo neurons** (or spindle neurons); the latter have so far only been found in cetaceans, primates, and elephants, and may be linked to higher cognitive faculties (Butti *et al.*, 2009; Eriksen and Pakkenberg, 2007; Montgomery *et al.*, 2013).

Taken together, all of these observations suggest that baleen whales may not be the simpletons that relative brain size would have us think.

As with mysticetes, the evolution of relative brain size in early odontocetes was mainly a result of changing body size, again calling into question the use of EQ as an indicator of cognitive ability. The same holds true for the proposed effects of echolocation: although biosonar likely evolved early in odontocete history (Geisler *et al.*, 2014), it seemingly did not result in an increase in absolute brain size anywhere along the odontocete stem lineage. It is still possible that cognitive or echolocation-related demands may have kept brain size stable, and thereby contributed to a rise in EQ, as odontocetes decreased in body size; however, neither of these factors seems to have been responsible for the emergence of the enlarged odontocete brain per se (Montgomery *et al.*, 2013). Nevertheless, both absolute and relative brain size did increase within individual lineages of stem odontocetes, such as xenorophids, which is particularly interesting in light of the seemingly independent elaboration of echolocation abilities within this family (section 5.2.2) (Geisler *et al.*, 2014).

The hypothesis that enlarged brains help cetaceans to generate body heat (the **thermogenesis hypothesis**) poses that the origin of neocetes coincided with a one-off, pronounced increase in brain size, especially through the enlargement of the cerebral cortex (Manger, 2006). This increase is thought to have been driven by cetaceans moving from the relatively warm, shallow waters of the ancient Tethys Sea, where they originated, into open-ocean environments. In addition, the thermogenesis hypothesis predicts that the EQ of living cetaceans correlates with the range of water temperatures they inhabit (Manger, 2006). Relative to the "cognitive" and "echolocation" hypotheses, this idea draws strength from (1) the absence of a clear idea as to which, if any, cognitive abilities could have driven the enlargement of the cerebral cortex of early neocetes; (2) the lack of an increase in absolute brain size associated with the emergence of echolocation; and (3) the absence of a clear correlation between EQ/cortex size and specialized sensory systems in other mammals (Manger, 2006, 2013). It furthermore offers an explanation for the seemingly simple architecture of the enlarged cetacean cerebral cortex, by interpreting the latter as a structure used mainly for generating heat, rather than computation. This interpretation of cortex structure is, however, disputed, with other studies arguing for considerably greater complexity of the cetacean cortex, or at least a functional equivalent to complexity in the form of a marked increase in size (Huggenberger, 2008; Marino, 2004).

As originally proposed, the thermogenesis hypothesis envisaged an allometric shift in the relationship between brain and body mass that affected both mysticetes and odontocetes (Manger, 2006). Later research detected such a shift but, curiously, associated it with the origin of odontocetes, rather than neocetes (Montgomery et al., 2013). This would seem to contradict the idea of the brain as a heat-generating organ, although it is important to note that the study by Montgomery et al. did not include any fossil mysticetes. Bigger problems for the thermogenesis hypothesis lie in a dispute about the proposed association between brain size and water temperature (Maximino, 2009a, 2009b), as well as in the proposed timing of its major predictions (Marino et al., 2008). The global distribution of archaeocetes demonstrates that cetaceans moved out of the Tethys long before the origin of neocetes (section 7.6), with basilosaurids reaching mid-southern latitudes by about 37 Ma, or maybe even earlier (Köhler and Fordyce, 1997; Reguero et al., 2012). This implies that cetaceans were able to adapt to a wide range of temperatures without an increase in relative brain size, and certainly well before the allometric shift that affected the relationship between brain and body mass in odontocetes.

It is also unclear why a heat-generating specialization of the brain should only have arisen with the origin of neocetes, rather than at the time when cetaceans first moved back into the water. The transition to an aquatic environment was completed by basilosaurids, and possibly even some of the more crownward protocetids, well before the origin of the large-brained neocetes (section 5.1). When the latter finally appeared, they did so at a time when cetacean body size decreased *despite* concurrent global cooling (Montgomery et al., 2013; Zachos et al., 2008). This contradicts the general expectation that

cooler environments lead to larger body sizes, also known as **Bergmann's rule** (Millien et al., 2006), and suggests that early neocetes were well above the minimum size needed for effective thermoregulation in a cool ocean (Downhower and Blumer, 1988).

In summary, it thus appears that all of the proposed explanations for the large brains of modern cetaceans have their particular difficulties, and hence are in need of further study: echolocation cannot explain the large brains of mysticetes, thermogenesis does not fit the fossil record, and the driving factors behind any heightened cognitive abilities, as well as the precise nature and extent of these abilities, remain to be established.

7.6 Paleobiogeography

Biogeography studies the present and past distribution, dispersal, and disappearance of species within a geographical context. Most living cetaceans occur over wide geographic areas. This is particularly true for large-bodied species, such as balaenopterids and giant sperm whales, which have genuinely worldwide, **cosmopolitan** distributions. By contrast, many smaller species, such as phocoenids and certain delphinids, are restricted (**endemic**) to a particular region. Nevertheless, body size is not always a good predictor of distribution, since there are both some large species that are endemic to a single ocean (e.g., *Balaena mysticetus* and *Hyperoodon ampullatus*) and several dolphins (e.g., *Grampus griseus*) that range almost as widely as rorquals.

Besides body size, there are a variety of other factors that govern both the range of a species and its ability to disperse into new areas. The most important of these include reproduction (e.g., breeding grounds), feeding strategy, and prey distribution (e.g., freshwater vs marine, coastal vs pelagic, and shallow vs deep water), geographical and environmental barriers (land masses and temperature gradients), and the innate ability of a species to disperse into new environments (e.g., constraints imposed by osmoregulation, diving ability or reproduction). In practice, their effects are difficult to observe in the fossil record, primarily because the latter is too patchy to establish the range of a particular taxon with any degree of confidence (section 2.2). Nevertheless, some generalizations, especially with regard to geographical barriers and innate dispersal ability, can still be made.

7.6.1 Initial dispersal from land

The distribution of the earliest cetaceans (all archaeocetes except basilosaurids) is mainly determined by their place of origin, as well as biological factors such as osmoregulation and reproduction that still tied them to land. Cetaceans originated along the shores of the ancient **Tethys Sea**, which originally separated Eurasia in the north from Africa and India to the south. Pakicetids, ambulocetids, and, with a single exception (Bebej *et al.*, 2012), remingtonocetids are exclusively known from modern-day India and Pakistan (Figure 7.14), with the oldest fossils likely dating to the Early Eocene (Ypresian), around 53 Ma (Bajpai and Gingerich, 1998; Williams, 1998). No other cetacean fossils are known from the Ypresian, with the possible exception of a single putative basilosaurid from the Early Eocene of Seymour Island, Antarctica (Reguero *et al.*, 2012). This claim is quite extraordinary, especially considering that no basilosaurids older than Bartonian (41 Ma) have ever been found anywhere else. Given the difficulty of dating the Antarctic material, the lack of clarity as to whether the specimen was found *in situ*, and the unlikelihood of such an early basilosaurid, this report remains to be verified.

By the end of the Lutetian, protocetids and at least some remingtonocetids had managed to disperse out of Indo-Pakistan to the western Tethys (Bebej *et al.*, 2012; Gingerich, 2010). Protocetids went even further and dispersed into the Atlantic, ultimately reaching the western coast of Africa and eastern North America (Gingerich and Cappetta, 2014; Uhen, 2014) (Plate 16b). No protocetids are yet known from Oceania, the Southern Ocean, the South Atlantic or the North Pacific (Figure 7.14). The lack of finds from these regions may reflect sampling bias, or else indicate that the still primarily coastal and partially land-bound protocetids had not yet managed to colonize all of the world's oceans (section 5.1). In either case, it is clear that archaeocetes eventually managed to attain a truly global distribution by the Late Eocene (ca 37 Ma), thanks to the dispersal of the fully aquatic basilosaurids. The latter have been reported from as far afield as Antarctica, Europe, New Zealand, North Africa, North America, and Peru, although, intriguingly, they have so far not been found around the North Pacific (e.g., Kellogg, 1936; Köhler and Fordyce, 1997; Reguero *et al.*, 2012). The last surviving archaeocetes were, possibly, the Late Oligocene kekenodontids (Clementz *et al.*, 2014; Fordyce, 2002). Although currently only known from New Zealand, this group may actually have been considerably more widespread, and potentially includes several undescribed specimens from the Oligocene of the Atlantic and Pacific coasts of North America.

7.6.2 Neoceti

Unlike most of their archaeocete forebears (except maybe basilosaurids and kekenodontids), the distribution of modern cetaceans is more determined by physical barriers and environmental change than innate factors limiting dispersal. One of the most striking patterns in extant cetacean biogeography is the presence of populations, or pairs of closely related species, that are separated from each other by the equator (Davies, 1963). Examples of such an **antitropical** distribution are found in balaenopterids, ziphiids, delphinids, and phocoenids, and suggest that the equator is, or until recently was, an oceanographic barrier. This led to **vicariance** (cessation of dispersal and gene flow) and, in some cases, **allopatric speciation** (i.e., the emergence of new species from geographically isolated populations).

Other important barriers in the modern ocean are the Isthmus of Panama, which prevents direct exchange between the central Pacific and Atlantic oceans; the Isthmus of Suez, which separates the Mediterranean from the Red Sea, and thus ultimately the Atlantic from the Indian Ocean; the Turkish Straits (the Bosporus and the Dardanelles), which leave only a narrow connection between the Mediterranean and the Black Sea; and the (seasonal) ice coverage of the Arctic Ocean, which prevents exchange between the North Atlantic and the North Pacific. Like the equator, these barriers are reflected in the presence of populations, or pairs of closely related species, tied to particular ocean basins (e.g., Northern Hemisphere species of the right whale *Eubalaena*, and some species of

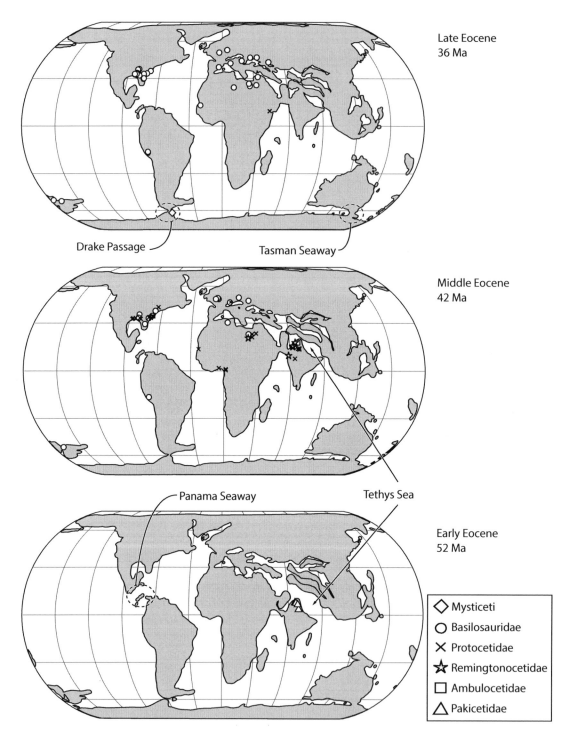

Figure 7.14 Dispersal of archaeocetes from the eastern part of the ancient Tethys Sea (Early Eocene) to the Western Tethys and across the North Atlantic (Middle Eocene) and, finally, the rest of the globe (Late Eocene). Maps were created with the plotting software available at fossilworks.org (© J. Alroy).

the delphinid *Lagenorhynchus*). By contrast, fewer subdivisions exist close to the poles, which are surrounded by the uninterrupted Arctic and Southern Oceans, respectively. As a result, many of the species inhabiting those waters have a more or less continuous, **circumpolar** distribution (e.g., monodontids, the bowhead whale *Balaena* and the pygmy right whale *Caperea*).

The very existence of closely related, yet geographically isolated, modern populations and species implies that the barriers that shape cetacean distribution today did not always exist in the past. Over the course of their evolution, cetaceans went through several periods of marked, tectonically driven ocean restructuring, the most important of which include: (1) the formation of the Southern Ocean via the gradual separation of Australia (**Tasman Seaway**) and South America (**Drake Passage**) from Antarctica (Figure 7.14); this process mostly took place during the Oligocene, but its exact timing remains a matter of debate (Hill *et al.*, 2013; Katz *et al.*, 2011); (2) the gradual closure of the ancient Tethys Sea, which became effective (i.e., it biogeographically separated the Mediterranean from the Indian Ocean) from around 22 Ma onward (Hamon *et al.*, 2013; Harzhauser *et al.*, 2009); (3) the fragmentation and eventual loss or isolation (around 6 Ma) of a northern arm of the Tethys known as the **Parathethys**, which originally encompassed the Black, Caspian, and Aral seas and extended all the way to Central Europe (Krijgsman *et al.*, 2010; Popov *et al.*, 2006); (4) the gradual closure of the Panama Seaway, which formerly connected the equatorial Pacific and Atlantic Oceans, between 8 and 4 Ma (Haug *et al.*, 2001); and (5) the opening of the **Bering Strait** between 5.5 and 3.5 Ma (Marincovich, 2000). These changes not only created or removed barriers to dispersal, but also affected the strength and direction of marine currents, the availability of shallow-water habitats and the geographical variation in seawater temperature and nutrient upwelling—all of which have themselves the power to shape cetacean distributions, whether directly or by proxy (e.g., Amaral *et al.*, 2012; Moreno *et al.*, 2005; Whitehead *et al.*, 2008).

The fossil record of mysticetes is generally too patchy to determine even an approximate range for most of the described taxa. Nevertheless, it appears that toothed mysticetes were much more geographically restricted than extant baleen whales, with aetiocetids occurring only in the North Pacific (Barnes *et al.*, 1995), mammalodontids only in Australia and New Zealand (Fordyce and Marx, 2011), and a third, as yet unnamed, group only along the south-eastern coast of the United States (Barnes and Sanders, 1996). A single find potentially referable to Aetiocetidae has been reported from Australia, and a potential mammalodontid is known from the Mediterranean (Bianucci *et al.*, 2011; Pledge, 2005); however, both of these specimens are fragmentary, and so far have not been backed up by further finds. If the Mediterranean individual indeed represents a mammalodontid, it might imply the existence of an Oligocene dispersal route connecting the Southwest Pacific with the Western Tethys and Proto-Mediterranean, via a still-open Tethyan Seaway.

Unlike toothed mysticetes, fossil chaeomysticetes resemble their modern relatives in being geographically widespread, at least at the level of the family. Thus, eomysticetids are known from the North and South Pacific, as well as the North Atlantic; cetotheriids from all of the former, as well as the Parathethys; and fossil (pre-Pleistocene) balaenids and balaenopterids from all oceans except the Indian, with their absence from the latter likely reflecting the generally poor sampling of this region (Uhen, 2015). Virtually nothing is known about where any of these clades originated, although a North Pacific origin has been suggested for cetotheriids (Bisconti, 2014). Besides these cosmopolitan families, there are three (sub)groups of chaeomysticetes with a more limited geographical distribution, namely: (1) a clade of cetotheriids closely related to *Cetotherium* (*Cetotherium*, *Brandtocetus*, *Kurdalagonus*, and *Zygiocetus*), which appears to have been restricted to the Eastern Parathethys (Gol'din and Startsev, 2014); (2) pygmy right whales, which—like the living *Caperea*—are so far only known from the Southern Hemisphere (Buono *et al.*, 2014); and (3) eschrichtiids.

Eschrichtiids are today represented by a single living species, the gray whale *Eschrichtius robustus*. Living *Eschrichtius* is found only along the fringes of the North Pacific, where it seems to date back to at least the Late Pliocene (Ichishima *et al.*, 2006). However, fossil eschrichtiids as old as the Late Miocene have been recovered from around the North Atlantic basin (Bisconti and Varola, 2006; Whitmore and Kaltenbach, 2008), and both subfossil finds and historical accounts demonstrate that even *Eschrichtius* inhabited the North Atlantic until as late as the 16th century (Mead and Mitchell, 1984). A Mediterranean or North Atlantic origin of gray whales has been proposed based on the occurrence of the oldest known fossils in this region (Bisconti and Varola, 2006). It is currently impossible to say whether eschrichtiids moved between the Atlantic and Pacific via the Panama Seaway, the Arctic Ocean, or both. Recent anomalous records of living gray whales from the Laptev Sea (north of Russia), the Mediterranean, and even the South Atlantic suggest, however, that at least extant *Eschrichtius* is capable of traversing polar seas during times of reduced ice cover (Meschersky *et al.*, 2015; Scheinin *et al.*, 2011).

Like that of mysticetes, the fossil record of odontocetes is diverse, but patchy. To avoid duplication, only marine taxa will be discussed here—for a description of the biogeography of fossil freshwater dolphins, see section 5.3. During the Oligocene to early Middle Miocene, a variety of relatively cosmopolitan taxa (e.g., physeteroids, squalodontids, eurhinodelphinids, and kentriodontids) coexisted with more geographically limited groups, such as xenorophids (western North Atlantic), allodelphinids (North Pacific), and eoplatanistids (Mediterranean). Squalodelphinids also may have been geographically widespread, but the contents of this clade are still under debate (Tanaka and Fordyce, 2014). Waipatiids are known with certainty only from the Late Oligocene of New Zealand, but they may additionally occur throughout the Late Oligocene and Early–Middle Miocene of the western North Pacific, the Parathethys, and the Mediterranean (Bianucci *et al.*, 2011; Tanaka and Fordyce, 2014). Like the potential mammalodontid from this region (as discussed in this section), this distribution supports the existence of an Oligocene dispersal route linking the Mediterranean with the Southwest Pacific, via the Tethyan Seaway.

During the second half of the Miocene and the Pliocene, fossil ziphiids, delphinids, and phocoenids appeared in all major ocean basins—although little can be said about the Indian Ocean, for which data are extremely sparse. There are generally not enough specimens to estimate the ranges of particular genera or species, but the occurrence of certain species pairs north and south of the equator (e.g., the pithanodelphinine *Atocetus*, the phocoenid *Piscolithax*, and the ziphiid *Messapicetus*) may suggest an antitropical distribution. Phocoenids, though widespread overall, are markedly concentrated in the equatorial and North Pacific, where they may also have originated (Barnes, 1985; Fajardo-Mellor *et al.*, 2006). Iterative dispersal events via the Bering Strait and the Arctic region may explain the Pliocene and modern distribution of phocoenids in the North Atlantic (Colpaert *et al.*, 2015; Lambert, 2008). Contrary to this interpretation, the origins of both phocoenids and delphinids have also been located in the North Atlantic (Banguera-Hinestroza *et al.*, 2014).

Monodontids once seem to have occurred far beyond their current Arctic range, with fossils known from as far south as Baja California, North Carolina, and Belgium (Barnes, 1984; Vélez-Juarbe and Pyenson, 2012). By contrast, the poorly diversified and now entirely extinct albireonids and odobenocetopsids were apparently restricted to the North and equatorial Pacific, respectively (Barnes, 2008; Pyenson *et al.*, 2013). In the case of *Odobenocetops*, its highly specialized cranial morphology and related ecology may well have constrained its habitat, and thus its spatial distribution.

7.7 Convergent evolution

Convergent evolution describes the independent appearance of superficially similar morphologies or molecular sequences in two or more unrelated taxa. Such resemblances often arise as a result of similar feeding strategies or other comparable functional needs (e.g., locomotion). A striking example of molecular convergence is the evolution of the protein Prestin, which is involved in the amplification of sounds entering the inner ear, and thus hearing sensitivity. Curiously, Prestin evolved along similar lines in several bats, beaked whales, porpoises, and dolphins, likely in response to their shared need for echolocation-related high-frequency hearing (Figure 7.15) (Liu *et al.*, 2010).

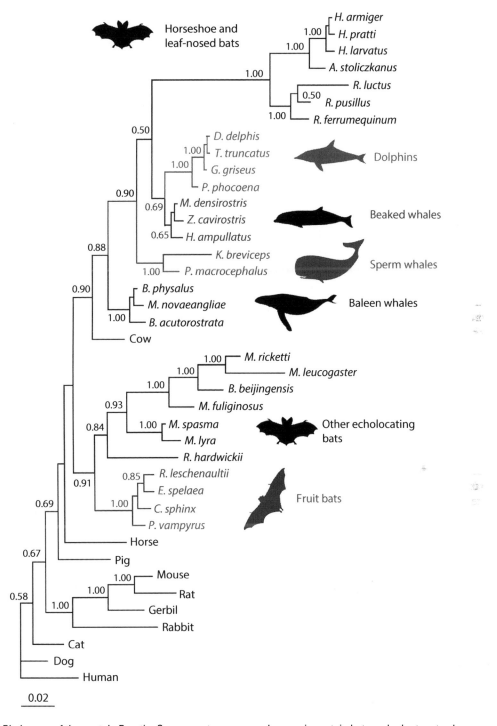

Figure 7.15 Phylogeny of the protein Prestin. Convergent sequence changes in certain bats and odontocetes have resulted in their proteins being more similar to each other than to those of other bats or cetaceans. Reproduced from Liu *et al.* (2010), with permission of Elsevier.

Nothing can be said about the DNA of fossil taxa, but there is good evidence for various degrees of morphological convergence. At the most basic level, cetaceans evolved a general body shape similar to that of several other aquatic vertebrates: fish, penguins, sirenians, and a group of extinct Mesozoic marine reptiles known as parvipelvian ichthyosaurs (Kelley and Pyenson, 2015). The resemblance with ichthyosaurs is particularly marked, and includes the evolution of a streamlined, tuna-shaped (**thunniform**) body shape, the transformation of limbs into flippers, a pronounced dorsal fin, and the presence of flukes attached to the end of the tail (Motani, 2002, 2005). Ichthyosaurs differ, however, in retaining external (albeit reduced) hind limbs, and in having their flukes oriented vertically, rather than horizontally. Whereas the first difference may reflect a real difference in swimming style, the second is easily explained with reference to the way the ancestors of cetaceans and ichthyosaurs moved on land: the spinal column of "reptiles," like that of fish, flexes from side to side as they move forward; by contrast, that of mammals flexes up and down. These movements are conserved in the swimming styles of the aquatic members of both groups, and thus necessitate differently oriented hydrofoils.

A similarly marked example of convergence concerns mysticetes and a group of large-bodied, Mesozoic teleost fishes known as pachycormids. Like baleen whales and a variety of extant filter-feeding sharks, at least some pachycormids were planktivorous. As the ecological equivalent of modern mysticetes, pachycormids may have prevented contemporary marine reptiles, which were otherwise highly diverse, from specializing on plankton as a main source of food (Friedman et al., 2010). Like mysticetes, filter-feeding pachycormids had long, slender jaws bearing a reduced dentition, or no teeth at all. Also like mysticetes and filter-feeding sharks, these fishes attained gigantic body sizes (up to 9 m), especially relative to living teleost filter feeders. In the same way that whales and ichthyosaurs converged, from rather different starting points, upon roughly the same locomotory style, whales and pachycormids thus independently moved toward similar feeding strategies (Friedman, 2012). The same can be said for rorquals and pelicans, both of which independently acquired elongate, kinetic jaws and an expandable throat pouch (Field et al., 2011).

Among odontocetes, convergent evolution has resulted in the emergence of particular skull and tooth morphologies resembling those of other marine tetrapods with similar feeding modes (Kelley and Motani, 2015). Convergent evolution is often also invoked to explain the markedly elongated rostrum, long and fused mandibular symphysis, flexible neck, and broad flippers characterizing the four extant "river dolphins" *Inia*, *Lipotes*, *Platanista*, and *Pontoporia*. All of these features may represent independent adaptations to freshwater habitats, but at least some of them were likely already present in the marine ancestors of the living species (section 5.3) (Geisler et al., 2011). Elongated rostra indicative of raptorial snap feeding also occur in many fossil odontocetes, such as allodelphinids, eurhinodelphinids, and pomatodelphinines. However, given their still-uncertain phylogenetic positions, it currently remains unclear whether this morphology is genuinely convergent or simply reflective of shared ancestry.

Much less ambiguous examples of odontocete convergent evolution also exist. The aberrant Pliocene delphinoid *Odobenocetops* shares striking similarities with the walrus *Odobenus*, including an extremely shortened, blunt, and highly vascularized rostrum, conspicuous muscle attachment sites indicative of a well-developed upper lip, a large, vaulted palate, and a pair of enlarged tusks (Figure 7.16). This peculiar combination of features suggests similar, benthic feeding strategies involving suction (de Muizon and Domning, 2002; de Muizon et al., 2002). Similarly, the bizarre Pliocene delphinid *Australodelphis* may have used suction to catch squid in a manner similar to that of extant ziphiids, and, as a result, evolved a cranial morphology nearly indistinguishable from that of beaked whales, such as *Mesoplodon* (Figure 7.17) (Fordyce et al., 2002).

A peculiar kind of evolutionary convergence is shown by dwarf and pygmy sperm whales (*Kogia*). Uniquely among odontocetes, they have acquired a somewhat shark-like appearance thanks to their squared head profile, underslung mandible, and pigmented false gill slits (Fordyce, 2009). This kind of superficial resemblance with another animal—in this case, a predator—is a case of **mimicry**, and presumably reduces the risk of being preyed upon. Although evolution has made *Kogia* look increasingly similar to a shark, mimicry is strictly not a case of convergent evolution, but of **coevolution**. Unlike convergent evolution, mimicry is not independent, but is specifically modeled on another species. Most of the characters involved in this particular example of the phenomenon unfortunately leave little or no traces on the bones, and thus are impossible to trace in the fossil record.

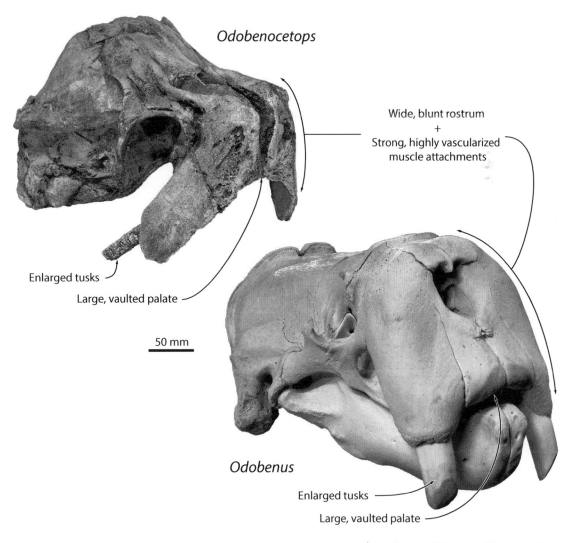

Odobenocetops

Wide, blunt rostrum
+
Strong, highly vascularized muscle attachments

Enlarged tusks

Large, vaulted palate

50 mm

Odobenus

Enlarged tusks

Large, vaulted palate

Figure 7.16 Comparison of the skull of the extinct delphinoid *Odobenocetops* (cast; National Museum of Nature and Science specimen PV 20741, Tsukuba, Japan) and the extant walrus *Odobenus* (National Museum of Nature and Science specimen M36186, Tsukuba, Japan). Note the considerable morphological similarities between these otherwise very different taxa.

Figure 7.17 Comparison of the skull of the extinct delphinid *Australodelphis* (cast of Australian Geological Survey Organisation, Commonwealth Palaeontological Collections specimen CPC 25730, held at the Geology Museum of the University of Otago, Dunedin, New Zealand) and the extant ziphiid *Mesoplodon* (National Museum of Nature and Science specimen M42042, Tsukuba, Japan). Note the considerable morphological similarities between these otherwise very different taxa. Photo of *Australodelphis* courtesy of R. E. Fordyce.

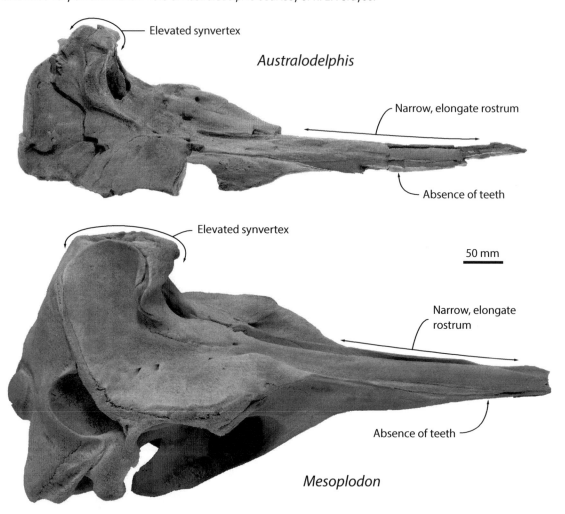

7.8 Suggested readings

Clauset, A. 2013. How large should whales be? PLoS One 8:e53967.

Kelley, N. P., and N. D. Pyenson. 2015. Evolutionary innovation and ecology in marine tetrapods from the Triassic to the Anthropocene. Science 348.

Marx, F. G., and R. E. Fordyce. 2015. Baleen boom and bust: a synthesis of mysticete phylogeny, diversity and disparity. Royal Society Open Science 2:140434.

Marx, F. G., and M. D. Uhen. 2010. Climate, critters, and cetaceans: Cenozoic drivers of the evolution of modern whales. Science 327:993–996.

Montgomery, S. H., J. H. Geisler, M. R. McGowen, C. Fox, L. Marino, and J. Gatesy. 2013. The evolutionary history of cetacean brain and body size. Evolution 67:3339–3353.

References

Acosta Hospitaleche, C., and M. Reguero. 2010. First articulated skeleton of *Palaeeudyptes gunnari* from the late Eocene of Isla Marambio (Seymour Island), Antarctica. Antarctic Science 22:289–298.

Amaral, A. R., L. B. Beheregaray, K. Bilgmann, D. Boutov, L. Freitas, K. M. Robertson, M. Sequeira, K. A. Stockin, M. M. Coelho, and L. M. Möller. 2012. Seascape genetics of a globally distributed, highly mobile marine mammal: the short-beaked common dolphin (genus *Delphinus*). PLoS One 7:e31482.

Atchley, W. R., B. Riska, L. A. P. Kohn, A. A. Plummer, and J. J. Rutledge. 1984. A quantitative genetic analysis of brain and body size associations, their origin and ontogeny: data from mice. Evolution 38:1165–1179.

Bajpai, S., and P. D. Gingerich. 1998. A new Eocene archaeocete (Mammalia, Cetacea) from India and the time of origin of whales. Proceedings of the National Academy of Sciences USA 95:15464–15468.

Banguera-Hinestroza, E., A. Hayano, E. Crespo, and A. R. Hoelzel. 2014. Delphinid systematics and biogeography with a focus on the current genus *Lagenorhynchus*: multiple pathways for antitropical and trans-oceanic radiation. Molecular Phylogenetics and Evolution 80:217–230.

Barnes, L. G. 1984. Fossil odontocetes (Mammalia: Cetacea) from the Almejas Formation, Isla Cedros, Mexico. PaleoBios 42:1–46.

Barnes, L. G. 1985. Evolution, taxonomy and antitropical distributions of the porpoises (Phocoenidae, Mammalia). Marine Mammal Science 1:149–165.

Barnes, L. G. 2008. Miocene and Pliocene Albireonidae (Cetacea, Odontoceti), rare and unusual fossil dolphins from the eastern North Pacific Ocean. Natural History Museum of Los Angeles County Science Series 41:99–152.

Barnes, L. G., J. L. Goedert, and H. Furusawa. 2001. The earliest known echolocating toothed whales (Mammalia; Odontoceti): preliminary observations of fossils from Washington State. Mesa Southwest Museum Bulletin 8:91–100.

Barnes, L. G., M. Kimura, H. Furusawa, and H. Sawamura. 1995. Classification and distribution of Oligocene Aetiocetidae (Mammalia; Cetacea; Mysticeti) from western North America and Japan. The Island Arc 3:392–431.

Barnes, L. G., and A. E. Sanders. 1996. The transition from archaeocetes to mysticetes: Late Oligocene toothed mysticetes from near Charleston, South Carolina. Paleontological Society Special Publication 8:24.

Bebej, R. M., I. S. Zalmout, A. A. Abed El-Aziz, M. S. M. Antar, and P. D. Gingerich. 2012. First evidence of Remingtonocetidae (Mammalia, Cetacea) outside Indo-Pakistan: new genus from the early Middle Eocene of Egypt. Journal of Vertebrate Paleontology: Program and Abstracts 32:62.

Beck, R. M. D., and M. S. Y. Lee. 2014. Ancient dates or accelerated rates? Morphological clocks and the antiquity of placental mammals. Proceedings of the Royal Society B 281:20141278.

Berger, W. H. 2007. Cenozoic cooling, Antarctic nutrient pump, and the evolution of whales. Deep Sea Research Part II: Topical Studies in Oceanography 54:2399–2421.

Bianucci, G. 2013. *Septidelphis morii*, n. gen. et sp., from the Pliocene of Italy: new evidence of the explosive radiation of true dolphins (Odontoceti, Delphinidae). Journal of Vertebrate Paleontology 33:722–740.

Bianucci, G., M. Gatt, R. Catanzariti, S. Sorbi, C. G. Bonavia, R. Curmi, and A. Varola. 2011. Systematics, biostratigraphy and evolutionary pattern of the Oligo-Miocene marine mammals from the Maltese Islands. Geobios 44:549–585.

Bisconti, M. 2014. Anatomy of a new cetotheriid genus and species from the Miocene of Herentals, Belgium, and the phylogenetic and palaeobiogeographical relationships of Cetotheriidae s.s. (Mammalia, Cetacea, Mysticeti). Journal of Systematic Palaeontology 13:377.

Bisconti, M., and A. Varola. 2006. The oldest eschrichtiid mysticete and a new morphological diagnosis of Eschrichtiidae (gray whales). Rivista Italiana di Paleontologia e Stratigrafia 112:447–457.

Boessenecker, R. W. 2013a. A new marine vertebrate assemblage from the Late Neogene Purisima Formation in Central California, part II: Pinnipeds and Cetaceans. Geodiversitas 35:815–940.

Boessenecker, R. W. 2013b. Pleistocene survival of an archaic dwarf baleen whale (Mysticeti: Cetotheriidae). Naturwissenschaften 100:365–371.

Buono, M. R., M. T. Dozo, F. G. Marx, and R. E. Fordyce. 2014. A Late Miocene potential neobalaenine mandible from Argentina sheds light on the origins of the living pygmy right whale. Acta Palaeontologica Polonica 59:787–793.

Butti, C., C. C. Sherwood, A. Y. Hakeem, J. M. Allman, and P. R. Hof. 2009. Total number and volume of Von Economo neurons in the cerebral cortex of cetaceans. The Journal of Comparative Neurology 515:243–259.

Chen, Z., S. Xu, K. Zhou, and G. Yang. 2011. Whale phylogeny and rapid radiation events revealed using novel retroposed elements and their flanking sequences. BMC Evolutionary Biology 11:314.

Clauset, A. 2013. How large should whales be? PLoS One 8:e53967.

Clauset, A., and D. H. Erwin. 2008. The evolution and distribution of species body size. Science 321:399–401.

Clementz, M. T., R. E. Fordyce, S. Peek, L., and D. L. Fox. 2014. Ancient marine isoscapes and isotopic evidence of bulk-feeding by Oligocene cetaceans. Palaeogeography, Palaeoclimatology, Palaeoecology 400:28–40.

Colpaert, W., M. Bosselaers, and O. Lambert. 2015. Out of the Pacific: a second fossil porpoise from the Pliocene of the North Sea Basin. Acta Palaeontologica Polonica 60:1–10.

Connor, R. C., J. Mann, P. L. Tyack, and H. Whitehead. 1998. Social evolution in toothed whales. Trends in Ecology & Evolution 13:228–232.

Davies, J. L. 1963. The antitropical factor in cetacean speciation. Evolution 17:107–116.

de Muizon, C. 1988. Les relations phylogénétiques des Delphinida (Cetacea, Mammalia). Annales de Paléontologie 74:159–227.

de Muizon, C., and D. P. Domning. 2002. The anatomy of Odobenocetops (Delphinoidea, Mammalia), the walrus-like dolphin from the Pliocene of Peru and its palaeobiological implications. Zoological Journal of the Linnean Society 134:423–452.

de Muizon, C., D. P. Domning, and D. R. Ketten. 2002. Odobenocetops peruvianus, the walrus-convergent delphinoid (Mammalia: Cetacea) from the Early Pliocene of Peru. Smithsonian Contributions to Paleobiology 93:223–261.

De Schepper, S., P. L. Gibbard, U. Salzmann, and J. Ehlers. 2014. A global synthesis of the marine and terrestrial evidence for glaciation during the Pliocene Epoch. Earth-Science Reviews 135:83–102.

DeConto, R. M., and D. Pollard. 2003. Rapid Cenozoic glaciation of Antarctica induced by declining atmospheric CO_2. Nature 421:245–249.

Dornburg, A., M. C. Brandley, M. R. McGowen, and T. J. Near. 2012. Relaxed clocks and inferences of heterogeneous patterns of nucleotide substitution and divergence time estimates across whales and dolphins (Mammalia: Cetacea). Molecular Biology and Evolution 29:721–736.

Downhower, J. F., and L. S. Blumer. 1988. Calculating just how small a whale can be. Nature 335:675–675.

Egan, K. E., R. E. M. Rickaby, K. R. Hendry, and A. N. Halliday. 2013. Opening the gateways for diatoms primes Earth for Antarctic glaciation. Earth and Planetary Science Letters 375:34–43.

Eriksen, N., and B. Pakkenberg. 2007. Total neocortical cell number in the mysticete brain. The Anatomical Record 290:83–95.

Evans, A. R., D. Jones, A. G. Boyer, J. H. Brown, D. P. Costa, S. K. M. Ernest, E. M. G. Fitzgerald, M. Fortelius, J. L. Gittleman, M. J. Hamilton, L. E. Harding, K. Lintulaakso, S. K. Lyons, J. G. Okie, J. J. Saarinen, R. M. Sibly, F. A. Smith, P. R. Stephens, J. M. Theodor, and M. D. Uhen. 2012. The maximum rate of mammal evolution. Proceedings of the National Academy of Sciences 109:4187–4190.

Fajardo-Mellor, L., A. Berta, R. L. Brownell, C. Boy, and R. N. Goodall. 2006. The phylogenetic relationships and biogeography of true porpoises (Mammalia: Phocoenidae) based on morphological data. Marine Mammal Science 22:910–932.

Field, D. J., S. C. Lin, M. Ben-Zvi, J. A. Goldbogen, and R. E. Shadwick. 2011. Convergent evolution driven by similar feeding mechanics in balaenopterid whales and pelicans. The Anatomical Record 294:1273–1282.

Fitzgerald, E. M. G. 2005. Pliocene marine mammals from the Whalers Bluff Formation of Portland, Victoria, Australia. Memoirs of Museum Victoria 62:67–89.

Fitzgerald, E. M. G. 2006. A bizarre new toothed mysticete (Cetacea) from Australia and the early evolution of baleen whales. Proceedings of the Royal Society B 273:2955–2963.

Fitzgerald, E. M. G. 2010. The morphology and systematics of *Mammalodon colliveri* (Cetacea: Mysticeti), a toothed mysticete from the Oligocene of Australia. Zoological Journal of the Linnean Society 158:367–476.

Fordyce, R. E. 1980. Whale evolution and Oligocene Southern Ocean environments. Palaeogeography, Palaeoclimatology, Palaeoecology 31:319–336.

Fordyce, R. E. 2002. Oligocene archaeocetes and toothed mysticetes: Cetacea from times of transition. Geological Society of New Zealand Miscellaneous Publication 114A:16–17.

Fordyce, R. E. 2003. Early crown-group Cetacea in the southern ocean: the toothed archaic mysticete *Llanocetus*. Journal of Vertebrate Paleontology 23 (Suppl. to 3):50A.

Fordyce, R. E. 2006. A southern perspective on cetacean evolution and zoogeography; pp. 755–782 in J. R. Merrick, M. Archer, G. M. Hickey, and M. S. Y. Lee (eds.), Evolution and Biogeography of Australasian Vertebrates. Auscipub, Oatlands, Australia.

Fordyce, R. E. 2009. Cetacean evolution; pp. 201–207 in J. G. M. Thewissen, W. F. Perrin, and B. Würsig (eds.), Encyclopedia of Marine Mammals. Academic Press, Burlington, MA.

Fordyce, R. E., and C. de Muizon. 2001. Evolutionary history of cetaceans: a review; pp. 169–223 in J.-M. Mazin, and V. de Buffrénil (eds.), Secondary Adaptation of Tetrapods to Life in Water. Verlag Dr. Friedrich Pfeil, München.

Fordyce, R. E., and F. G. Marx. 2011. Toothed mysticetes and ecological structuring of Oligocene whales and dolphins from New Zealand. Geological Survey of Western Australia Annual Record 2011/9:33.

Fordyce, R. E., P. G. Quilty, and J. Daniels. 2002. *Australodelphis mirus*, a bizarre new toothless ziphiid-like fossil dolphin (Cetacea: Delphinidae) from the Pliocene of Vestfold Hills, East Antarctica. Antarctic Science 14:37–54.

Freckleton, R. P., and P. H. Harvey. 2006. Detecting non-Brownian trait evolution in adaptive radiations. PLoS Biol 4:e373.

Friedman, M. 2012. Parallel evolutionary trajectories underlie the origin of giant suspension-feeding whales and bony fishes. Proceedings of the Royal Society B 279:944–951.

Friedman, M., K. Shimada, L. D. Martin, M. J. Everhart, J. Liston, A. Maltese, and M. Triebold. 2010. 100-million-year dynasty of giant planktivorous bony fishes in the Mesozoic seas. Science 327:990–993.

Geisler, J. H., M. W. Colbert, and J. L. Carew. 2014. A new fossil species supports an early origin for toothed whale echolocation. Nature 508:383–386.

Geisler, J. H., M. R. McGowen, G. Yang, and J. Gatesy. 2011. A supermatrix analysis of genomic, morphological, and paleontological data from crown Cetacea. BMC Evolutionary Biology 11:1–33.

Geisler, J. H., and A. E. Sanders. 2003. Morphological evidence for the phylogeny of Cetacea. Journal of Mammalian Evolution 10:23–129.

Gingerich, P. 2015. Body weight and relative brain size (encephalization) in Eocene Archaeoceti (Cetacea). Journal of Mammalian Evolution:1–15.

Gingerich, P. D. 2010. Cetacea; pp. 873–899 in L. Werdelin, and W. J. Sanders (eds.), Cenozoic Mammals of Africa. University of California Press, Berkeley.

Gingerich, P. D., M. S. M. Antar, and I. Zalmout. 2014. Skeleton of new protocetid (Cetacea,

Archaeoceti) from the lower Gehannam Formation of Wadi al Hitan in Egypt: survival of a protocetid into the Priabonian Late Eocene. Journal of Vertebrate Paleontology: Program and Abstracts 138.

Gingerich, P. D., and H. Cappetta. 2014. A new archaeocete and other marine mammals (Cetacea and Sirenia) from lower Middle Eocene phosphate deposits of Togo. Journal of Paleontology 88:109–129.

Goedert, J. L., L. G. Barnes, and H. Furusawa. 2007. The diversity and stratigraphic distribution of cetaceans in early Cenozoic strata of Washington State, U.S.A. Geological Society of Australia Abstracts 85:44.

Gol'din, P., and D. Startsev. 2014. *Brandtocetus*, a new genus of baleen whales (Cetacea, Cetotheriidae) from the Late Miocene of Crimea, Ukraine. Journal of Vertebrate Paleontology 34:419–433.

Goldbogen, J. A., J. Calambokidis, D. A. Croll, M. F. McKenna, E. Oleson, J. Potvin, N. D. Pyenson, G. Schorr, R. E. Shadwick, and B. R. Tershy. 2012. Scaling of lunge-feeding performance in rorqual whales: mass-specific energy expenditure increases with body size and progressively limits diving capacity. Functional Ecology 26:216–226.

Güntürkün, O. 2014. Is dolphin cognition special? Commentary on Manger PR (2013): Questioning the interpretations of behavioral observations of cetaceans: is there really support for a special intellectual status for this mammalian order? Neuroscience 250:664–696. Brain, Behavior and Evolution 83:177–180.

Hamon, N., P. Sepulchre, V. Lefebvre, and G. Ramstein. 2013. The role of eastern Tethys seaway closure in the Middle Miocene Climatic Transition (ca. 14 Ma). Climate of the Past 9:2687–2702.

Hampe, O., and S. Baszio. 2010. Relative warps meet cladistics: a contribution to the phylogenetic relationships ofbaleen whales based on landmark analyses of mysticete crania. Bulletin of Geosciences 85:199–218.

Harzhauser, M., M. Reuter, W. E. Piller, B. Berning, A. Kroh, and O. Mandic. 2009. Oligocene and Early Miocene gastropods from Kutch (NW India) document an early

biogeographic switch from Western Tethys to Indo-Pacific. Paläontologische Zeitschrift 83:333–372.

Haug, G. H., R. Tiedemann, R. Zahn, and A. C. Ravelo. 2001. Role of Panama uplift on oceanic freshwater balance. Geology 29:207–210.

Hill, D. J., A. M. Haywood, P. J. Valdes, J. E. Francis, D. J. Lunt, B. S. Wade, and V. C. Bowman. 2013. Paleogeographic controls on the onset of the Antarctic circumpolar current. Geophysical Research Letters 40:2013GL057439.

Hof, P. R., R. Chanis, and L. Marino. 2005. Cortical complexity in cetacean brains. The Anatomical Record 287A:1142–1152.

Huggenberger, S. 2008. The size and complexity of dolphin brains—a paradox? Journal of the Marine Biological Association of the United Kingdom 88:1103–1108.

Hyeong, K., J. Lee, I. Seo, M. J. Lee, C. M. Yoo, and B.-K. Khim. 2014. Southward shift of the Intertropical Convergence Zone due to Northern Hemisphere cooling at the Oligocene-Miocene boundary. Geology 42:667–670.

Ichishima, H., E. Sato, T. Sagayama, and M. Kimura. 2006. The oldest record of Eschrichtiidae (Cetacea: Mysticeti) from the Late Pliocene, Hokkaido, Japan. Journal of Paleontology 80:367–379.

Jerison, H. J. 1973. Evolution of the Brain and Intelligence. Academic Press, New York.

Katz, M. E., B. S. Cramer, J. R. Toggweiler, G. Esmay, C. Liu, K. G. Miller, Y. Rosenthal, B. S. Wade, and J. D. Wright. 2011. Impact of Antarctic Circumpolar Current development on Late Paleogene ocean structure. Science 332:1076–1079.

Kelley, N. P., and R. Motani. 2015. Trophic convergence drives morphological convergence in marine tetrapods. Biology Letters 11:20140709.

Kelley, N. P., and N. D. Pyenson. 2015. Evolutionary innovation and ecology in marine tetrapods from the Triassic to the Anthropocene. Science 348.

Kellogg, R. 1936. A Review of the Archaeoceti. Carnegie Institution of Washington

Publication 482. Carnegie Institution, Washington, DC.

Kemper, C. M. 2009. Pygmy right whale *Caperea marginata*; pp. 939–941 in W. F. Perrin, B. Würsig, and J. G. M. Thewissen (eds.), Encyclopedia of Marine Mammals. Academic Press, Burlington, MA.

Kiel, S., and J. L. Goedert. 2006. Deep-sea food bonanzas: early Cenozoic whale-fall communities resemble wood-fall rather than seep communities. Proceedings of the Royal Society of London B 273:2625–2632.

Köhler, R., and R. E. Fordyce. 1997. An archaeocete whale (Cetacea: Archaeoceti) from the Eocene Waihao Greensand, New Zealand. Journal of Vertebrate Paleontology 17:574–583.

Krijgsman, W., M. Stoica, I. Vasiliev, and V. V. Popov. 2010. Rise and fall of the Paratethys Sea during the Messinian Salinity Crisis. Earth and Planetary Science Letters 290:183–191.

Lambert, O. 2008. A new porpoise (Cetacea, Odontoceti, Phocoenidae) from the Pliocene of the North Sea. Journal of Vertebrate Paleontology 28:863–872.

Liu, Y., S. J. Rossiter, X. Han, J. A. Cotton, and S. Zhang. 2010. Cetaceans on a molecular fast track to ultrasonic hearing. Current Biology 20:1834–1839.

Manger, P. R. 2006. An examination of cetacean brain structure with a novel hypothesis correlating thermogenesis to the evolution of a big brain. Biological Reviews 81:293–338.

Manger, P. R. 2013. Questioning the interpretations of behavioral observations of cetaceans: Is there really support for a special intellectual status for this mammalian order? Neuroscience 250:664–696.

Marincovich, L. 2000. Central American paleogeography controlled Pliocene Arctic Ocean molluscan migrations. Geology 28:551–554.

Marino, L. 1998. A comparison of encephalization between odontocete cetaceans and anthropoid primates. Brain, Behavior and Evolution 51:230–238.

Marino, L. 2004. Cetacean brain evolution: multiplication generates complexity. International Journal of Comparative Psychology 17:1–16.

Marino, L. 2005. Big brains do matter in new environments. Proceedings of the National Academy of Sciences USA 102:5306–5307.

Marino, L., C. Butti, R. C. Connor, R. E. Fordyce, L. M. Herman, P. R. Hof, L. Lefebvre, D. Lusseau, B. McCowan, E. A. Nimchinsky, A. A. Pack, J. S. Reidenberg, D. Reiss, L. Rendell, M. D. Uhen, E. Van der Gucht, and H. Whitehead. 2008. A claim in search of evidence: reply to Manger's thermogenesis hypothesis of cetacean brain structure. Biological Reviews 83:417–440.

Marino, L., R. C. Connor, R. E. Fordyce, L. M. Herman, P. R. Hof, L. Lefebvre, D. Lusseau, B. McCowan, E. A. Nimchinsky, A. A. Pack, L. Rendell, J. S. Reidenberg, D. Reiss, M. D. Uhen, E. Van der Gucht, and H. Whitehead. 2007. Cetaceans have complex brains for complex cognition. PLoS Biology 5:e139.

Marino, L., D. W. McShea, and M. D. Uhen. 2004. Origin and evolution of large brains in toothed whales. The Anatomical Record 281A:1247–1255.

Marino, L., M. D. Uhen, B. Frohlich, J. M. Aldag, C. Blane, D. Bohaska, and F. C. Whitmore, Jr. 2000. Endocranial volume of Mid-Late Eocene archaeocetes (Order: Cetacea) revealed by computed tomography: implications for cetacean brain evolution. Journal of Mammalian Evolution 7:81–94.

Martínez Cáceres, M., C. de Muizon, O. Lambert, G. Bianucci, R. Salas-Gismondi, and M. Urbina Schmidt. 2011. A toothed mysticete from the Middle Eocene to Lower Oligocene of the Pisco Basin, Peru: new data on the origin and feeding evolution of Mysticeti. Sixth Triennial Conference on Secondary Adaptation of Tetrapods to Life in Water, San Diego, 56–57.

Martínez-García, A., D. M. Sigman, H. Ren, R. F. Anderson, M. Straub, D. A. Hodell, S. L. Jaccard, T. I. Eglinton, and G. H. Haug. 2014. Iron fertilization of the Subantarctic Ocean during the last ice age. Science 343:1347–1350.

Martínez-García, A., and G. Winckler. 2014. Iron fertilization in the glacial ocean. Past Global Changes Magazine 22:82–83.

Marx, F. G. 2011. The more the merrier? A large cladistic analysis of mysticetes, and comments on the transition from teeth to baleen. Journal of Mammalian Evolution 18:77–100.

Marx, F. G., and R. E. Fordyce. 2015. Baleen boom and bust: a synthesis of mysticete phylogeny, diversity and disparity. Royal Society Open Science 2:140434.

Marx, F. G., and M. D. Uhen. 2010. Climate, critters, and cetaceans: Cenozoic drivers of the evolution of modern whales. Science 327:993–996.

Maximino, C. 2009a. A quantitative test of the thermogenesis hypothesis of cetacean brain evolution, using phylogenetic comparative methods. Marine and Freshwater Behaviour and Physiology 42:1–17.

Maximino, C. 2009b. Reply to Manger's Commentary on "A quantitative test of the thermogenesis hypothesis of cetacean brain evolution, using phylogenetic comparative methods." Marine and Freshwater Behaviour and Physiology 42:363–372.

McGowen, M. R., M. Spaulding, and J. Gatesy. 2009. Divergence date estimation and a comprehensive molecular tree of extant cetaceans. Molecular Phylogenetics and Evolution 53:891–906.

Mead, J. G., and E. D. Mitchell. 1984. Atlantic gray whales; pp. 33–53 in M. L. Jones, S. L. Swartz, and S. Leatherwood (eds.), The Gray Whale, *Eschrichtius robustus*. Academic Press, Orlando, FL.

Meschersky, I. G., M. A. Kuleshova, D. I. Litovka, V. N. Burkanov, R. D. Andrews, G. A. Tsidulko, V. V. Rozhnov, and V. Y. Ilyashenko. 2015. Occurrence and distribution of mitochondrial lineages of gray whales (*Eschrichtius robustus*) in Russian Far Eastern seas. Biology Bulletin 42:34–42.

Millar, J. S., and G. J. Hickling. 1990. Fasting endurance and the evolution of mammalian body size. Functional Ecology 4:5–12.

Millien, V., S. Kathleen Lyons, L. Olson, F. A. Smith, A. B. Wilson, and Y. Yom-Tov. 2006. Ecotypic variation in the context of global climate change: revisiting the rules. Ecology Letters 9:853–869.

Mitchell, E. D. 1989. A new cetacean from the late Eocene La Meseta Formation, Seymour Island, Antarctic Peninsula. Canadian Journal of Fisheries and Aquatic Science 46:2219–2235.

Montgomery, S. H., J. H. Geisler, M. R. McGowen, C. Fox, L. Marino, and J. Gatesy. 2013. The evolutionary history of cetacean brain and body size. Evolution 67:3339–3353.

Moreno, I. B., A. N. Zerbini, D. Danilewicz, M. C. de O. Santos, P. C. Simões-Lopes, J. Lailson-Brito Jr., and A. F. Azevedo. 2005. Distribution and habitat characteristics of dolphins of the genus *Stenella* (Cetacea: Delphinidae) in the southwest Atlantic Ocean. Marine Ecology Progress Series 300:229–240.

Motani, R. 2002. Scaling effects in caudal fin propulsion and the speed of ichthyosaurs. Nature 415:309–312.

Motani, R. 2005. Evolution of fish-shaped reptiles (Reptilia: Ichthyopterygia) in their physical environment and constraints. Annual Review of Earth and Planetary Sciences 33:395–420.

Murakami, M., C. Shimada, Y. Hikida, Y. Soeda, and H. Hirano. 2014. *Eodelphis kabatensis*, a new name for the oldest true dolphin *Stenella kabatensis* Horikawa, 1977 (Cetacea, Odontoceti, Delphinidae), from the upper Miocene of Japan, and the phylogeny and paleobiogeography of Delphinoidea. Journal of Vertebrate Paleontology 34:491–511.

Pfuhl, H. A., and I. N. McCave. 2005. Evidence for late Oligocene establishment of the Antarctic Circumpolar Current. Earth and Planetary Science Letters 235:715–728.

Pledge, N. S. 2005. A new species of early Oligocene Cetacean from Port Willunga, South Australia. Memoirs of the Queensland Museum 51:121–133.

Popov, S. V., I. G. Shcherba, L. B. Ilyina, L. A. Nevesskaya, N. P. Paramonova, S. O. Khondkarian, and I. Magyar. 2006. Late Miocene to Pliocene palaeogeography of the Paratethys and its relation to the Mediterranean. Palaeogeography, Palaeoclimatology, Palaeoecology 238:91–106.

Potvin, J., J. A. Goldbogen, and R. E. Shadwick. 2012. Metabolic expenditures of lunge feeding rorquals across scale: implications for the evolution of filter feeding and the limits to maximum body size. PLoS One 7:e44854.

Pyenson, N. D., C. S. Gutstein, M. A. Cozzuol, J. Velez-Juarbe, and M. Suárez. 2013. New material of *Odobenocetops* form the late Miocene of Chile clarifies the systematics and paleobiology of walrus-convergent odontocetes. Journal of Vertebrate Paleontology: Program and Abstracts 195.

Pyenson, N. D., and D. R. Lindberg. 2011. What happened to gray whales during the Pleistocene? The ecological impact of sea-level change on benthic feeding areas in the North Pacific Ocean. PLoS One 6:e21295.

Pyenson, N. D., and S. N. Sponberg. 2011. Reconstructing body size in extinct crown Cetacea (Neoceti) using allometry, phylogenetic methods and tests from the fossil record. Journal of Mammalian Evolution 18:269–288.

Pyron, R. A. 2011. Divergence time estimation using fossils as terminal taxa and the origins of Lissamphibia. Systematic Biology 60:466–481.

Rabosky, D. L. 2014. Automatic detection of key innovations, rate shifts, and diversity-dependence on phylogenetic trees. PLoS One 9:e89543.

Reguero, M. A., S. Marenssi, A., and S. N. Santillana. 2012. Weddellian marine/coastal vertebrates diversity from a basal horizon (Ypresian, Eocene) of the Cucullaea I Allomember, La Meseta formation, Seymour (Marambio) Island, Antarctica. Revista Peruana de Biología 19:275–284.

Rivin, M. A. 2010. Early Miocene cetacean diversity in the Vaqueros Formation, Laguna Canyon, Orange County, California: In *Geological Sciences*, Vol. Master of Science, pp. 105. California State University, Fullerton, Fullerton, CA.

Sarmiento, J. L., N. Gruber, M. A. Brzezinski, and J. P. Dunne. 2004. High-latitude controls of thermocline nutrients and low latitude biological productivity. Nature 427:56–60.

Scheinin, A. P., D. Kerem, C. D. MacLeod, M. Gazo, C. A. Chicote, and M. Castellote. 2011. Gray whale (*Eschrichtius robustus*) in the Mediterranean Sea: anomalous event or early sign of climate-driven distribution change? Marine Biodiversity Records 4:e28.

Schreer, J. F., and K. M. Kovacs. 1997. Allometry of diving capacity in air-breathing vertebrates. Canadian Journal of Zoology 75:339–358.

Slater, G. J., S. A. Price, F. Santini, and M. E. Alfaro. 2010. Diversity versus disparity and the radiation of modern cetaceans. Proceedings of the Royal Society B 277:3097–3104.

Smith, A. B. 2001. Large-scale heterogeneity of the fossil record: implications for Phanerozoic biodiversity studies. Philosophical Transactions of the Royal Society of London B 356:351–367.

Steeman, M. E., M. B. Hebsgaard, R. E. Fordyce, S. Y. W. Ho, D. L. Rabosky, R. Nielsen, C. Rhabek, H. Glenner, M. V. Sørensen, and E. Willerslev. 2009. Radiation of extant cetaceans driven by restructuring of the oceans. Systematic Biology 58:573–585.

Tanaka, Y., and R. E. Fordyce. 2014. Fossil dolphin *Otekaikea marplesi* (Latest Oligocene, New Zealand) expands the morphological and taxonomic diversity of Oligocene cetaceans. PLoS One 9:e107972.

Tsai, C.-H., and T. Ando. 2015. Niche partitioning in Oligocene toothed mysticetes (Mysticeti: Aetiocetidae). Journal of Mammalian Evolution 23:33–41.

Uhen, M. D. 2014. New material of *Natchitochia jonesi* and a comparison of the innominata and locomotor capabilities of Protocetidae. Marine Mammal Science 30:1029–1066.

Uhen, M. D. 2015. Systematics Archive 9–Cetacea. Paleobiology Database, https://paleobiodb.org

Uhen, M. D., A. G. Boyer, J. H. Brown, D. P. Costa, S. K. M. Ernest, A. R. Evans, J. L. Gittleman, M. J. Hamilton, L. E. Harding, K. Lintulaakso, S. K. Lyons, J. G. Okie, J. J. Saarinen, R. M. Sibly, F. A. Smith, P. R. Stephens, and J. M. Theodor. in preparation. The trajectory of marine mammal body mass over the Cenozoic.

Uhen, M. D., and N. D. Pyenson. 2007. Diversity estimates, biases and historiographic effects: resolving cetacean diversity in the Tertiary. Palaeontologia Electronica 10:10.2.10A.

Valenzuela-Toro, A. M., C. S. Gutstein, R. M. Varas-Malca, M. E. Suarez, and N. D. Pyenson. 2013. Pinniped turnover in the South Pacific Ocean: new evidence from the Plio-Pleistocene of the Atacama Desert, Chile. Journal of Vertebrate Paleontology 33:216–233.

Vélez-Juarbe, J., and N. D. Pyenson. 2012. *Bohaskaia monodontoides*, a new monodontid (Cetacea, Odontoceti, Delphinoidea) from the Pliocene of the western North Atlantic Ocean. Journal of Vertebrate Paleontology 32:476–484.

Whitehead, H., B. McGill, and B. Worm. 2008. Diversity of deep-water cetaceans in relation to temperature: implications for ocean warming. Ecology Letters 11:1198–1207.

Whitmore, F. C., Jr. 1994. Neogene climatic change and the emergence of the modern whale fauna of the North Atlantic Ocean. Proceedings of the San Diego Society of Natural History 29:223–227.

Whitmore, F. C., Jr., and J. A. Kaltenbach. 2008. Neogene Cetacea of the Lee Creek Phosphate Mine, North Carolina. Virginia Museum of Natural History Special Publication 14:181–269.

Williams, E. M. 1998. Synopsis of the earliest cetaceans; pp. 1–28 in J. G. M. Thewissen (ed.), The Emergence of Whales. Plenum Press, New York.

Worthy, G. A. J., and J. P. Hickie. 1986. Relative brain size in marine mammals. The American Naturalist 128:445–459.

Zachos, J. C., G. R. Dickens, and R. E. Zeebe. 2008. An early Cenozoic perspective on greenhouse warming and carbon-cycle dynamics. Nature 451:279–283.

8 Paleontological Insights into Evolution and Development

8.1 Limb morphology and development

The transformations of the fore- and hind limbs are among the most striking morphological features distinguishing cetaceans from terrestrial mammals. While the forelimbs have turned into flippers, the hind limbs are almost completely missing and no longer visible externally (Abel, 1908). The evolution and development of both limbs have been studied through the fossil record, as well as embryological and gene expression techniques.

8.1.1 Forelimb

In contrast to most terrestrial mammals, the forelimb of modern cetaceans is not used as a weight-bearing driver of locomotion. Instead, its main functions are to steer and stabilize the body during swimming, with the propulsive force being generated by the flukes (Cooper *et al.*, 2008). Externally, the cetacean flipper is dorsoventrally flattened, shows no signs of separate fingers, and completely lacks claws. Its overall shape varies between species, likely as a result of differences in habitat use, social interactions, and swimming style. For example, elongated, narrow flippers are associated with fast swimmers, whereas broad, triangular flippers

are more typical of slow cruisers and highly maneuvrable taxa (Sanchez and Berta, 2010; Woodward *et al.*, 2006).

Despite its markedly different function, the basic osteological structure of the cetacean forelimb is the same as in any other mammal. Thus, it is anchored to the body via the scapula and, from proximal to distal, consists of the humerus, radius and ulna, carpals, metacarpals, and phalanges (section 3.3.2). Raoellids and pakicetids have forelimbs closely resembling those of terrestrial artiodactyls, with a narrow scapula, gracile long bones, a mobile wrist, and a reduced (or no) ability to **pronate** and **supinate** (i.e., axially rotate) the forearm (Figure 8.1) (Cooper *et al.*, 2012; Madar, 2007). The manus is incompletely known, but likely was **pentadactyl** (five-fingered) and **mesaxonic** (i.e., with the axis of symmetry passing through the third digit). The retention of the first digit in cetaceans is interesting, given its absence in extant artiodactyls, including hippopotamids (Boisserie *et al.*, 2005). It is possible that cetaceans underwent an evolutionary reversal, which led them to reacquire digit I (Wang *et al.*, 2009). More plausibly from a morphological perspective, however, they may simply have retained the ancestral condition of a five-fingered manus, as seen in the early artiodactyl *Diacodexis*

Cetacean Paleobiology, First Edition. Felix G. Marx, Olivier Lambert, and Mark D. Uhen.
© 2016 John Wiley & Sons, Ltd. Published 2016 by John Wiley & Sons, Ltd.

Figure 8.1 Cetacean forelimb evolution illustrated by the pakicetid *Pakicetus*, the protocetid *Maiacetus*, the basilosaurid *Dorudon*, and the extant mysticetes *Balaena* and *Balaenoptera*. Note the accelerated evolution of particular *Hoxb9* and *Hoxd12* along the cetacean stem lineage and the lineage leading to tetradactyl mysticetes. Drawings of *Pakicetus*, *Maiacetus*, and *Dorudon* adapted from Thewissen *et al.* (2001), Gingerich *et al.* (2009), and Uhen (2004), respectively.

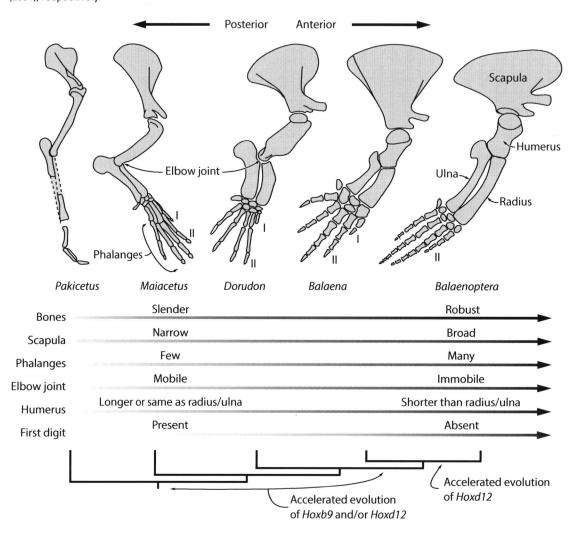

(Thewissen and Hussain, 1990). In either case, the retention of the first digit long after it was lost in artiodactyls is striking, and likely explained by the need to maintain a broad, webbed hand for swimming (Gingerich *et al.*, 2001).

The forelimb of *Ambulocetus* resembles that of pakicetids in its basic construction, but is considerably more robust and ends in a proportionally much larger hand. Distally, the digits of *Ambulocetus*—and most likely also those of raoellids and pakicetids—bore small hooves, reflecting their artiodactyl ancestry (Thewissen *et al.*, 1994, 1996). The forelimb morphology of protocetids is mainly known from relatively complete and well-preserved specimens belonging to *Rodhocetus* and *Maiacetus* (Gingerich *et al.*,

2001, 2009). These animals retain a relatively narrow scapula, but the major long bones are more robust than those of *Ambulocetus*, and the forearm in particular is somewhat flattened (Gingerich *et al.*, 2009). Alternating rows of carpal bones with saddle-shaped articular surfaces indicate that they retained the ability to extend the wrist. Unlike in *Ambulocetus*, the protocetid carpus does not include a centrale. The metacarpals are robust, and there is no suggestion of hyperphalangy (Figure 8.1). Like their forebears, protocetids seem to have borne small, nail-like hooves on the tips of their fingers (Gingerich *et al.*, 2001, 2009).

Basilosaurid forelimb morphology is well known, based on a wide range of specimens belonging to *Ancalecetus*, *Basilosaurus*, *Chrysocetus*, *Cynthiacetus*, *Dorudon*, and *Zygorhiza* (Gingerich and Uhen, 1996; Kellogg, 1936; Martínez Cáceres and de Muizon, 2011; Uhen, 2004; Uhen and Gingerich, 2001). With the exception of *Ancalecetus* (discussed further in this chapter), basilosaurids have a relatively broad, fan-shaped scapula compared to other archaeocetes. The humerus is about as long as the forearm and bears a well-developed ridge for the attachment of a powerful deltoid muscle along its anterior margin. Like in *Maiacetus*, the radius and ulna are locked into position relative to one another, and together articulate with the distal portion of the humerus via a rounded elbow joint. Flexion and extension of the elbow are possible but, judging from the size of the articular trochlea, restricted to a range of about 30–40° (Uhen, 2004). The carpals are still organized in alternating rows, but the magnum and trapezoid are fused into a single unit. Small, flat articular surfaces between individual carpal elements mean that the wrist was incapable of much extension, and thus unable to support the body on land (Uhen, 1998, 2004). Like those of protocetids, the metacarpals are robust. The presence or absence of distal hooves remains unclear, as there are currently no unequivocal examples of distal phalanges.

In terms of its forelimb, *Ancalecetus simonsi*, which is otherwise very similar to the contemporaneous *Dorudon atrox*, is a major exception to the general basilosaurid pattern, and indeed markedly different from all other cetaceans (Gingerich and Uhen, 1996). The scapula of this species is much narrower than that of other basilosaurids, and the acromion process is folded back on itself and projects ventrally. The glenoid cavity is very shallow and projects posteriorly, rather than ventrally as in other cetaceans. The shaft of the humerus is transversely flattened and topped by a relatively small head. In addition, there seems to have been an unusual ligamentous connection between the scapula and the humeral head, which presumably restricted motion at the shoulder. The radius and ulna are oriented perpendicularly to the humerus and connected to the latter via an immovable, variably fused elbow joint. The ulna lacks a true olecranon process. The wrist of *Ancalecetus* is similar to that of *Dorudon atrox*, but some of the carpals are fused to one another. *Ancalecetus simonsi* is known from just a single specimen, and could plausibly represent an aberrant individual of *Dorudon atrox*. For now, however, it is considered a separate species in its own genus (Gingerich and Uhen, 1996).

Neocetes inherit and, in many cases, further exaggerate the broad, fan-shaped scapula of basilosaurids (Figure 8.1). One of the most striking features characterizing this group is the complete loss of mobility of the elbow, so that the forelimb can only be moved at the shoulder. In most lineages, the humerus is shortened, especially relative to the radius and ulna. In living mysticetes and, to a lesser degree, odontocetes, carpals ossify relatively late during ontogeny and often remain partially or entirely cartilaginous (Flower, 1885). In addition, individual carpals frequently fuse, which makes their identification difficult in some cases. Both of these features likely reflect the effective immobilization of the wrist, and hence a reduced need to maintain functional carpal articulations.

Most neocetes deviate from the primitive mammalian pattern in the number of phalanges comprising each digit (typically two in the thumb and three in all other digits). In digits II–IV, this frequently leads to the development of additional phalanges, or **hyperphalangy**, whereas the opposite (i.e., a reduction in the number of phalanges) often affects digits I and V (Cooper *et al.*, 2007a). In pygmy right whales, gray whales, and rorquals, digit reduction has proceeded even further: in these lineages, one finger has been lost completely, and the flipper has thus become **tetradactyl**. Exactly which of the digits was lost has long been a matter of debate (Kükenthal, 1889–1893), but recent work provided strong evidence for the reduction of digital ray I, based on the pattern of articulation between the metacarpals and the carpals (Figure 8.2) (Cooper *et al.*, 2007a; Howell, 1930). In line with the immobilization of the elbow and wrist, the digits of extant cetaceans are held in place by thick connective tissue and thus cannot move individually. Delphinoids furthermore have extremely reduced forelimb muscles, which precludes any active movement of the flipper distal to the elbow. By contrast, most living mysticetes and early diverging odontocetes (physeteroids and ziphiids) retain reasonably developed digital flexors and extensors, which may serve to stiffen the flipper or, in the case of balaenopterids, might even allow a slight degree of palmar flexion (Cooper *et al.*, 2007b).

Developmentally, the renewed importance of the first digit in archaeocetes, hyperphalangy, and the loss of digit I in some mysticetes may be apparent in the expression of a particular family of genes responsible for body plan patterning

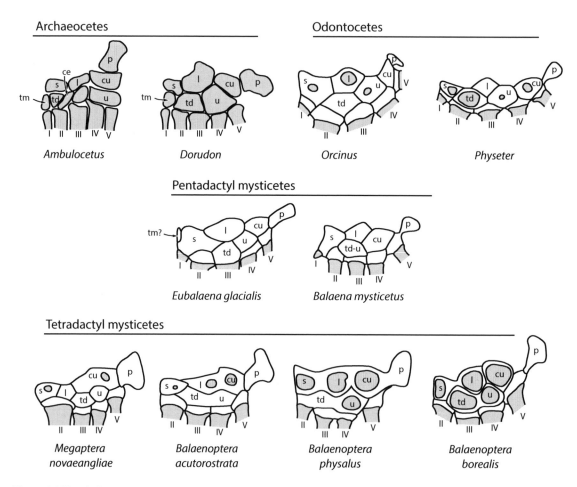

Figure 8.2 Morphology and ossification of the carpus of various extinct and extant cetaceans. Roman numerals denote metacarpals. Cartilage is shown in white, bone in gray; cu, cuneiform; ce, centrale; l, lunate; m, magnum; p, pisiform; s, scaphoid; td, trapezoid; tm, trapezium; u, unciform. Adapted from Cooper *et al.* (2007a).

(**homeobox**, or **Hox** genes). *Hox* genes encode **transcription factors** (proteins regulating DNA expression) which function near the base of a chain of regulatory genes controlling the differentiation and development of body segments in the embryo. The role played by *Hox* genes is so fundamental that their DNA sequences are highly conserved across vertebrates as a whole. Nevertheless, the origin of (modern) cetaceans appears to have been accompanied by accelerated evolution of the genes *Hoxd12* and/or *Hoxb9*, both of which are involved in forelimb development. In addition, faster than normal evolution of *Hoxd12* also marked the origins of tetradactyl mysticetes or, at least, that of balaenopterids (Liang *et al.*, 2013; Wang *et al.*, 2009).

Hoxd12 and *Hoxb9* ultimately help to control the expression of *Sonic hedgehog* (*Shh*), a signaling molecule acting directly on cells and implicated in the formation and patterning (e.g., digit number and identity) of individual limb elements (Rabinowitz and Vokes, 2012; Xu and Wellik, 2011). In the case of cetaceans, prolonged expression of *Shh* or related signaling factors may have been responsible for the emergence of hyperphalangy by causing the distalmost region of the flipper to continue growing for longer than is normal (Richardson *et al.*, 2004; Richardson and Oelschläger, 2002; Sanz-Ezquerro and Tickle, 2003). Similarly, changes in the (*Hoxd12*-regulated) expression of *Shh* may have resulted in the loss of digit I in tetradactyl mysticetes (Shapiro *et al.*, 2003; Wang *et al.*, 2009). Given the highly conservative nature of *Hox* genes, the question arises whether the comparatively strong deviations of cetaceans from the usual vertebrate bauplan (e.g., hyperphalangy, digit loss, and loss of hind limb) may have been favored by relaxed selective pressures. In this context, it is interesting to note that *Hoxd12* (unlike *Hox9b*) also contributes to the patterning of the hind limb, whose almost complete loss (section 8.1.2) plausibly may have relaxed the developmental constraints usually acting on this gene (Wang *et al.*, 2009; Xu and Wellik, 2011). However, this process could not have become important until well after the origin of cetaceans, since all archaeocetes except basilosaurids retain well-developed, functional legs and feet.

8.1.2 Hind limb

Overall, the evolution of the cetacean hind limb is a history of reduction, from a fully functioning, weight-bearing limb to a small, internal remnant that only serves as a support structure for the reproductive system (Figure 8.3) (Tajima *et al.*, 2004). Like many terrestrial mammals, pakicetids and raoellids have a well-developed pelvis firmly anchored to the sacral portion of the vertebral column, as well as an elongate femur, a similarly sized tibia, and gracile metatarsals (section 5.1.1). The astragalus resembles that of artiodactyls in bearing trochleated proximal and distal articular surfaces, which constrain the ankle to anteroposterior, hinge-like movements (Cooper *et al.*, 2012; Madar, 2007). The hind limb of *Ambulocetus* is generally more robust than that of pakicetids and ends in an elongate foot used to generate propulsion (Thewissen *et al.*, 1994). This trend is seemingly continued in remingtonocetids, although the morphology of their hind limb is still rather poorly known (Thewissen and Bajpai, 2009).

Hind limb elements have been described for a range of protocetids, including *Rodhocetus*, *Maiacetus*, *Qaisracetus*, *Natchitochia*, and *Georgiacetus* (Uhen, 2014). These taxa show a progression from a pelvis consisting of a large ischium and a relatively short pubis and ilium in *Rodhocetus* and *Maiacetus* (as well as more basal archaeocetes) to exactly the opposite condition in the more crownward *Qaisracetus*, *Natchitochia*, and *Georgiacetus* (Figure 8.4). The articular surface of the pubis is more complex in the more crownward taxa, which likely means that part or all of the articulation between the pelves was cartilaginous. Together with the enlarged pubis, this looser connection likely evolved to accommodate the increasingly wide transverse processes of the sacral vertebrae (Hulbert, 1998). At the level of the vertebral column, the transformation of the pubic symphysis was accompanied by a reduction in the stability of the sacrum: whereas *Rodhocetus* and *Maiacetus* have four sacral vertebrae, this number is smaller (two or three) in *Qaisracetus* and *Natchitochia*, and effectively zero in *Georgiacetus* and an undescribed protocetid from the Late Eocene of Egypt (section 8.2) (Gingerich *et al.*, 2014; Uhen, 2014). The lack of a bony connection between the pelvis and

Figure 8.3 Cetacean hind limb evolution illustrated by the pakicetid *Pakicetus*, the protocetid *Maiacetus*, the basilosaurid *Basilosaurus*, and two extant taxa: the mysticete *Balaena* and the odontocete *Tursiops*. Non-basilosaurid archaeocetes (*Pakicetus* and *Maiacetus*) have weight-bearing hind limbs. Note the changes in *Sonic Hedgehog* (*Shh*) expression along the lineages leading to Pelagiceti (*Basilosaurus* and neocetes) and Neoceti. Drawings of *Pakicetus*, *Maiacetus* and *Basilosaurus* adapted from Thewissen *et al.* (2001), Gingerich *et al.* (2009), and Gingerich *et al.* (1990), respectively.

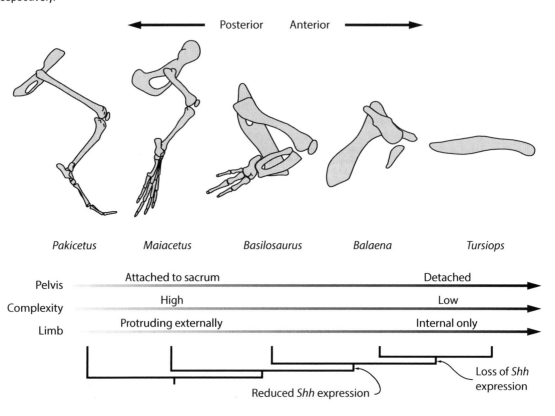

the vertebral column means that *Georgiacetus* was likely unable to support its weight on land. Nevertheless, *Georgiacetus* retains a large acetabulum, which is suggestive of an equally large femur (Hulbert *et al.*, 1998).

As in *Georgiacetus*, the pelvis is detached from the vertebral column in basilosaurids and all neocetes. There is currently only a single described fragment of the pelvis of the basal basilosaurid *Basilotritus*, which stands out mainly for its extremely shallow, presumably nonfunctional acetabulum (Uhen, 1999). By contrast, the pelves of *Basilosaurus* and *Chrysocetus* are better known and generally strap-like, with a variably developed acetabulum (Gingerich *et al.*, 1990; Uhen and

Gingerich, 2001). The unusual shape of the basilosaurid pelvis may reflect an enlarged pubis, which, unlike in living cetaceans, remains in articulation with its counterpart (Gingerich *et al.*, 1990; Uhen and Gingerich, 2001). However, this arrangement has been questioned, and the enlarged, strap-like portion of the pelvis may instead represent the ischium (Bajpai *et al.*, 2009; Gol'din, 2014). Unlike in earlier archaeocetes, the hind limb of basilosaurids is extremely small relative to the size of the body, but most likely it remained externally visible. Most of the tarsals are fused, and both the total number of digits and the number of phalanges per digit are reduced compared to protocetids. Nevertheless, the hip and knee joints

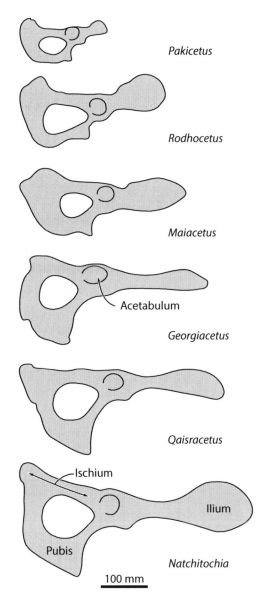

Figure 8.4 Evolution of the pelvis in non-basilosaurid archaeocetes. Adapted from Uhen (2014).

are well formed and mobile, implying that the limb may have remained functional in some sense (e.g., for mating purposes; Gingerich *et al.*, 1990).

Hind limbs are reduced to just a small, rod-like remnant of the pelvis in most extant odontocetes, and a tripartite pelvis plus, in some species, the femur in mysticetes. As a result of this marked reduction in size, the leg has become entirely enclosed by the soft tissues forming the body wall. Even in the earliest fossil neocetes, the hind limb appears so small that it was likely internal (Fordyce *et al.*, 2000). However, data are lacking for most taxa, and little can be said about the exact position of the bones relative to the nonpreserved soft tissues. Among extant cetaceans, the best developed hind limbs occur in balaenids, which besides a relatively elaborate bony pelvis and femur also preserve a cartilaginous tibia (Abel, 1908; Lönnberg, 1911). In addition, atavistic individuals of extant species may preserve additional limb segments and, rarely, even develop external hind limbs (Andrews, 1921; Ohsumi and Kato, 2008). Although the pelvis of modern cetaceans has clearly lost its ancestral locomotory function, it still serves as an attachment site for muscles that support the reproductive system and, in several species, appears to be sexually dimorphic (Dines *et al.*, 2014; Simões-Lopes and Gutstein, 2004; Tajima *et al.*, 2004).

Like that of the forelimb, the development of the hind limb is partially driven by the expression of the signaling factor *Sonic hedgehog* (*Shh*). Modern cetacean embryos still express an external limb bud, which in most mammals would continue to grow and differentiate under the influence of *Shh*. In cetaceans, however, *Shh* is no longer expressed in this part of the body, which results in the gradual degradation of the hind limb bud from around the fifth gestational week onward (Thewissen *et al.*, 2006). The regulatory mechanisms controlling the production of *Shh* differ between the fore- and hind limbs: although both are partly under the control of certain *HoxA/D* genes (e.g., *Hoxd12*), the development of the hind limb is not dependent on *Hox9* (Xu and Wellik, 2011). Instead, *Shh* expression is regulated by the transcription factor *Islet1*, which also plays a role in the formation of the pelvis and thus is a prime candidate for a developmental mechanism underlying hind limb loss (Itou *et al.*, 2012). However, the evolution of this protein in cetaceans has not yet been investigated.

Judging from the fossil record, it seems likely that disappearance of *Shh* from the hind limb bud was, at least initially, a gradual process (Bejder and Hall, 2002). Thus, reduced expression of *Shh* in basilosaurids may plausibly have led to the loss of the distal phalanges and the first two digits, while still ensuring the development of a functional, mobile appendage (Figure 8.3) (Thewissen et al., 2006). Subsequently, limb patterning was lost entirely in neocetes, although the current data are insufficient to say whether this happened suddenly or over a protracted period of time. In snakes, a similar, stepwise loss of the hind limb was accompanied by a concurrent loss of vertebral patterning (Cohn and Tickle, 1999). Cetaceans seem to have evolved along similar lines, with the reduction of the hind limb coinciding with the homogenization of the lumbar, sacral, and anterior caudal vertebrae. If these two processes are indeed linked, then the point in time when the sacral vertebrae became morphologically indistinguishable might also mark the loss of *Shh* expression in the hind limb bud (Thewissen et al., 2006).

8.2 Regionalization of the vertebral column

The mammalian vertebral column is generally divided into five functional regions: cervical, thoracic, lumbar, sacral, and caudal (section 3.3.1). Each region differs in the size, shape, and function (e.g., neck rotation or support of the chest) of the vertebrae it contains, as well as its role in locomotion. Terrestrial mammals load the vertebral column at two points corresponding to the attachment of the fore- and hind limbs. This loading is reflected in the differentiation of the column—in particular, the presence of a stable, fused sacrum capable of supporting the hind limbs. Unlike their ancestors, modern cetaceans are fully aquatic and approximately neutrally buoyant in water. Their return to the sea not only freed cetaceans from the point loads that constrain vertebral organization in terrestrial mammals, but also exposed them to new requirements related mainly to swimming. Adapting to this new environment necessitated a wide-ranging reorganization of the vertebral column, which ultimately became reflected in the fundamental pattern underlying its development.

Artiodactyls typically have a vertebral formula of 7C 13T 6L 4S (i.e., 7 cervical vertebrae, 13

thoracics, etc.), with a variable number of caudals (Buchholtz, 1998). The same formula appears in *Remingtonocetus domandaensis*, the oldest cetacean for which the pre-caudal vertebral count is known with certainty (Bebej et al., 2012). It therefore seems likely that pakicetids and ambulocetids also retained the ancestral artiodactyl pattern, although alternative, larger vertebral counts have also been proposed (Madar, 2007; Madar et al., 2002). All protocetids for which sufficient material is known likewise have a presacral vertebral count identical to that of typical artiodactyls (Uhen, 2014), except for an as-yet-undescribed protocetid from the Priabonian of Egypt (Gingerich et al., 2014). This single outlier has 15 thoracic and 4 lumbar vertebrae, suggesting "thoracic capture" of the first two lumbars, given that the thoracic plus lumbar vertebral count is the same as in all other protocetids.

Like the vast majority of mammals, all cetaceans retain seven cervical vertebrae, although the latter are frequently shortened or even fused to each other in order to stabilize the head during swimming (Figure 5.7) (Buchholtz, 2001). By contrast, all other vertebral counts change in post-protocetid cetaceans (i.e., Pelagiceti). Thus, within Basilosauridae the thoracics number 12 in *Basilotritus*, 16 in *Basilosaurus*, 17 in *Dorudon* and *Zygorhiza*, and 20 in *Cynthiacetus* (Bebej et al., 2012; Martínez Cáceres and de Muizon, 2011; Uhen, 2001, 2004). In general, basilosaurids (except for, perhaps, the basal *Basilotritus*) also have an increased number of lumbars (13+), whereas the number of caudals seems to be relatively stable at around 21. In all cases, the additional vertebrae may have been added at a single point in the vertebral column, likely near the intersection of the thorax and lumbus (Uhen, 2014). Neocetes show a similar degree of variability, but with no apparent restriction on the number of caudals and a relatively stable number of thoracics (Buchholtz and Schur, 2004).

In both basilosaurids and neocetes, increased thoracic and lumbar counts mostly reflect **meristic** changes, which are additions to the vertebral count by way of intensified **somitogenesis** (subdivision of the embryonic tissue into discrete developmental units). In basilosaurids, this change affects both the thoracic and the lumbar regions at the same time, which suggests that the development of these two units is linked. A similar pattern occurs in terrestrial mammals and the undescribed Late Eocene protocetid from Egypt (Gingerich *et al.*, 2014), in which an increase in one of these units is mirrored by a reciprocal decrease in the other. This developmental link between the thoracic and lumbar regions is known as **terrestrial modularity** (Buchholtz, 2007). By contrast, the thoracic count is relatively stable in neocetes, and increases in the number of lumbar vertebrae are instead associated with concurrent increases in the anterior caudal region. This suggests (1) a novel developmental link between the lumbar and the anterior caudal regions, a phenomenon known as **neocete modularity**; and (2) the dissociation of the caudal vertebrae into an **anterior tail** unit (linked to the lumbars) and a **fluke** section, whose overall count remains relatively stable (Figure 8.5). A separate

fluke section can also be identified in basilosaurids, based on the shape of the vertebrae; however, it is currently unclear whether it was already developmentally independent from the remainder of the caudal series (Buchholtz, 2007).

Besides coupled increases in number, the lumbar and anterior caudal vertebrae of neocetes and, to a lesser degree, basilosaurids also display similarities in morphology (e.g., centrum length). This is accompanied by the reduction and eventual disappearance of the sacral vertebrae, thus creating a single developmental and functional unit extending from the thorax to the fluke. This new functional unit, the **torso**, allows a single undulatory wave to travel along the central and posterior vertebral column, thus enabling basilosaurids and later cetaceans to swim by caudal undulation or oscillation (section 5.1.1) (Buchholtz, 2001;

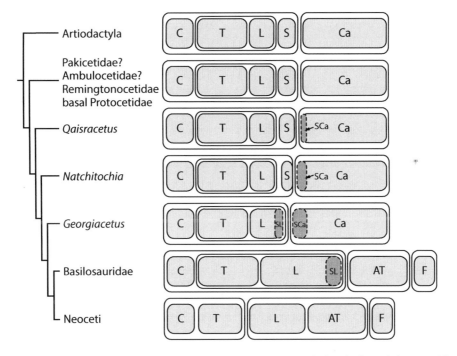

Figure 8.5 Modular organization of the vertebral column in early cetacean evolution. As far as is known with certainty, all early cetaceans (Remingtonocetidae and basal Protocetidae) retain the ancestral artiodactyl pattern. Note how the sacral vertebrae first become caudalized in more crownward protocetids, and then lumbarized in basilosaurids and neocetes. In artiodactyls and archaeocetes, the thoracic and lumbar vertebrae form a unit within a bigger module also including the cervicals and the sacrum. In neocetes, these units are dissolved and the lumbars instead become linked to the anterior portion of the tail. AT, anterior tail; C, cervical; Ca, caudal; F, fluke; L, lumbar; S, Sacral; SCa, sacral caudal; SL, sacral lumbar; T, thoracic. Adapted from Uhen (2014).

Thewissen and Fish, 1997). The reduction of the sacrum is already apparent in several derived protocetids, in which varying numbers of posterior sacral vertebrae are no longer connected to the pelvis (via the first sacral vertebra). Instead, these former sacral vertebrae assume the appearance of anterior caudals (**sacral caudal vertebrae**) by developing hemal processes reminiscent of chevron bones (Uhen, 2014).

In the most crownward protocetids and basilosaurids, this pattern reverses. Thus, *Georgiacetus* retains three sacral caudal vertebrae, but the first sacral vertebra has instead turned into a posterior lumbar (**sacral lumbar vertebra**) (Buchholtz, 1998). In basilosaurids, the posterior three sacrals lose their caudalized identity, and join the former S1 as part of the lumbar column. Finally, the last morphological vestiges which set the sacral vertebrae apart (e.g., thickened transverse processes) disappear with the origin of neocetes, thus resulting in the complete lumbarization of the sacrum (Figure 8.5). In modern cetaceans, the sacral region can only be identified based on circumstantial evidence, such as the position of associated soft tissues (e.g., the pudendal nerve plexus) or the position of the hind limb bud in the embryo (Moran *et al.*, 2015; Slijper, 1936; Thewissen *et al.*, 2006). This kind of **homeotic** change, in which particular vertebrae change their morphological identity, likely reflects underlying changes in the expression domains of genes responsible for axial patterning (e.g., *Hox* genes). Reciprocal changes between the thoracic and lumbar counts, as seen in terrestrial mammals and the undescribed Late Eocene protocetid from Egypt (as discussed in this section), likely also fall into this category (Buchholtz, 2007). Little is known about exactly which genes underlie the particular patterning of the cetacean vertebral column. However, modified regulation of *Hoxc8*, which is expressed along the posterior region of the developing spine, may be involved in this process in mysticetes (Shashikant *et al.*, 1998).

In neocetes, the emergence of a functional torso and the complete absence of the sacrum are accompanied by another innovation: an inverse relationship between the number of vertebrae in the torso and the length of each individual vertebra. Whereas in terrestrial mammals and

archaeocetes higher vertebral counts are usually accompanied by longer centra, and thus an overall longer vertebral column, neocetes either have a large number of short vertebrae in the torso, or a small number of long ones. A large number of short vertebrae may increase the proportion of intervertebral disk space, thus allowing elastic energy storage during swimming, and also makes the column more rigid by restricting intervertebral flexion (section 5.1.3) (Buchholtz, 1998, 2007).

8.3 The origins of homodonty, polydonty, and monophyodonty

Teeth are partly composed of enamel, the hardest vertebrate substance known. As a result, teeth tend to preserve well after death, and much of the mammalian fossil record is based on them. Mammals differ from most other vertebrates in (1) being **heterodont**, or having different types of teeth in each jaw, with each tooth class—incisors, canines, premolars, and molars—serving a different function (e.g., cutting, slicing, piercing, crushing, and grinding); and (2) having upper and lower teeth that precisely match (**occlude**) and work against each other, thus forming an effective food-processing apparatus. To maintain occlusion, mammals generally have a restricted number of teeth that are replaced only once during the lifetime of the individual—a condition known as **diphyodonty**. Thus, the **deciduous** milk teeth developed in juveniles are ultimately shed and followed by the **permanent** dentition, which in placental mammals ancestrally comprises three incisors, a single canine, four premolars, and three molars in each quadrant. Modern cetaceans are among the few groups of mammals that have fundamentally diverged from this pattern, either by giving up on precise occlusion and increasing the number of teeth, or by losing their dentition altogether.

8.3.1 Archaeocetes
Archaic cetaceans largely follow the ancestral mammalian condition in retaining a relatively small number of clearly heterodont teeth (Figure 8.6). All non-basilosaurid archaeocetes furthermore retain the ancestral placental tooth count of 11

Figure 8.6 Simplified overview of cetacean dental evolution. Note the relatively simple tooth shape of extant (and some extinct) odontocetes. Adapted from Uhen (2009), with permission of Elsevier.

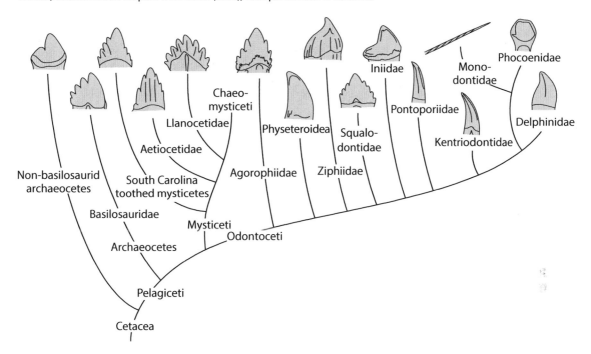

teeth per quadrant (see section 8.3), except for the protocetid *Makaracetus*, which is thought to lack the third upper incisor (Gingerich *et al.*, 2005). The incisors are anteroposteriorly aligned with the cheek teeth and separated from each other by large diastemata. Like the more posterior canine, they have conical crowns and are single-rooted. Premolar shape changes from being nearly conical and single- or double-rooted anteriorly, to equilaterally triangular and double- or triple-rooted posteriorly (Cooper *et al.*, 2009). The third or fourth premolar is often the largest upper or lower tooth. Molar teeth lack deciduous antecedents and are not separated by diastemata. The lower molars lack a crushing basin but, except in remingtonocetids, retain a distinct protoconid and hypoconid. The upper molars variably retain a paracone and a metacone on the buccal margin, as well as a protocone projecting lingually (this is less developed or absent in remingtonocetids and protocetids). Unlike other non-basilosaurid archaeocetes, remingtonocetids have crenulations on the cutting edges of the cheek teeth,

which may or may not be homologous with incipiently developed accessory denticles on some of the cheek teeth of the protocetid *Georgiacetus* (Thewissen and Bajpai, 2001).

The cheek teeth of basilosaurids differ from those of other archaeocetes in lacking a protocone and in being uniformly double-rooted, with the distal root often showing evidence of fusion with a former third root. The crown consists of a central main cusp flanked by a series of well-developed accessory denticles, which are likely homologous with the incipient denticles seen in *Georgiacetus*. The third upper molar is absent in all members of this family, resulting in a total upper tooth count of 10. Some individuals of *Dorudon* and *Zygorhiza* preserve various stages of tooth replacement, including—unusually for a derived mammal—the first premolar (Figure 8.7) (Kellogg, 1936; Uhen, 2000). Judging from the relatively crownward position of these species, it is likely that most, if not all, archaeocetes were diphyodont. *Chrysocetus* is a possible exception, as it appears to have attained an adult

Figure 8.7 Fossil evidence for tooth replacement in the basilosaurid *Zygorhiza kochii*. Adapted from Kellogg (1936).

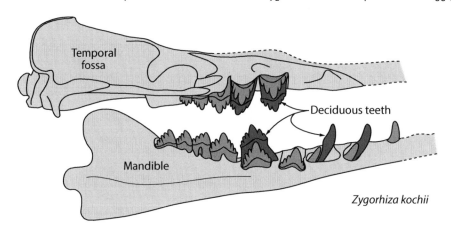

dentition at an unusually young ontogenetic age. This either means that the deciduous dentition was replaced more rapidly in this species than in other basilosaurids, or that tooth development was restricted to just a single set of permanent teeth (Uhen and Gingerich, 2001).

8.3.2 Neoceti

Archaic neocetes, such as toothed mysticetes, agorophiids, and squalodontids, inherited the basilosaurid condition of widely spaced anterior teeth, followed posteriorly by a limited number of denticulate premolars and molars. The cheek teeth tend to be broadly triangular and morphologically distinct in both basal mysticetes and odontocetes. In later forms, parallel trends toward increasingly conical, single-rooted teeth started to blur the boundaries between different tooth classes (Figure 8.8) (Armfield *et al.*, 2013). Ultimately, this process led to the emergence of genuinely **homodont** (i.e., morphologically homogeneous) dentitions in all living odontocetes (except *Inia*), and nearly homodont teeth in at least one genus of toothed mysticete (*Aetiocetus*).

The simplification of neocete crown shape was accompanied by a similar reduction in the complexity of the enamel ultrastructure (Loch *et al.*, 2015). In general, the hydroxyapatite crystals that make up mammalian enamel are organized into bundles, or **prisms**. In heterodont cetaceans, these prisms tend to form an inner layer of decussating **Hunter–Schreger bands**, followed by an outer layer of **radial enamel** (i.e., with prisms oriented perpendicular to the enamel–dentine junction). By contrast, homodont cetaceans often have a simpler enamel structure, consisting of an inner layer of radial enamel covered by a thin outer layer of prismless enamel (Figure 8.9). Hunter–Schreger bands are common in large-bodied mammals and are thought to guard against enamel cracking, whereas radial enamel is relatively wear-resistant (e.g., in the face of teeth sliding against each other). Together, these features likely made the teeth of heterodont cetaceans more resistant to damage arising from shearing and mastication. Prismless enamel, on the other hand, is best suited to withstanding attrition and abrasion, and thus piercing and grasping actions (Loch *et al.*, 2015).

Extant mysticetes lack teeth as adults, but still develop a series of embryonic **tooth primordia**. The latter progress through the **bud, cap**, and **bell** stages of development and even start to mineralize, but never erupt, and are ultimately resorbed (Karlsen, 1962). Morphologically, the incipient teeth are usually simple and mostly conical. However, groups of two or three may fuse prior to resorption, thus creating more complex structures reminiscent of premolars or molars (Karlsen, 1962; Slijper, 1936). Whether the latter are really homologous with cheek teeth, and hence might indicate that extant mysticetes arose from heterodont ancestors—rather than incipiently homodont ones similar to *Aetiocetus*—currently remains unclear. Thus, studies on mice have shown that fusion of

Figure 8.8 Evolution of tooth morphology and number of teeth per jaw quadrant. Pointedness index was calculated as the height of the main cusp divided by the total mesiodistal length of the tooth crown. Tooth class refers to main mammalian tooth types (incisor, canine, premolar, and molar). Note the concurrent emergence of homodonty (higher pointedness index and fewer tooth classes) and polydonty (number of teeth >11 per quadrant). Adapted from Armfield (2013) under a Creative Common Attribution license.

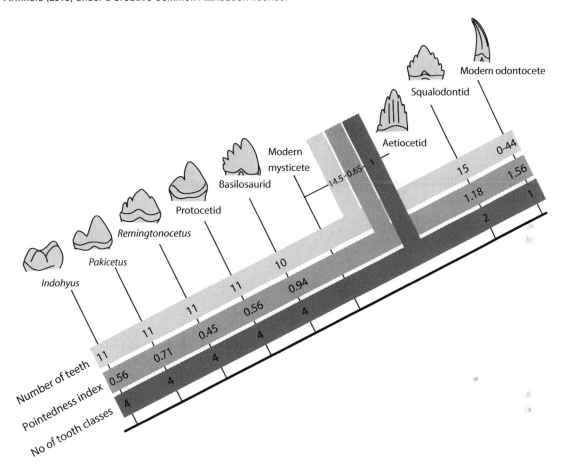

ancestral tooth primordia may indeed result in morphologically more complex adult teeth (Peterkova *et al.*, 2014). In general, however, cusp formation and tooth complexity appear to be controlled by a string of localized signaling centers known as the primary and secondary **enamel knots**, which ultimately relate the development of a multicusped tooth back to a single primordium (Jernvall and Thesleff, 2000).

In line with the simple morphology of the tooth buds, cusp formation in the bowhead whale *Balaena* appears to be driven by a single signaling center per tooth bud, which may be reflective of the simplified dental morphology of certain archaic mysticetes, such as aetiocetids (Thewissen *et al.*, 2014). Nevertheless, both extant mysticetes and odontocetes retain some of the developmental patterns that underlie heterodonty in terrestrial mammals. In the latter, the development of simple and complex tooth shapes is regulated by the expression of *Bone morphogenetic protein* (*Bmp*) 4 and *Fibroblast growth factor* (*Fgf*) 8, respectively: *Bmp4* is mainly restricted to the mesial portion (incisor field) of the jaw, whereas *Fgf8*

Figure 8.9 Enamel ultrastructure in (a) an undescribed kekenodontid (Otago University Geology Museum specimen 22023); (b) *Prosqualodon* (Museo Paleontológico Egidio Feruglio, Argentina, specimen PV 1868); (c) an undescribed squalodontid (Otago University Geology Museum 22457); and (d) *Otekaikea* sp. (Otago University Geology Museum specimen 22306). All specimens are shown in longitudinal section. EDJ, enamel–dentine junction; HSB, Hunter–Schreger bands; PL, prismless enamel; R, radial enamel. Adapted from Loch *et al.* (2015) under a Creative Common Attribution license.

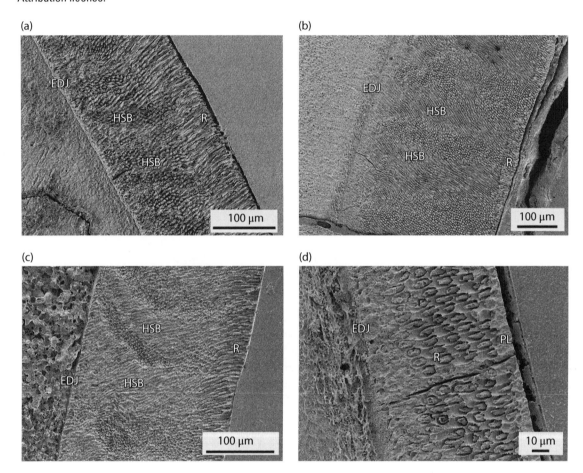

occurs distally, in the molar field (Figure 8.10) (Tucker and Sharpe, 2004). Cetaceans retain both of these molecular signals, and thus the potential for heterodonty; however, unlike terrestrial mammals, they express *Bmp4* along the entire tooth row. As a result, *Bmp4* overlaps with *Fgf8* along the distal portion of the jaw, where it may contribute to the development of simplified teeth (Armfield *et al.*, 2013).

Some of the most archaic neocetes (e.g., mammalodontids, early aetiocetids, and simocetids) retain the ancestral placental tooth count or resemble basilosaurids in lacking the third upper molar. Most, however, show a clear tendency to increase the number of teeth implanted in the maxilla (but not the premaxilla), from at least 9 per quadrant in *Agorophius pygmaeus* and *Aetiocetus polydentatus*, to 16 in *Waipatia*, and up to 80 in some fossil platanistids (Fordyce, 1981, 1994; Lambert, 2006). An elevated number of teeth, or **polydonty**, is typical of all extant odontocetes, although some groups (e.g., ziphiids and certain delphinids) secondarily

Figure 8.10 Expression patterns of *Bone morphogenetic protein* (*Bmp*) 4 and *Fibroblast growth factor* (*Fgf*) 8 across the jaws of various mammals. Dotted lines indicate hypothesized major changes in dental patterning. Note the overlap of *Bmp4* and *Fgf8* along the distal part of the tooth row in the dolphin. Adapted from Armfield (2013) under a Creative Common Attribution license.

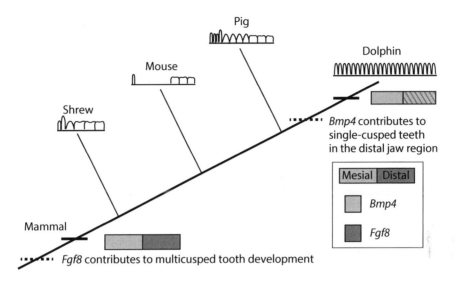

lost most or all of their dentition. Based on the number of embryonic tooth buds (more than 30 in *Balaena* and *Balaenoptera*), extant mysticetes also appear to be ancestrally polydont (Ishikawa and Amasaki, 1995; Thewissen *et al.*, 2014).

Developmentally, polydonty may have been driven by the same, distal extension of the *Bmp4* expression domain that underlies the emergence of homodonty (Figure 8.10) (Armfield *et al.*, 2013). Specifically, extra teeth could have arisen (1) via initial intercalation and retention of the deciduous and permanent teeth, followed by the addition of supernumerary teeth; or (2) via direct addition of supernumerary teeth, such as in the form of extra molars at the end of the tooth row. Under the first scenario, it would be difficult to explain why there are no taxa in which the premaxilla bears more than three teeth, given that the latter likely correspond to a single generation of incisors (Fordyce, 1982). By contrast, the second scenario may be supported by the presence of extra cheek teeth in several archaic mysticetes and odontocetes, as well as developmental evidence for a single wave of tooth development in extant mysticetes (Karlsen, 1962; Thewissen *et al.*, 2014). The latter observation also

bears on the question of tooth replacement: whereas archaeocetes are clearly diphyodont, there is no fossil, morphological, or developmental evidence for more than one tooth generation in any extant or extinct neocete. All mysticetes and odontocetes are therefore presumed to be **monophyodont**. Whether the teeth actually present in extant neocetes represent the original deciduous or permanent dentition currently remains unclear (Karlsen, 1962; Uhen and Gingerich, 2001).

8.4 Heterochrony: aged youngsters, juvenile adults

Heterochrony describes a change in the relative timing or rate of individual development, which can contribute to, but does not necessarily dominate, marked shifts in morphology (Hanken, 2015; Koyabu *et al.*, 2014; Werneburg and Sánchez-Villagra, 2015). In general, heterochrony comes in two guises: truncated development (**pedomorphosis**), or the retention of "juvenile" traits in sexually mature adults of a given species, relative to its immediate ancestor; and extended development (**peramorphosis**), that is, delayed maturation of traits beyond the adult level of the immediate ancestor (Alberch *et al.*, 1979; Reilly *et al.*, 1997).

Pedomorphosis can be produced by slower rates of development (**deceleration,** also known as neoteny), a late onset of development (**post-displacement**) or an early termination of development (**hypomorphosis**, also known as progenesis). Exactly the opposite holds true for peramorphosis, which is the result of **acceleration**, **pre-displacement** or **hypermorphosis**. Most of these processes are not mutually exclusive, and they can affect both somatic and reproductive traits in different ways. Thus, for example, hypomorphosis on its own generally causes a given trait to be less developed in the descendant species, relative to its immediate ancestor; however, the trait in question may remain the same in both the ancestor and the descendant if hypomorphosis is combined with proportional acceleration. Because the shape of the trait itself remains constant in this case, the change in its ontogenetic trajectory is also known as **isomorphosis** (Reilly *et al.*, 1997).

At a general level, heterochrony may have shaped cetacean evolution by being involved in the emergence of hyperphalangy (peramorphosis, possibly hypermorphosis) (section 8.1.1), the loss of the hind limb (pedomorphosis, possibly hypo-morphosis) (section 8.1.2), and variations in bone density and thickness (i.e., osteosclerosis, pachy-ostosis, and osteoporosis; both pedo- and per-amorphosis) (Richardson and Oelschläger, 2002; Thewissen *et al.*, 2006). Besides these large-scale effects, heterochrony has also affected particu-lar species, or groups of taxa, in more specific ways. Of these, phocoenids (porpoises) are the most obvious and best studied example.

Compared to other modern delphinoids, extant phocoenids stand out for their (1) small size; (2) short rostrum; (3) rounded braincase; (3) grac-ile cranium marked by relatively inconspicuous processes and bony crests; (4) lesser degree of cranial asymmetry (Figure 8.11); (5) lesser degree of fusion of sutures and epiphyses; and (6) lesser degree of ossification in the manus (Barnes, 1985; de Ricqlès and de Buffrénil, 2001; Galatius *et al.*,

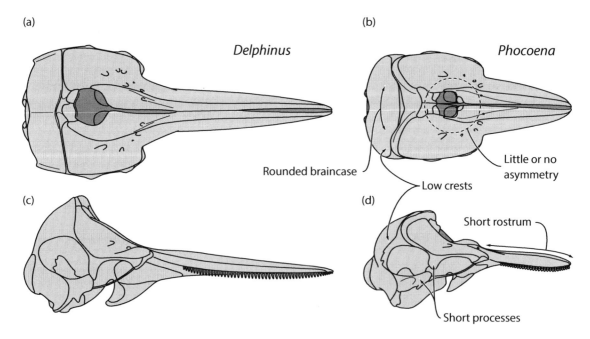

Figure 8.11 Skulls of the delphinid *Delphinus* and the phocoenid *Phocoena* in (a, b) dorsal and (c, d) right lateral views, with morphological differences resulting from pedomorphosis highlighted. The skulls are scaled so that the distance between the antorbital process and the occipital condyle is the same in all figures.

Figure 8.12 Effects of pedomorphosis on porpoises (*Phocoena phocoena*), the delphinid *Cephalorhynchus* (*C. commersonii*), and other delphinids (*Lagenorhynchus albirostris*), as evident in (a) the degree of sutural fusion in the skull with increasing age; and (b) the change in skull shape (relative to neonates) with increasing body size. In (a), higher values of the cranial suture score denote a higher degree of sutural fusion. Note how fusion in *P. phocoena* and *C. commersonii* stays well below that of *L. albirostris* even in fully adult individuals, implying the presence of hypomorphosis. In (b), "Procrustes distance to neonate shape" is a measure of cranial shape calculated from three-dimensional landmarks (Figure 1.4), whereas "Centroid size" measures body size. Note how cranial shape changes markedly faster in *P. phocoena* and *C. commersonii* compared to *L. albirostris*, demonstrating the effects of acceleration. Reproduced from Galatius (2010), with permission of the Linnean Society of London.

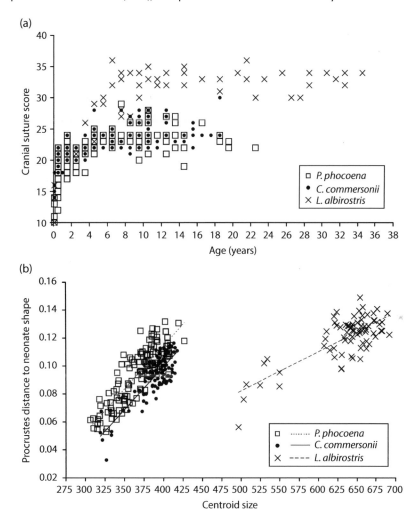

2006; Mellor *et al.*, 2009). A similar set of features, all of which are reminiscent of juvenile individuals of other delphinoids, also characterizes species of the delphinid *Cephalorhynchus* (Figure 8.12). Together with the relatively early onset of sexual maturity, this suggests that extant phocoenids and *Cephalorhynchus* became pedomorphic via hypomorphosis—a phenomenon that is otherwise rare among mammals (Galatius, 2010; Ichishima and Kimura, 2005).

The emergence of pedomorphosis in these two groups was almost certainly convergent, as both other delphinids and most pre-Pleistocene fossil phocoenids show either fewer or no signs of heterochrony (e.g., Lambert, 2008; Murakami *et al.*, 2012a, 2012b).

Interestingly, both porpoises and *Cephalorhynchus* also show evidence of acceleration, which offsets some of the effects of hypomorphosis on their overall morphology. Thus, the shape changes brought on by hypomorphosis were likely not adaptive in this case. Rather, their truncated development

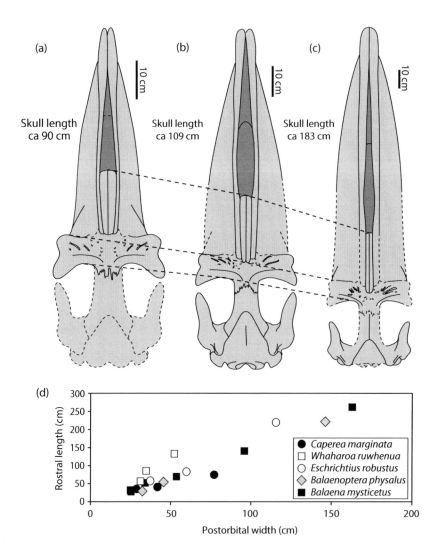

Figure 8.13 Comparison of the relative proportions of the rostrum and braincase in (a) a young juvenile, (b) a large juvenile, and (c) an adult of the eomysticetid *Waharoa ruwhenua*. Note the pronounced elongation of the rostrum relative to the rest of the skull. Graph (d) compares the rostral ontogeny of *W. ruwhenua* with that of extant species, including the pygmy right whale (*Caperea marginata*), the gray whale (*Eschrichtius robustus*), the fin whale (*Balaenoptera physalus*), and the bowhead whale (*Balaena mysticetus*). In *W. ruwhenua*, peramorphism causes the relative length of the rostrum to increase considerably faster than in any of the other species. Adapted from Boessenecker and Fordyce (2015) under a Creative Commons Attribution license.

may have been driven by selective pressures favoring smaller body size and/or shorter generation times, which in turn could have been triggered by unstable environmental conditions or the colonization of small-scale, yet reliably productive shelf habitats (Galatius, 2010; Ichishima and Kimura, 2005). The second option may be more likely, since their long lifespans, constant energetic needs, and limited capacity for energy storage put small cetaceans in a poor position to deal with highly unstable food supplies. Support for the shelf colonization idea furthermore comes from the observation that coastal porpoises tend to be both smaller and more obviously pedomorphic than phocoenids inhabiting less productive, comparatively unpredictable offshore waters (Galatius, 2010).

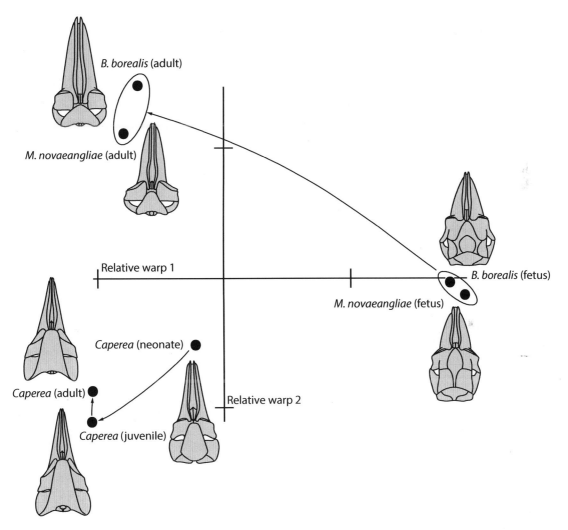

Figure 8.14 Change in skull shape through ontogeny in the extant pygmy right whale *Caperea*, as well as two balaenopterids: the sei (*Balaenoptera borealis*) and humpback (*Megaptera novaeangliae*) whales. The "relative warp" axes record deviation from the average shape of all of the skulls shown here. Note how all specimens of *Caperea* cluster within the same area, reflecting the relatively constant overall skull shape (e.g., far anteriorly projected supraoccipital) at various ontogenetic stages. By contrast, balaenopterids undergo marked changes in the relative size and shape of the rostrum, temporal fossa, neurocranium, and supraoccipital. Adapted from Tsai and Fordyce (2014).

Among mysticetes, heterochrony may have played a role in the appearance of the Oligocene mammalodontids, aetiocetids, and eomysticetids: whereas the former two are characterized by their small size, short rostra, and large eyes—features typical of juvenile mammals, as well as pedomorphic odontocetes (Fitzgerald, 2010; Sanders and Barnes, 2002)—the latter likely acquired their extremely elongated rostra via peramorphic acceleration (Figure 8.13) (Boessenecker and Fordyce, 2015). More recently, heterochrony became involved in the development of neobalaenids and balaenopterids (Tsai and Fordyce, 2014). Thus, the skull shape of *Caperea* not only changes little between the fetal and adult stages (Figure 8.14), but also closely resembles that of its only known fossil relative, *Miocaperea*, suggesting that deceleration may have constrained the development of pygmy right whales and resulted in long-term structural conservatism. By contrast, balaenopterids undergo marked changes over the course of their ontogeny that seem to reflect acceleration, especially as regards the telescoping of the supraoccipital and rostral bones, the shape of the supraorbital process of the frontal, and the length of the rostrum (Tsai and Fordyce, 2014).

8.5   Suggested readings

Buchholtz, E. A. 2007. Modular evolution of the cetacean vertebral column. Evolution & Development 9:278–289.

Thewissen, J., M. Cohn, L. Stevens, S. Bajpai, J. Heyning, and W. Horton. 2006. Developmental basis for hind-limb loss in dolphins and origin of the cetacean bodyplan. Proceedings of the National Academy of Sciences 103:8414–8418.

Thewissen, J. G. M., L. N. Cooper, and R. R. Behringer. 2012. Developmental biology enriches paleontology. Journal of Vertebrate Paleontology 32:1223–1234.

References

Abel, O. 1908. Die Morphologie der Hüftrudimente der Cetaceen. Denkschriften der Kaiserlichen Akademie der Wissenschaften 81:139–195.

Alberch, P., S. J. Gould, G. F. Oster, and D. B. Wake. 1979. Size and shape in ontogeny and phylogeny. Paleobiology 5:296–317.

Andrews, R. C. 1921. A remarkable case of external hind limbs in a humpback whale. American Museum Novitates 9:1–6.

Armfield, B. A., Z. Zheng, S. Bajpai, Christopher J. Vinyard, and J. G. M. Thewissen. 2013. Development and evolution of the unique cetacean dentition. PeerJ 1:e24.

Bajpai, S., J. G. M. Thewissen, and A. Sahni. 2009. The origin and early evolution of whales: macroevolution documented on the Indian Subcontinent. Journal of Biosciences 34:673–686.

Barnes, L. G. 1985. Evolution, taxonomy and antitropical distributions of the porpoises (Phocoenidae, Mammalia). Marine Mammal Science 1:149–165.

Bebej, R. M., M. Ul-Haq, I. S. Zalmout, and P. D. Gingerich. 2012. Morphology and function of the vertebral column in *Remingtonocetus domandaensis* (Mammalia, Cetacea) from the Middle Eocene Domanda Formation of Pakistan. Journal of Mammalian Evolution 19:77–104.

Bejder, L., and B. K. Hall. 2002. Limbs in whales and limblessness in other vertebrates: mechanisms of evolutionary and developmental transformation and loss. Evolution & Development 4:445–458.

Boessenecker, R. W., and R. E. Fordyce. 2015. Anatomy, feeding ecology, and ontogeny of a transitional baleen whale: a new genus and species of Eomysticetidae (Mammalia: Cetacea) from the Oligocene of New Zealand. PeerJ 3:e1129.

Boisserie, J.-R., F. Lihoreau, and M. Brunet. 2005. The position of Hippopotamidae within Cetartiodactyla. Proceedings of the National Academy of Sciences of the United States of America 102:1537–1541.

Buchholtz, E. 1998. Implications of vertebral morphology for locomotor evolution in early Cetacea; pp. 325–351 in J. G. M. Thewissen (ed.), The Emergence of Whales. Plenum Press, New York.

Buchholtz, E. A. 2001. Vertebral osteology and swimming style in living and fossil whales (Order: Cetacea). Journal of Zoology 253:175–190.

Buchholtz, E. A. 2007. Modular evolution of the cetacean vertebral column. Evolution & Development 9:278–289.

Buchholtz, E. A., and S. A. Schur. 2004. Vertebral osteology in Delphinidae (Cetacea). Zoological Journal of the Linnean Society 140:383–401.

Cohn, M. J., and C. Tickle. 1999. Developmental basis of limblessness and axial patterning in snakes. Nature 399:474–479.

Cooper, L. N., A. Berta, S. D. Dawson, and J. S. Reidenberg. 2007a. Evolution of hyperphalangy and digit reduction in the cetacean manus. The Anatomical Record 290:654–672.

Cooper, L. N., S. D. Dawson, J. S. Reidenberg, and A. Berta. 2007b. Neuromuscular anatomy and evolution of the cetacean forelimb. The Anatomical Record 290:1121–1137.

Cooper, L. N., N. Sedano, S. Johansson, B. May, J. D. Brown, C. M. Holliday, B. W. Kot, and F. E. Fish. 2008. Hydrodynamic performance of the minke whale (*Balaenoptera acutorostrata*) flipper. Journal of Experimental Biology 211:1859–1867.

Cooper, L. N., J. G. M. Thewissen, S. Bajpai, and B. N. Tiwari. 2012. Postcranial morphology and locomotion of the Eocene raoellid *Indohyus* (Artiodactyla: Mammalia). Historical Biology 24:279–310.

Cooper, L. N., J. G. M. Thewissen, and S. T. Hussain. 2009. New middle Eocene archaeocetes (Cetacea: Mammalia) from the Kuldana Formation of northern Pakistan. Journal of Vertebrate Paleontology 29:1289–1299.

de Ricqlès, A., and V. de Buffrénil. 2001. Bone histology, heterochronies and the return of tetrapods to life in water: were are we?; pp. 289–310 in J.-M. Mazin, and V. de Buffrénil (eds.), Secondary Adaptation of Tetrapods to Life in Water. Verlag Dr. Friedrich Pfeil, München.

Dines, J. P., E. Otárola-Castillo, P. Ralph, J. Alas, T. Daley, A. D. Smith, and M. D. Dean. 2014. Sexual selection targets cetacean pelvic bones. Evolution 68:3296–3306.

Fitzgerald, E. M. G. 2010. The morphology and systematics of *Mammalodon colliveri* (Cetacea: Mysticeti), a toothed mysticete from the Oligocene of Australia. Zoological Journal of the Linnean Society 158:367–476.

Flower, W. H. 1885. An Introduction to the Osteology of the Mammalia. Macmillan and Co., London.

Fordyce, R. E. 1981. Systematics of the odontocete whale *Agorophius pygmaeus* and the family Agorophidae (Mammalia: Cetacea). Journal of Paleontology 55:1028–1045.

Fordyce, R. E. 1982. Dental anomaly in a fossil squalodont dolphin from New Zealand, and the evolution of polydonty in whales. New Zealand Journal of Zoology 9:419–426.

Fordyce, R. E. 1994. *Waipatia maerewhenua*, new genus and new species (Waipatiidae, new family), an archaic Late Oligocene dolphin (Cetacea: Odontoceti: Platanistoidea) from New Zealand. Proceedings of the San Diego Society of Natural History 29:147–176.

Fordyce, R. E., L. G. Barnes, J. L. Goedert, and B. J. Crowley. 2000. Pelvic girdle elements of Oligocene and Miocene Mysticeti: Whale hind legs in transition. Journal of Vertebrate Paleontology 20 (Suppl. to 3):41A.

Galatius, A. 2010. Paedomorphosis in two small species of toothed whales (Odontoceti): how and why? Biological Journal of the Linnean Society 99:278–295.

Galatius, A., M.-B. E. R. Andersen, B. Haugan, H. E. Langhoff, and Å. Jespersen. 2006. Timing of epiphyseal development in the flipper skeleton of the harbour porpoise (*Phocoena phocoena*) as an indicator of paedomorphosis. Acta Zoologica 87:77–82.

Gingerich, P. D., M. S. M. Antar, and I. Zalmout. 2014. Skeleton of new protocetid (Cetacea, Archaeoceti) from the lower Gehannam Formation of Wadi al Hitan in Egypt: survival of a protocetid into the Priabonian Late Eocene. Journal of Vertebrate Paleontology:Program and Abstracts, 138.

Gingerich, P. D., M. u. Haq, I. S. Zalmout, I. H. Khan, and M. S. Malakani. 2001. Origin of whales from early artiodactyls: hands and feet of Eocene Protocetidae from Pakistan. Science 293:2239–2242.

Gingerich, P. D., B. H. Smith, and E. L. Simons. 1990. Hind limbs of Eocene *Basilosaurus*: evidence of feet in whales. Science 249:154–157.

Gingerich, P. D., and M. D. Uhen. 1996. *Ancalecetus simonsi*, a new dorudontine archaeocete (Mammalia, Cetacea) from the early Late Eocene of Wadi Hitan, Egypt. Contributions from the Museum of Paleontology, University of Michigan 29:359–401.

Gingerich, P. D., M. Ul-Haq, W. von Koenigswald, W. J. Sanders, B. H. Smith, and I. S. Zalmout. 2009. New protocetid whale from the Middle Eocene of Pakistan: birth on land, precocial development, and sexual dimorphism. PLoS One 4:1–20.

Gingerich, P. D., I. S. Zalmout, M. Ul-Haq, and M. A. Bhatti. 2005. *Makaracetus bidens*, a new protocetid archaeocete (Mammalia, Cetacea) from the early Middle Eocene of Balochistan (Pakistan). Contributions from the Museum of Paleontology, University of Michigan 31:197–210.

Gol'din, P. 2014. Naming an innominate: pelvis and hindlimbs of Miocene whales give an insight into evolution and homology of cetacean pelvic girdle. Evolutionary Biology 41:473–479.

Hanken, J. 2015. Is heterochrony still an effective paradigm for contemporary studies of evo-devo?; pp. 97–110 in A. C. Love (ed.), Conceptual Change in Biology. Springer, Dordrecht.

Howell, A. B. 1930. Aquatic Mammals: Their Adaptations to Life in the Water. Charles C. Thomas, Springfield, IL.

Hulbert, R. C., Jr. 1998. Postcranial osteology of the North American Middle Eocene protocetid *Georgiacetus*; pp. 235–268 in J. G. M. Thewissen (ed.), The Emergence of Whales. Plenum Press, New York.

Hulbert, R. C., Jr., R. M. Petkewich, G. A. Bishop, D. Bukry, and D. P. Aleshire. 1998. A new middle Eocene protocetid whale (Mammalia: Cetacea: Archaeoceti) and associated biota from Georgia. Journal of Paleontology 72:907–927.

Ichishima, H., and M. Kimura. 2005. *Haborophocoena toyoshimai*, a new early Pliocene porpoise (Cetacea; Phocoenidae) from Hokkaido, Japan. Journal of Vertebrate Paleontology 25:655–664.

Ishikawa, H., and H. Amasaki. 1995. Development and physiological degradation of tooth buds and development of rudiment of baleen plate in southern minke whale, *Balaenoptera acutorostrata*. Journal of Veterinary Medical Science 57:665–670.

Itou, J., H. Kawakami, T. Quach, M. Osterwalder, S. M. Evans, R. Zeller, and Y. Kawakami. 2012. Islet1 regulates establishment of the posterior hindlimb field upstream of the Hand2-Shh morphoregulatory gene network in mouse embryos. Development 139:1620–1629.

Jernvall, J., and I. Thesleff. 2000. Reiterative signaling and patterning during mammalian tooth morphogenesis. Mechanisms of Development 92:19–29.

Karlsen, K. 1962. Development of tooth germs and adjacent structures in the whalebone whale (*Balaenoptera physalus* (L.)). Hvalrådets Skrifter 45:1–56.

Kellogg, R. 1936. A review of the Archaeoceti. Carnegie Institution of Washington Publication 482:1–366.

Koyabu, D., I. Werneburg, N. Morimoto, C. P. E. Zollikofer, A. M. Forasiepi, H. Endo, J. Kimura, S. D. Ohdachi, N. Truong Son, and M. R. Sánchez-Villagra. 2014. Mammalian skull heterochrony reveals modular evolution and a link between cranial development and brain size. Nature Communications 5:3625.

Kükenthal, W. 1889–1893. Vergleichend-anatomische und entwicklungsgeschichtliche Untersuchung an Walthieren. Denkschriften der Medicinisch-Naturwissenschaftlichen Gesellschaft zu Jena 3:1–448.

Lambert, O. 2006. First record of a platanistid (Cetacea, Odontoceti) in the North Sea Basin: a review of *Cyrtodelphis* Abel, 1899 from the Miocene of Belgium. Oryctos 6:69–79.

Lambert, O. 2008. A new porpoise (Cetacea, Odontoceti, Phocoenidae) from the Pliocene of the North Sea. Journal of Vertebrate Paleontology 28:863–872.

Liang, L., Y.-Y. Shen, X.-W. Pan, T.-C. Zhou, C. Yang, D. M. Irwin, and Y.-P. Zhang. 2013. Adaptive evolution of the *Hox* gene family for development in bats and dolphins. PLoS One 8:e65944.

Loch, C., J. A. Kieser, and R. E. Fordyce. 2015. Enamel ultrastructure in fossil cetaceans (Cetacea: Archaeoceti and Odontoceti). PLoS One 10:e0116557.

Lönnberg, E. 1911. The pelvic bones of some Cetacea. Arkiv för Zoologi 7:1–15.

Madar, S. I. 2007. The postcranial skeleton of Early Eocene pakicetid cetaceans. Journal of Paleontology 81:176–200.

Madar, S. I., J. G. M. Thewissen, and S. T. Hussain. 2002. Additional holotype remains of *Ambulocetus natans* (Cetacea, Ambulocetidae), and their implications for locomotion in early whales. Journal of Vertebrate Paleontology 22:405–422.

Martínez Cáceres, M., and C. de Muizon. 2011. A new basilosaurid (Cetacea, Pelagiceti) from the Late Eocene to Early Oligocene Otuma Formation of Peru. Comptes Rendus Palevol 10:517–526.

Mellor, L., L. N. Cooper, J. Torre, and R. L. J. Brownell. 2009. Paedomorphic ossification in porpoises with an emphasis on the vaquita (*Phocoena sinus*). Aquatic Mammals 35:193–202.

Moran, M., S. Bajpai, J. C. George, R. Suydam, S. Usip, and J. G. M. Thewissen. 2015. Intervertebral and epiphyseal fusion in the postnatal ontogeny of cetaceans and terrestrial mammals. Journal of Mammalian Evolution 22:93–109.

Murakami, M., C. Shimada, Y. Hikida, and H. Hirano. 2012a. A new basal porpoise, *Pterophocaena nishinoi* (Cetacea, Odontoceti, Delphinoidea), from the upper Miocene of Japan and its phylogenetic relationships. Journal of Vertebrate Paleontology 32:1157–1171.

Murakami, M., C. Shimada, Y. Hikida, and H. Hirano. 2012b. Two new extinct basal phocoenids (Cetacea, Odontoceti, Delphinoidea), from the upper Miocene Koetoi Formation of Japan and their phylogenetic significance. Journal of Vertebrate Paleontology 32:1172–1185.

Ohsumi, S., and H. Kato. 2008. A bottlenose dolphin (*Tursiops truncatus*) with fin-shaped hind appendages. Marine Mammal Science 24:743–745.

Peterkova, R., M. Hovorakova, M. Peterka, and H. Lesot. 2014. Three-dimensional analysis of the early development of the dentition. Australian Dental Journal 59:55–80.

Rabinowitz, A. H., and S. A. Vokes. 2012. Integration of the transcriptional networks regulating limb morphogenesis. Developmental Biology 368:165–180.

Reilly, S. M., E. O. Wiley, and D. J. Meinhardt. 1997. An integrative approach to heterochrony: the distinction between interspecific and intraspecific phenomena. Biological Journal of the Linnean Society 60:119–143.

Richardson, M. K., J. E. Jeffery, and C. J. Tabin. 2004. Proximodistal patterning of the limb: insights from evolutionary morphology. Evolution & Development 6:1–5.

Richardson, M. K., and H. H. A. Oelschläger. 2002. Time, pattern, and heterochrony: a study of hyperphalangy in the dolphin embryo flipper. Evolution & Development 4:435–444.

Sanchez, J. A., and A. Berta. 2010. Comparative anatomy and evolution of the odontocete forelimb. Marine Mammal Science 26:140–160.

Sanders, A. E., and L. G. Barnes. 2002. Paleontology of the Late Oligocene Ashley and Chandler Bridge Formations of South Carolina, 3: Eomysticetidae, a new family of primitive mysticetes (Mammalia: Cetacea). Smithsonian Contributions to Paleobiology 93:313–356.

Sanz-Ezquerro, J. J., and C. Tickle. 2003. Fgf signaling controls the number of phalanges and tip formation in developing digits. Current Biology 13:1830–1836.

Shapiro, M. D., J. Hanken, and N. Rosenthal. 2003. Developmental basis of evolutionary digit loss in the Australian lizard Hemiergis. Journal of Experimental Zoology Part B: Molecular and Developmental Evolution 297B:48–56.

Shashikant, C. S., C. B. Kim, M. A. Borbély, W. C. H. Wang, and F. H. Ruddle. 1998. Comparative studies on mammalian *Hoxc8* early enhancer sequence reveal a baleen whale-specific deletion of a cis-acting element. Proceedings of the National Academy of Sciences of the United States of America 95:15446–15451.

Simões-Lopes, P. C., and C. S. Gutstein. 2004. Notes on the anatomy, positioning and homology of the pelvic bones in small cetaceans (Cetacea, Delphinidae, Pontoporiidae). Latin American Journal of Aquatic Mammals 3:157–162.

Slijper, E. J. 1936. Die Cetaceen, vergleichend-anatomisch und systematisch. M. Nijhoff, Amsterdam.

Tajima, Y., Y. Hayashi, and T. K. Yamada. 2004. Comparative anatomical study on the relationships between the vestigial pelvic bones and the surrounding structures of finless porpoises (*Neophocaena phocaenoides*). Journal of Veterinary Medical Science 66:761–766.

Thewissen, J. G. M., and S. Bajpai. 2001. Dental morphology of Remingtonocetidae (Cetacea, Mammalia). Journal of Paleontology 75:463–465.

Thewissen, J. G. M., and S. Bajpai. 2009. New skeletal material of *Andrewsiphius* and *Kutchicetus*, two Eocene cetaceans from India. Journal of Paleontology 83:635–663.

Thewissen, J. G. M., M. J. Cohn, L. S. Stevens, S. Bajpai, J. Heyning, and W. E. Horton. 2006. Developmental basis for hind-limb loss in dolphins and origin of the cetacean bodyplan. Proceedings of the National Academy of Sciences of the United States of America 103:8414–8418.

Thewissen, J. G. M., and F. E. Fish. 1997. Locomotor evolution in the earliest cetaceans: functional model, modern analogues, and paleontological evidence. Paleobiology 23:482–490.

Thewissen, J. G. M., and S. T. Hussain. 1990. Postcranial osteology of the most primitive artiodactyl: *Diacodexis pakistanensis* (Dichobunidae). Anatomia, Histologia, Embryologia 19:37–48.

Thewissen, J. G. M., S. T. Hussain, and M. Arif. 1994. Fossil evidence for the origin of aquatic locomotion in archaeocete whales. Science 263:210–212.

Thewissen, J. G. M., S. I. Madar, and S. T. Hussain. 1996. *Ambulocetus natans*, an Eocene cetacean (Mammalia) from Pakistan. Courier Forschungsinstitut Senckenberg 191:1–86.

Thewissen, J. G. M., D. McBurney, J. C. George, and R. Suydam. 2014. Tooth and baleen development in the bowhead whale. Seventh Triennial Conference on Secondary Adaptation of Tetrapods to Life in Water: Abstracts, Fairfax, VA.

Thewissen, J. G. M., E. M. Williams, L. J. Roe, and S. T. Hussain. 2001. Skeletons of terrestrial cetaceans and the relationship of whales to artiodactyls. Nature 413:277–281.

Tsai, C.-H., and R. E. Fordyce. 2014. Disparate heterochronic processes in baleen whale evolution. Evolutionary Biology 41:299–307.

Tucker, A., and P. Sharpe. 2004. The cutting-edge of mammalian development; how the embryo makes teeth. Nature Reviews Genetics 5:499–508.

Uhen, M. D. 1998. Middle to Late Eocene basilosaurines and dorudontines; pp. 29–61 in J. G. M. Thewissen (ed.), The Emergence of Whales. Plenum Press, New York.

Uhen, M. D. 1999. New species of protocetid archaeocete whale, *Eocetus wardii* (Mammalia, Cetacea), from the Middle Eocene of North Carolina. Journal of Paleontology 73:512–528.

Uhen, M. D. 2000. Replacement of deciduous first premolars and dental eruption in archaeocete whales. Journal of Mammalogy 81:123–133.

Uhen, M. D. 2001. New material of *Eocetus wardii* (Mammalia, Cetacea) from the Middle Eocene of North Carolina. Southeastern Geology 40:135–148.

Uhen, M. D. 2004. Form, function, and anatomy of *Dorudon atrox* (Mammalia: Cetacea): An archaeocete from the Middle to Late Eocene of Egypt. University of Michigan Papers on Paleontology 34:1–222.

Uhen, M. D. 2009. Evolution of dental morphology; pp. 302–307 in J. G. M. Thewissen, W. F. Perrin, and B. Würsig (eds.), Encyclopedia of Marine Mammals. Academic Press, Burlington.

Uhen, M. D. 2014. New material of *Natchitochia jonesi* and a comparison of the innominata and locomotor capabilities of Protocetidae. Marine Mammal Science 30:1029–1066.

Uhen, M. D., and P. D. Gingerich. 2001. New genus of dorudontine archaeocete (Cetacea) from the middle-to-late Eocene of South Carolina. Marine Mammal Science 17:1–34.

Wang, Z., L. Yuan, S. J. Rossiter, X. Zuo, B. Ru, H. Zhong, N. Han, G. Jones, P. D. Jepson, and S. Zhang. 2009. Adaptive evolution of 5′*HoxD* genes in the origin and diversification of the cetacean flipper. Molecular Biology and Evolution 26:613–622.

Werneburg, I., and M. R. Sánchez-Villagra. 2015. Skeletal heterochrony is associated with the anatomical specializations of snakes among squamate reptiles. Evolution 69:254–263.

Woodward, B. L., J. P. Winn, and F. E. Fish. 2006. Morphological specializations of baleen whales associated with hydrodynamic performance and ecological niche. Journal of Morphology 267:1284–1294.

Xu, B., and D. M. Wellik. 2011. Axial *Hox9* activity establishes the posterior field in the developing forelimb. Proceedings of the National Academy of Sciences 108:4888–4891.

9 Living Cetaceans in an Evolutionary Context

9.1 A modern view of cetacean evolution

Seventy years after George Gaylord Simpson's (1945) statement of abject uncertainty as to the origins of cetaceans, where does our knowledge of whale evolution stand today? Much has changed since Simpson seemingly admitted defeat on the subject, although he unfortunately did not live to see most of the major breakthroughs that finally cracked the cetacean conundrum. Any list attempting to summarize these breakthroughs must necessarily be subjective and incomplete. Nevertheless, it likely would include:

1. The first molecular evidence grouping cetaceans with artiodactyls (Boyden and Gemeroy, 1950), which in turn confirmed the much earlier proposals of Hunter (1787) and Flower (1883);
2. The recognition of *Gandakasia*, *Ichthyolestes*, and *Pakicetus* as some of the most archaic cetaceans, which triggered a series of further discoveries from the crucial region of Indo-Pakistan (Gingerich and Russell, 1981; Gingerich *et al.*, 1983);
3. The discovery of the archaic mysticete *Llanocetus denticrenatus*, which pushed back the origin of neocetes to at least the Late Eocene (Fordyce, 2003; Mitchell, 1989);
4. The description of the "walking, swimming" whale, *Ambulocetus natans*, which provided some of the best evidence for both terrestrial *and* aquatic locomotion in early whales (Thewissen *et al.*, 1994);
5. The first discoveries of well-preserved cetacean astragali, whose double-pulley shape clearly demonstrated the artiodactyl affinities of cetaceans, and reconciled conflicting molecular and morphological hypotheses (Gingerich *et al.*, 2001; Thewissen *et al.*, 2001);
6. The recognition of hippopotamids and raoellids as the closest living and extinct relatives of whales, respectively, thus cementing the nesting of cetaceans within Artiodactyla (Thewissen *et al.*, 2007; Waddell *et al.*, 1999);
7. The recognition of the transitional nature of the aetiocetid feeding apparatus, which combines both teeth and palatal nutrient foramina, and thus closes the morphological gap between toothless mysticetes and their toothed, archaeocete ancestors (Deméré *et al.*, 2008); and
8. The realization that echolocation has ancient roots and may be as old as odontocetes themselves (Fordyce, 2002; Geisler *et al.*, 2014).

With these major steps in mind, it is now finally possible to tell the grand story of how cetaceans left the land to become the largest, and arguably most bizarre, mammals of all time. The story began 53 million years ago, when small, terrestrial artiodactyls gave up

Cetacean Paleobiology, First Edition. Felix G. Marx, Olivier Lambert, and Mark D. Uhen.
© 2016 John Wiley & Sons, Ltd. Published 2016 by John Wiley & Sons, Ltd.

the running habit of their ancestors, and instead took to nearby rivers or lakes to escape from predators. Small as it may have seemed at the time, this change in behavior opened up a whole new world of possibilities, which cetaceans were quick to exploit. In virtually no time at all, they adjusted their diet, teeth, and hearing to the new environment (section 5.1.1).

Soon after, this was followed by further, even more profound changes to the morphology of the spine, the limbs, and most of the sense organs, which enabled cetaceans to move from freshwater out to sea—an environment that had been virtually vacant of large vertebrate predators since the time of the dinosaurs (section 5.1.2). Initially, the newly marine whales remained close to the coast and even returned to shore to drink and reproduce. However, as more and more aquatic forms appeared, their link to the land became ever more tenuous (section 5.1.3). Ultimately, the connection became severed entirely when cetaceans managed to clear the last hurdle to a truly pelagic existence: the ability to mate and give birth at sea (section 6.3). From their first, tentative steps into the water to this point, a mere 12–13 million years had passed—in geological terms, little more than the blink of an eye.

Once firmly in the water, cetaceans initially kept chasing after fish and invertebrates in the manner of their ancestors (section 6.1.1). Soon, however, novel forms of feeding appeared, in tandem with dramatic changes in global ocean circulation that greatly increased food availability (e.g., the onset of the Antarctic Circumpolar Current; section 7.1.1). By circa 33 Ma, some species had found ways to filter large quantities of small prey directly from the water, whereas others had started to detect their prey using sounds, enabling them to hunt at depth or during the night (sections 5.2, 6.1.2, and 6.1.3). Both of these strategies were so successful that they went on to become defining characteristics. A great schism occurred, which gave rise to the two lineages of whales that still dominate the ocean today: the filter-feeding mysticetes and the echolocating odontocetes.

Armed with their new "key" adaptations, mysticetes and odontocetes quickly diversified into a colorful range of forms (section 7.3). Some of these went on to spawn long, successful lineages,

whereas others, such as aetiocetids, eomysticetids, and xenorophids, had their heyday and then soon went extinct—mostly by about 23 Ma. Also caught in this early turnover event were some archaic-looking animals (kekenodontids and at least some mammalodontids) that had not gone down the route of echolocation or mass feeding, and instead represented the last gasp of the traditional, archaeocete way of life (sections 7.1.1, 7.2.1, and 7.2.2).

Over the next 15 Ma, those lineages that made it past the Oligocene continued to diversify and gave rise to the earliest members of all of the living families. The Early–Middle Miocene seas were buzzing with a disparate cetacean assemblage, including heterodont squalodontids and waipatiids, other platanistoids, physeterids, kogiids, long-snouted eurhinodelphinids, eoplatanistids, ziphiids, pontoporiids, balaenids, and stem balaenopteroids. Lording it over everyone were several, sometimes gigantic killer sperm whales, including what may be the largest predator ever: the aptly named *Livyatan melvillei* (sections 5.4.3 and 6.1.3). By the late Middle to early Late Miocene (ca 9–14 Ma), some of these groups had already started to disappear. Most, however, combined with the then-emerging cetotheriids, crown balaenopteroids, and crown delphinoids to form the most diverse cetacean fauna ever (section 7.1.2).

Following the Middle–Late Miocene diversity peak, the luck of most cetacean families started to change. Some, like balaenopterids, delphinoids, and some ziphiids, thrived and laid the foundations for much of the modern fauna; most, however, went extinct. An increasingly cool climate and the gradual onset of Northern Hemisphere glaciation likely were bad news for many small species, which were faced with more unpredictable food supplies and stood to lose access to rapidly changing, or even disappearing, shallow-water habitats (sections 7.1.2 and 7.2.3). Overall cetacean diversity dramatically declined, from more than 70 genera in the Late Miocene to the roughly 40 genera alive today. Nevertheless, this time also saw the emergence of some of the best known cetacean extremes. As delphinids diversified, their complex social structure and/or echolocation apparatus seemingly drove them to acquire the largest brains of any animal except humans (section 7.5). At the same time, mysticetes took the evolution of body size to new heights and became the largest animals the world has ever seen (section 7.4).

9.2 Cetacea—quo vadis?

Today, it appears that the fate of cetaceans and their future evolution is intimately intertwined with humans and our activities regarding the oceans and the atmosphere. Some researchers have warned that we are in the midst of the sixth great mass extinction of life on Earth, which is caused directly by human hunting, but more significantly by habitat destruction and climate change (Barnosky *et al.*, 2011, 2012). We are certain that we have caused, or at least contributed to, the local extirpation of species from entire ocean basins, such as the Atlantic population of the gray whale, *Eschrichtius robustus* (Mead and Mitchell, 1984). We are relatively certain that we have caused the extinction of one of the rarest of the freshwater dolphins, *Lipotes vexillifer* (Pyenson, 2009). And we are almost certainly going to lose the vaquita, *Phocoena sinus*, because of poor fishing practices and the destruction of the Colorado River delta (Jaramillo-Legorreta *et al.*, 2007; Rodríguez-Quiroz *et al.*, 2012). As humans, we have the power not only to drive these amazing animals to extinction, but also to save them by protecting their environment, and thus our own.

Information provided by evolutionary biology and the fossil record can help us to understand the nature, adaptability, and distribution of living animals, and thus ultimately assist in guiding decisions related to their conservation. For example, phylogenies combining molecular, morphological, and fossil data have shown that all of the extant "river dolphins," *Caperea*, *Eschrichtius*, *Kogia*, and *Physeter*, are **relict** taxa, and as such the last survivors of ancient and formerly much more diverse lineages. Although not all of these taxa are necessarily endangered (for most, there are not enough data to tell), putting them into an evolutionary context nevertheless highlights the immense loss in genetic and morphological distinctiveness that their extinction would entail (Pyenson, 2009). Such insights, in turn, might justify additional, targeted efforts at protecting both them and the ecosystem on which they depend.

In some cases, fossils provide the only means of establishing what the "normal" range of an animal, prior to human interference, ought to be. Somewhat ironically, gray whales were first described based on a subfossil skeleton from Sweden, even though the species today only inhabits the North Pacific (Mead and Mitchell, 1984). In addition, older finds confirm that gray whales have been present in the North Atlantic since at least the Late Miocene (Bisconti and Varola, 2006). In this light, the recent sighting of stray individuals of *Eschrichtius* in the Mediterranean might represent the first steps of a possible recolonization event, rather than the invasion of a habitat by a foreign species (Scheinin *et al.*, 2011).

Finally, on a broader scale, the study of cetacean evolution has revealed that relatively recent events in Earth's geological history, such as global cooling and the onset of Northern Hemisphere glaciation, may have driven an overall decline in cetacean diversity, yet *also* correlate with the radiation of delphinids and, possibly, balaenopterids (e.g., Bianucci, 2013; Marx and Fordyce, 2015; McGowen *et al.*, 2009). In this light, the question of how future global warming and associated changes in ice volume and ocean currents might impact extant cetaceans deserves further study.

References

Barnosky, A. D., E. A. Hadly, J. Bascompte, E. L. Berlow, J. H. Brown, M. Fortelius, W. M. Getz, J. Harte, A. Hastings, P. A. Marquet, N. D. Martinez, A. Mooers, P. Roopnarine, G. Vermeij, J. W. Williams, R. Gillespie, J. Kitzes, C. Marshall, N. Matzke, D. P. Mindell, E. Revilla, and A. B. Smith. 2012. Approaching a state shift in Earth's biosphere. Nature 486:52–58.

Barnosky, A. D., N. Matzke, S. Tomiya, G. O. U. Wogan, B. Swartz, T. B. Quental, C. Marshall, J. L. McGuire, E. L. Lindsey, K. C. Maguire, B. Mersey, and E. A. Ferrer. 2011. Has the Earth's sixth mass extinction already arrived? Nature 471:51–57.

Bianucci, G. 2013. *Septidelphis morii*, n. gen. et sp., from the Pliocene of Italy: new evidence of the explosive radiation of true dolphins (Odontoceti, Delphinidae). Journal of Vertebrate Paleontology 33: 722–740.

Bisconti, M., and A. Varola. 2006. The oldest eschrichtiid mysticete and a new morphological diagnosis of Eschrichtiidae (gray whales). Rivista Italiana di Paleontologia e Stratigrafia 112:447–457.

Boyden, A., and D. Gemeroy. 1950. The relative position of the Cetacea among the orders of Mammalia as indicated by precipitin tests. Zoologica 35:145–151.

Deméré, T. A., M. R. McGowen, A. Berta, and J. Gatesy. 2008. Morphological and molecular evidence for a stepwise evolutionary transition from teeth to baleen in mysticete whales. Systematic Biology 57:15–37.

Flower, W. H. 1883. On whales, past and present, and their probable origin. Nature 28:199–202, 226–230.

Fordyce, R. E. 2002. *Simocetus rayi* (Odontoceti: Simocetidae, new family): a bizarre new archaic Oligocene dolphin from the eastern North Pacific. Smithsonian Contributions to Paleobiology 93:185–222.

Fordyce, R. E. 2003. Early crown-group Cetacea in the southern ocean: the toothed archaic mysticete *Llanocetus*. Journal of Vertebrate Paleontology 23 (Suppl. 3):50A.

Geisler, J. H., M. W. Colbert, and J. L. Carew. 2014. A new fossil species supports an early origin for toothed whale echolocation. Nature 508:383–386.

Gingerich, P. D., M. U. Haq, I. S. Zalmout, I. H. Khan, and M. S. Malakani. 2001. Origin of whales from early artiodactyls: hands and feet of Eocene Protocetidae from Pakistan. Science 293:2239–2242.

Gingerich, P. D., and D. E. Russell. 1981. *Pakicetus inachus*, a new archaeocete (Mammalia, Cetacea) from the Early-Middle Eocene Kuldana Formation of Kohat (Pakistan). Contributions from the Museum of Paleontology, University of Michigan 25:235–246.

Gingerich, P. D., N. A. Wells, D. E. Russell, and S. M. I. Shah. 1983. Origin of whales in epicontinental remnant seas: new evidence from the early Eocene of Pakistan. Science 220:403–406.

Hunter, J. 1787. Observations on the structure and oeconomy of whales. Philosophical Transactions of the Royal Society of London B 77:371–450.

Jaramillo-Legorreta, A., L. Rojas-Bracho, R. L. Brownell, A. J. Read, R. R. Reeves, K. Ralls, and B. L. Taylor. 2007. Saving the vaquita: immediate action, not more data. Conservation Biology 21:1653–1655.

Marx, F. G., and R. E. Fordyce. 2015. Baleen boom and bust: a synthesis of mysticete phylogeny, diversity and disparity. Royal Society Open Science 2:140434.

McGowen, M. R., M. Spaulding, and J. Gatesy. 2009. Divergence date estimation and a comprehensive molecular tree of extant cetaceans. Molecular Phylogenetics and Evolution 53:891–906.

Mead, J. G., and E. D. Mitchell. 1984. Atlantic gray whales; pp. 33–53 in M. L. Jones, S. L. Swartz, and S. Leatherwood (eds.), The Gray Whale, *Eschrichtius robustus*. Academic Press, Orlando.

Mitchell, E. D. 1989. A new cetacean from the late Eocene La Meseta Formation, Seymour Island, Antarctic Peninsula. Canadian Journal of Fisheries and Aquatic Science 46:2219–2235.

Pyenson, N. D. 2009. Requiem for *Lipotes*: an evolutionary perspective on marine mammal extinction. Marine Mammal Science 25:714–724.

Rodríguez-Quiroz, G., E. A. Aragón-Noriega, D. M. Ruiz-López, and H. A. González-Ocampo. 2012. The vaquita will be the second cetacean species driven to extinction by humans. Endangered Species UPDATE Wnter 2012:84–89.

Scheinin, A. P., D. Kerem, C. D. MacLeod, M. Gazo, C. A. Chicote, and M. Castellote. 2011. Gray whale (*Eschrichtius robustus*) in the Mediterranean Sea: anomalous event or early sign of climate-driven distribution change? Marine Biodiversity Records 4:e28.

Simpson, G. G. 1945. The principles of classification and a classification of mammals. Bulletin of the American Museum of Natural History 85:1–350.

Thewissen, J. G. M., L. N. Cooper, M. T. Clementz, S. Bajpai, and B. N. Tiwari. 2007. Whales originated form aquatic artiodactyls in the Eocene epoch of India. Nature 450:1190–1195.

Thewissen, J. G. M., S. T. Hussain, and M. Arif. 1994. Fossil evidence for the origin of aquatic locomotion in archaeocete whales. Science 263:210–212.

Thewissen, J. G. M., E. M. Williams, L. J. Roe, and S. T. Hussain. 2001. Skeletons of terrestrial cetaceans and the relationship of whales to artiodactyls. Nature 413:277–281.

Waddell, P. J., N. Okada, and M. Hasegawa. 1999. Towards resolving the interordinal relationships of placental mammals. Systematic Biology 48:1–5.

Index

Note: Page numbers in *italics* refer figures.

Cetacean Paleobiology, First Edition. Felix G. Marx, Olivier Lambert, and Mark D. Uhen.
© 2016 John Wiley & Sons, Ltd. Published 2016 by John Wiley & Sons, Ltd.